THE RECORD DECIDES

THE RECORD DECIDES

Safety and Industrial Hygiene Expert Services in Workplace Injury, Occupational Disease, and OSHA Litigation

Logan F. Martin, PhD, CIH, CSP, CHMM

LF Martin LLC

THE RECORD DECIDES

Safety and Industrial Hygiene Expert Services in Workplace Injury, Occupational Disease, and OSHA Litigation

Logan F. Martin, PhD, CIH, CSP, CHMM

ISBN 979-8-9972897-0-6 (hardcover)

ISBN 979-8-9972897-1-3 (paperback)

Library of Congress Control Number: 2026919563

Published by LF Martin LLC

www.lfmartinllc.com

Cover design by Logan F. Martin

First edition, 2026

Chapter 28 describes an enforcement landscape that is moving quickly. Corrections, and updates to the directives, dockets, and rulemakings that chapter tracks, are posted at www.lfmartinllc.com.

For my mother and my grandmother,

who gave me my love of learning,

and who never let me stop reaching for what came next.

And for my daughters, Maddox and Emery,

for their love, their laughter,

and their unwavering support.

With love and gratitude, always.

Contents

Foreword

My early career was as an executive in the oil and gas industry. As one might imagine, I spent a considerable amount of time with attorneys. I remember an old country lawyer one time saying to me, "facts are stubborn things." I have also found it to be true in my almost 20 years as a practicing safety professional. When issues arise, be they in workplace safety compliance, industrial hygiene, occupational disease, or an OSHA enforcement action, an even more precise truth emerges: The record decides.

Too often, discussions about workplace incidents become debates shaped by hindsight, emotion, or competing narratives. One side has a version of events. The other side has a different version. Lawyers advocate. Experts testify. Organizations defend their decisions. Injured workers seek answers. Yet, as this book convincingly demonstrates, outcomes are rarely determined by the loudest argument or the strongest personality. They are determined by the quality, completeness, and credibility of the record that existed before the dispute ever began.

What makes *The Record Decides* distinctive is that it does not approach workplace injury, occupational disease, or OSHA enforcement defense and litigation from only one perspective. It is not a plaintiff's manual. It is not a defense handbook. It is not merely a safety text, nor is it solely a legal reference. Instead, it occupies the important space where safety practice, industrial hygiene, regulatory enforcement, and litigation intersect.

Throughout these chapters, you will be reminded that technical questions often drive legal outcomes. What hazard existed? Who controlled it? What exposure actually occurred? What evidence supports that conclusion? These are questions that cannot be answered through assumption or advocacy alone. They require disciplined analysis, documented facts, reliable methodology, and a willingness to follow the evidence wherever it leads.

This book also delivers an important message to safety professionals. Many practitioners will never serve as expert witnesses, never sit through a deposition, and never testify before a jury. Yet every safety professional creates records that may someday be scrutinized in precisely those settings. Training documentation, inspection reports, exposure assessments, audit findings, corrective-action records, and incident investigations are often created as routine business documents.

Years later, those same documents may become critical evidence. This book challenges professionals to view their work through that lens and to consider how today's records may be interpreted tomorrow.

For attorneys, the text provides a practical guide to understanding technical evidence and evaluating expert opinions. For safety and industrial hygiene professionals, it offers valuable insight into how courts, regulators, and opposing counsel examine the work they perform every day. For employers and risk managers, it serves as a reminder that effective safety management is not simply about compliance but rather it is about creating reliable systems and credible records that accurately reflect reality.

The most compelling aspect of this work is its consistent commitment to intellectual honesty. Repeatedly, as you read each chapter, you will encounter the same theme: a trustworthy expert is not loyal to a party, but to a method. The record does not choose sides, and neither should the professional charged with interpreting it. That idea runs throughout the book, which gives it both credibility and lasting value.

Whether you are handling an active case, responding to an OSHA inspection, evaluating an occupational exposure, building a safety program, or simply striving to become a more effective professional, this book offers a thoughtful and practical framework for understanding how technical evidence shapes real-world outcomes.

James A. Junkin, CSP, MSP, SMP, ASP, CSHO
Chief Executive Officer
Mariner-Gulf Consulting and Services
August 2026

Preface

This book began as a search for a book that did not exist.

The work described in these pages sits between two literatures that do not read each other. There are excellent legal treatises on expert admissibility, and there are excellent industrial hygiene texts on sampling strategy, exposure assessment, and the hierarchy of controls. What there was not — not in one volume, and not written for both of the people who need it — was an account of the place where the two meet: where an air sample becomes a number in a citation, where a training roster becomes proof of employer knowledge, where a calibration log that nobody thought to keep decides a case about a man's lungs. That account had to be assembled out of directives, decisions, standards, consensus guidance, and case files, in the margins of the practice itself, because the practice kept demanding an answer the shelf would not supply.

The necessity is not academic, and I want to be plain about it before the reader reaches a single chapter. Behind every technical failure catalogued in this book is a document that somebody actually wrote. Somebody drafted the exposure assessment that was never revisited when the process changed. Somebody signed the audit finding that was never closed out. Somebody entered the case on the 300 log, or decided not to. Somebody stood beneath the load. Years later a jury will be read those documents aloud, and the professional who created them — in good faith, on an ordinary afternoon, with nothing yet at stake — will discover that the record he made without thinking about it has become the whole of the case. That gap between how these records are created and how they are eventually read is the reason this book exists. It is not a gap the literature was closing.

I care about that gap more than a technical subject usually warrants, and the reason is not complicated. This is a field in which the consequence of careless work is not an embarrassment or a lost motion. It is a hand, a lung, a family that receives an explanation instead of an answer. The methods in this book are demanding because the stakes are, and the discipline it argues for — the primary record over the recollection, the published method over the confident intuition, the assumption stated rather than buried — is not a professional courtesy. It is the only thing that reliably protects a worker before the injury and the truth after it.

A word about how the book was made, because it bears on what the reader should expect of it. It was self-published, which means the functions a publishing house distributes across a department were performed by one person, in the evenings and between engagements. Structural editing, technical review, citation checking, copyediting, proofreading, the topic finder, the typesetting, the cover, and every decision about what belonged in the book and what did not: all of it was mine. There was no acquiring editor to argue with about scope, which is a freedom, and no citation department to catch me, which is not. Twenty-nine chapters across five parts, seven appendices, roughly four hundred and fifty sources in thirty-two chapter reference lists, two citation systems reconciled into one alphabetical sequence, and several hundred internal cross-references that must each survive every revision made to the chapter it points at. A single renumbered section can quietly falsify a dozen pointers elsewhere in the manuscript. I do not know a way to do that work at scale except slowly, by hand, more than once.

The law did not hold still while it was being done. The amended Rule 702 took effect while the chapters were in draft and required every opinion-side argument in Part I to be rewritten against it rather than merely annotated with it. The silica docket that Part III leans on moved every quarter. Chapter 28 describes an enforcement landscape that was changing on the day it was finished and is changing as you read this, which is why its corrections and updates are posted at www.lfmartinllc.com rather than pretended into permanence between covers.

The manuscript has been read through several times, front to back, for exactly the failures that matter in a book of this kind. Every citation was verified against the source document rather than against another source's description of it — the same discipline Chapter 29 demands of any expert who lets a machine near a report. Cross-references were re-walked after the final structural pass. Regulatory limits, standard numbers, directive designations, and reported holdings were checked, and the ones that could not be confirmed were removed rather than softened.

And a book of this length, produced this way, will still contain errors. That is not modesty; it is arithmetic. If you find one, it is mine, and I apologize for it in advance. I would also ask that you tell me. Corrections are posted at www.lfmartinllc.com and folded into subsequent printings, and I would rather be corrected by a careful reader than quoted by a careless one. In the meantime, the distinction worth holding on to is the one this book makes about every other technical document: a typographical error costs you a moment, but a transposed exposure limit or a holding stated one degree too broadly could cost you a motion. Verify the governing authority in your jurisdiction before you rely on it. That instruction appears on the copyright page as a legal necessity. I mean it as professional advice.

What I hope the difficulty of assembling this book does not obscure is how ordinary the argument at the center of it is. Nothing here is clever. There is no technique in these pages that rescues a bad record, and I have tried hard not to imply otherwise, because that implication is the one thing a book like this could do that would make matters worse. The cases described here are won and lost long before anyone is retained, in the quality of work performed by people who had no idea a case was coming.

I would rather this book were unnecessary. It is not. Records are being created today — in good faith, in a hurry, by professionals doing their best with what they were given — that will be read aloud years from now to determine what an employer knew and when. Some of those records will hold. The ones that do will not hold because someone was clever after the fact. They will hold because someone did the work properly on an ordinary Tuesday, when nothing was at stake yet and no one was watching.

That is the argument of this book. It is why I wrote it, and why I would write it again.

Logan F. Martin, PhD, CIH, CSP, CHMM
August 2026

Acronyms

Terms are defined at greater length in Appendix B. This key covers the abbreviations used most often in the text.

ACGIH	American Conference of Governmental Industrial Hygienists
AIHA	American Industrial Hygiene Association
ANSI	American National Standards Institute
APF	Assigned Protection Factor
ASSP	American Society of Safety Professionals
ASTM	ASTM International (formerly American Society for Testing and Materials)
BNA OSHC	Bureau of National Affairs, Occupational Safety and Health Cases — the reporter in which Review Commission decisions are cited (e.g., 16 BNA OSHC 2131)
C.F.R.	Code of Federal Regulations
CHMM	Certified Hazardous Materials Manager
CIH	Certified Industrial Hygienist
CPE	Certified Professional Ergonomist
CPL	OSHA compliance directive (CPL series)
CSHO	Compliance Safety and Health Officer
CSP	Certified Safety Professional
dBA	A-weighted decibel
EHS	Environment, Health, and Safety
FELA	Federal Employers Liability Act
FOIA	Freedom of Information Act
FOM	OSHA Field Operations Manual
FRE	Federal Rules of Evidence

GSD	Geometric Standard Deviation
IEC	International Electrotechnical Commission
IH	Industrial Hygiene; Industrial Hygienist
ISEA	International Safety Equipment Association
ISO	International Organization for Standardization
ITA	Injury Tracking Application
JHA	Job Hazard Analysis
JSA	Job Safety Analysis
LEV	Local Exhaust Ventilation
LHWCA	Longshore and Harbor Workers' Compensation Act
LMS	Learning Management System
MSD	Musculoskeletal Disorder
MSHA	Mine Safety and Health Administration
NEP	National Emphasis Program
NFPA	National Fire Protection Association
NIOSH	National Institute for Occupational Safety and Health
OEL	Occupational Exposure Limit
OIICS	Occupational Injury and Illness Classification System
OSH Act	Occupational Safety and Health Act of 1970
OSHA	Occupational Safety and Health Administration
OSHRC	Occupational Safety and Health Review Commission
PEL	Permissible Exposure Limit
PFAS	Polyfluoroalkyl Substances
PPE	Personal Protective Equipment
REL	Recommended Exposure Limit
SAE	Sampling and Analytical Error
SEG	Similar Exposure Group
SHMS	Safety and Health Management System
STEL	Short-Term Exposure Limit

TED OSHA Training and Education Directive (TED series), which includes the OSHA Technical Manual, TED 01-00-015

TLV Threshold Limit Value

TWA Time-Weighted Average

UTL Upper Tolerance Limit

A Note on Citations

Legal authorities are cited in Bluebook form; scientific, technical, and regulatory-guidance sources are cited in APA form. Both appear in a single alphabetical list at the end of each chapter. Regulatory provisions are cited to the Code of Federal Regulations, and OSHA directives by their CPL, TED, or other instruction number. Decisions of the Occupational Safety and Health Review Commission are cited to the Bureau of National Affairs' Occupational Safety and Health Cases — abbreviated BNA OSHC, in the form volume, reporter, page (16 BNA OSHC 2131) — because the Commission has no official reporter and that is the citation counsel and the Commission itself use. Readers who have not met the abbreviation before will find it, and the rest of the working shorthand, in the Acronyms key.

Because the two systems disagree about typography, this book resolves the conflict in one direction and states the resolution here. In running text, names are italicized, in full-name and short-form references alike, and in headings, tables, and the index as well as in the body — *Daubert*, *Kumho Tire*, *Sommerville v. Union Carbide Corp.* In the reference list at the end of each chapter, entries are set in roman throughout, following the flat, unitalicized convention this book uses for its scientific and regulatory sources, so that a reader scanning a reference list is not asked to parse two typographic systems in one alphabetical run.

Rules of court are cited in the abbreviated Bluebook form on first and every subsequent reference — Fed. R. Civ. P. 26(a)(2)(B), Fed. R. Evid. 702 — rather than spelled out. Where a rule is discussed at length and no ambiguity can arise, the short form ("Rule 702," "Rule 26") is used in the surrounding prose; the abbreviated full citation appears at the first mention in each chapter and in every reference list. Subdivision letters are given whenever the argument turns on a particular subdivision, because in this area of practice it usually does.

Case citations, verdicts, regulatory limits, agency procedures, and decisions of the Occupational Safety and Health Review Commission are stated as of the date given on the copyright page and remain subject to appeal, remand, settlement, amendment, or revision.

How to Use This Book

The chapters that follow were written to be read in order, and the reader who does that will find an argument rather than a reference work. But five different readers open a book like this one, usually with a deadline, and each of them wants a different path through it. This page is that map.

1. Read it as an argument. Start at the Introduction and read forward. The parts are sequenced the way these matters actually unfold — the expert and the court, the OSHA case, the science of exposure, the injury case file, the professional's own practice — and the same four principles compound across them: the primacy of the primary record, the published method over the confident intuition, the assumption stated rather than buried, and the discovery request framed by someone who knows what the records are called.

2. Consult it by chapter for a live matter. The chapter titles and the epigraphs are written to let you find the chapter you need without reading the ones you don't. Three entry points cover most emergencies. An incident that happened this week: Chapter 22, then Chapter 21. A citation that arrived this week: Chapter 7, then Chapter 8, then Chapter 9. An expert report that arrived this week: Chapter 17 if it is an exposure opinion, Chapter 2 and Chapter 4 in every case.

3. Use it as a self-audit. The safety professional who wants to know how her own files will read when they are exhibits should take Chapter 25 first, then Chapter 24, then Chapter 23, and then Appendix G. Those four, run against an actual program, will produce a longer list of findings than most internal audits do.

4. Use it as a discovery checklist. Counsel building demands will find inventories of what each system generates, and what its absences mean, in Chapter 6 §III, Chapter 10, Chapter 18 §VI, Chapter 19 §III, and Chapter 22 §IV. Appendix E is the playbook for the OSHA inspection file specifically, and it supports a practice this book recommends roughly a dozen times: get the file, not the citation summary.

5. Use it as a reference. Appendix B is the vocabulary. Appendix C is the case table, with one-line holdings and chapter cross-references. Appendix D is the standards and directives matrix. Appendix A is the ten questions to ask before retaining an expert, and it is meant to be photocopied. Appendix F is a model Rule 26 report and Appendix G a model exposure assessment report, both annotated; they are the artifacts the rest of the book describes. Appendix G is collected with the other appendices, though it belongs with the exposure chapters, rather than in the back matter.

A note on jurisdiction. The examples in this book lean on the jurisdictions that generate the case law — New York for statutory strict liability in construction, California for retained control, exclusivity, and state-plan divergence — because that is where the decided cases are. They are illustrations of a category, not a national rule. Wherever a chapter turns on a state-law question, the localizing questions are the same four: Is this a *Daubert* jurisdiction or a *Frye* jurisdiction, and what does its expert-disclosure regime require (Chapter 2 §IV)? Is there a state plan, and does it supply a prescriptive standard where federal OSHA has only the General Duty Clause (Chapter 12)? What are this state's exceptions to workers'-compensation exclusivity, and may fault be apportioned to the immune employer (Chapter 19 §§II, VI)? And does this state recognize negligence per se from an OSHA violation, evidence of the standard of care, or neither (Chapters 11, 12, 23)? Answer those four at intake and the rest of the book localizes itself.

Introduction: The Record Decides

There is a moment, common to nearly every matter described in this book, when the case stops being about what happened and becomes about what can be proven to have happened. A worker's hand is gone, or his lungs are scarred, or a trench has closed over him, and everyone in the room — the plaintiff's lawyer, the defense lawyer, the claims professional, the safety director who now cannot sleep — holds a version of events. The versions do not agree, and none of them will decide anything. What decides the case is the record: the sampling data and the calibration logs behind it, the energy-control procedure that was or was not written for that machine, the soil that was or was not classified, the training file, the inspection log, the OSHA 300 entry that was made or conspicuously not made, the citation and the technical chain beneath it. The record was created long before anyone chose sides, and it will say what it says for whichever party reads it with discipline.

This book is about reading it with discipline.

Its thesis can be stated in three sentences. First, workplace injury, occupational disease, and OSHA enforcement cases are decided on technical questions — what hazard existed, what standard governed it, who controlled it, what exposure the worker actually sustained, and whether the measurement or the reconstruction behind the claim can survive scrutiny — before they are decided on legal ones. Second, those technical questions are answered from records that a qualified safety professional or industrial hygienist can audit, reconstruct, and test, using published methods the courts can independently check. Third, this discipline serves both sides of the "v." equally, because the record does not pick a side, and an expert worth retaining does not either.

I. Who This Book Is For

I have written this book for two audiences who rarely read the same literature, and that separation is part of the problem.

The first audience is legal: the plaintiff's attorney building a third-party case out of a construction site fatality, the defense attorney answering an overexposure citation or a $400 million dropped-object complaint, the in-house counsel deciding whether to contest a willful classification, and the paralegals who will actually assemble the discovery requests, chase the sampling records, and index the training files that these chapters describe. For this audience, the book is a technical field guide to territory the law schools do not cover: how an air sample becomes a number and where that number breaks; how an eight-hour time-weighted average is constructed and how it is taken apart; what the Field Operations Manual requires of a compliance officer and what happens when the case file shows he did not do it; what a soil classification is, why it matters more than the photographs, and why the "competent person" on the organizational chart may not have been one; what the OSHA 300 log establishes about knowledge and what its silences reveal.

The second audience is the profession the first audience will retain, depose, and cross-examine: the certified industrial hygienists, certified safety professionals, safety engineers, and EHS managers whose measurements, audits, programs, and reports become the record these cases are built on. Most safety professionals will never sit in the expert's chair. Nearly all of them will someday find their work sitting in the evidence binder: the exposure assessment they wrote three years ago, the audit finding that was or was not closed out, the email in which they recommended the guard that was never purchased. For this audience, the book is a mirror held up at an unfamiliar angle: this is how your ordinary professional work reads when litigation counsel reads it, and this is what it means to do that work in a manner that will hold.

The two audiences need each other more than either usually admits. Counsel without a technical guide will litigate the citation's legal elements while the measurement underneath them, the element that actually decides an exposure case, goes unexamined. The safety professional without an understanding of the litigation that follows an incident will write reports, keep logs, and close out findings in ways that a competent discovery request will later turn against the very employer or worker the professional meant to protect. Each chapter in this book is written so that both readers can use it: the attorney will find the discovery targets, the governing standards, and the decided cases; the safety professional will find the technical substance and, between the lines, a standing answer to the question every practitioner should ask of every document she creates — *how will this read when it is Exhibit 14?* Chapter 25 makes that question, and its answers, the explicit subject.

II. What an Expert in This Field Actually Does

Because the expert witness is the connective figure between these two audiences, it is worth being precise at the outset about what the role is and is not.

The safety or industrial hygiene expert serves in one of two capacities, and the distinction has procedural teeth. A consulting expert is retained to educate counsel: to audit the government's case file or the opposing expert's report, to identify the technical vulnerabilities and the discovery that will develop them, to tell counsel — candidly and early, while the information can still shape strategy — whether the exposure reconstruction holds, whether the citation's measurement chain is intact, whether the client's own records will survive the reading opposing counsel intends to give them. In most jurisdictions the consulting expert's work is shielded from discovery, which is precisely what makes the role valuable: it is the one setting in which counsel can afford an unvarnished technical answer. A testifying expert performs the same analysis under opposite conditions — every note discoverable, every assumption subject to cross-examination, every opinion required to satisfy Rule 702 or its state analogue — and adds the further obligation of explaining the analysis to a trier of fact who has never worn a sampling pump or walked a trench.

The same professional may serve either role, but not casually both in the same matter, and counsel should decide which is needed before the retention letter is drafted rather than after the disclosure deadline has made the choice irrevocable. When to retain is equally consequential and the answer is nearly always *earlier than you think*: the expert retained after the close of discovery can only work with the record counsel happened to assemble, while the expert retained at intake shapes what the record contains — which is the difference, in the cases this book examines, between an opinion and an admissible one.

One further point about the role, because it frames everything that follows. The expert's obligations run to the analysis, not to the retention letter. The industrial hygienist who re-runs the Secretary's time-weighted average computation owes counsel the result of the arithmetic whether it helps or hurts; the safety professional reconstructing a dropped-object incident owes counsel the physics whether the trajectory implicates the plaintiff's employer or the defendant's crane. This is not merely an ethical posture: it is the load-bearing wall of admissibility. An opinion assembled to fit a litigation position announces itself: in the assumptions that all lean one direction, in the alternative causes never addressed, in the methodology adopted for the case rather than brought to it. Courts have become steadily better at recognizing the pattern, and since December 2023 they have been under instruction to look for it.

III. Why Now

Three developments make this book timely rather than merely useful.

The first is the amendment to Fed. R. Evid. 702 that took effect on December 1, 2023. The amendment did not change the standard for expert testimony so much as it repaired the enforcement of one: it clarified that the proponent must demonstrate, by a preponderance of the evidence, that the expert's opinion reflects a reliable application of a reliable methodology to sufficient facts, and that this demonstration is for the court, not the jury, and cannot be waved through as a matter going "to weight rather than admissibility." For safety and industrial hygiene opinions, which live or die on methodology, the practical effect has been pronounced. The expert who eyeballed an exposure, adopted the client's narrative as the factual record, or reasoned backward from the injury to the breach could once count on reaching the jury and absorbing the criticism on cross. That expert is now being excluded, and the cases discussed throughout this book — opinions struck, verdicts vacated because the opinion should have been struck — are the evidence. Every chapter here is written against the amended Rule, because every opinion in this field will be tested against it.

The second development is the transformation of exposure litigation, with respirable crystalline silica as its leading edge. The 2016 silica standards armed OSHA with a modern enforcement framework at the same time that engineered-stone fabrication was manufacturing a population of young, catastrophically injured workers, and the result — a decade on — is the most active occupational exposure docket in the country, running simultaneously in the enforcement forum and the civil courts. Its real significance is as a template: the same convergence of a measurable agent, a prescriptive standard, a discoverable paper trail, and a retrospective exposure question defines the litigation now forming around other agents, and the technical methods that decide the silica cases will decide those as well.

The third development is quieter but reaches more readers: the steady migration of safety and industrial hygiene documentation into discoverable electronic systems. The exposure database, the learning-management system's training records, the electronically maintained injury log, the audit platform with its time-stamped findings and closure records: these have made the safety professional's work product more durable, more searchable, and more mercilessly complete than the paper files of a generation ago. The record that decides these cases has never been richer. Neither has the record of what an employer knew and when.

IV. How the Book Is Organized

The twenty-nine chapters that follow are arranged in five parts, in an order that tracks how these matters actually unfold.

Part I, The Expert and the Court, takes up the figure on whom everything else depends and follows the opinion from retention to verdict: how to select a safety or industrial hygiene expert and run the engagement (Chapter 1); how to build —

and how to attack — an opinion under the amended Rule 702 (Chapter 2); how to write the Rule 26 report so the opinion survives what comes next (Chapter 3); how to depose and cross-examine the opposing expert, and how to armor your own against the identical campaign (Chapter 4); how the opinion is delivered at the *Daubert* hearing and to the jury (Chapter 5); and what disciplined, side-neutral expert work looks like in practice, illustrated through the dropped-object cases where the same record must serve whichever party reads it honestly (Chapter 6).

Part II, The Anatomy of an OSHA Case, follows federal enforcement from the first knock at the gate through the contest before the Review Commission: the inspection as the Field Operations Manual scripts it and how to prepare a workforce for it; the affirmative defenses, the contest, and the settlement calculus that gets citations struck, downgraded, or grouped; the three recurring technical fractures in the Secretary's prima facie case; the technical audit of OSHA's own sampling methodology; the multi-employer worksite and the question of who really controlled the hazard; the state-plan overlay that can change the governing standard, the reporting deadline, and the settlement value; and the whistleblower and retaliation claims that increasingly travel alongside the citation.

Part III, The Science of Exposure, is the industrial hygiene core of the book: how exposure becomes injury in the law's two-part causation framework, and where the hygienist's opinion must yield to the physician's; what a handful of air samples can and cannot support; how an exposure is reconstructed years or decades after the fact in a manner that survives challenge; how to audit an opposing expert's exposure opinion — the sampling design, the borrowed data, the misapplied method, and the buried uncertainty — and route the findings into the motion, the deposition, and the rebuttal report; and where the workplace boundary itself is drawn, since the alternative cause that decides so many of these cases must be reconstructed to the same standard as the occupational one rather than merely named.

Part IV, The Injury Case File, turns to the incident types that fill the third-party docket — the workers'-compensation escape routes that make an employer's jobsite litigable even when the employer is immune; the exposure that leaves the workplace altogether and is sued upon by a household member or a neighbor whom the exclusive remedy never touched; the preservation obligations that attach the day of the incident, before anyone knows which side of the caption they will occupy; the investigation that must be conducted, correctly and completely, while the scene still exists — and closes with the single most under-used document in workplace injury litigation, the OSHA 300 log.

Part V, The Professional's Practice, ends the book where the well-advised employer begins and where every safety professional eventually arrives: the mock regulatory audit, which is the same examination the government will eventually conduct, performed on the employer's schedule and closed out in corrective action rather than citations; the drafting disciplines — wording, completeness, compliance, and the resisted over-promise — that decide how the program's own documents will read when they become exhibits; the day the professional's own file — and the professional — are called to testify; how a professional deliberately builds an expert practice — the engagements, the ethics, and the career record that make neutrality believable; the enforcement frontier that will shape the next five years of this docket; and the arrival of machine methods on the four surfaces where they touch this work — the model, the report, the demonstrative, and the employer's own monitoring systems.

Each chapter is written to be usable on its own, and the reader with an incident to investigate tomorrow morning should go directly to Chapter 22 without apology; the page preceding this Introduction maps four other entry points. But the parts are sequenced as an argument, and where a later chapter depends on an earlier one it says so and does not repeat it. Chapter 19, for example, assumes the method Chapter 6 sets out and takes up only what the third-party posture adds; Chapter 9 assumes the procedural framework Chapter 8 supplies and confines itself to the technical fractures. The reader who takes the parts in order will find the same principles compounding: the primacy of the primary record, the published method over the confident intuition, the assumption stated and tested rather than buried, and the discovery request framed by someone who knows what the records are called and where they live.

V. A Note on Sides

It will be apparent by now that this book declines to be a plaintiff's book or a defense book, and the refusal is deliberate. It is written for both sides of the "v." because the technical analysis it describes is the same in either chair. The hierarchy of controls does not rearrange itself for the defense. The calibration record is missing or it is not. The exposure reconstruction is built on measured data and stated assumptions, or it is built on advocacy, and the difference is visible to anyone trained to look, which increasingly includes the court. A book that taught one side to find the truth and the other to bury it would not be a book about expert work; it would be a partisan manual, and it would fail the reader on whichever side the next case placed them. The methods here are offered to plaintiff and defense counsel alike, and to the safety and industrial hygiene professionals both will retain, precisely because a method worth trusting is one that reaches the same answer no matter who asks the question.

Counsel sometimes hears this as naïveté about the adversarial process. It is the opposite. The adversarial process is precisely why the disciplined, record-bound opinion wins: it is the only kind that survives the process. The plaintiff's expert who overreaches hands the defense a Rule 702 motion; the defense expert who explains away the record hands the plaintiff a cross-examination; and in either event the client pays for the indulgence. The chapters that follow contain, I hope, a good deal of useful technique for both sides. But the deepest strategy they teach is not a technique at all. It is the habit of finding out what the record actually shows, early enough to act on it, from someone with no stake in the answer.

That is what an expert is for. The rest of this book is the particulars.

A Note on Scope

This book addresses land-based employment governed by the Occupational Safety and Health Act and by state workers' compensation law. Maritime employment under the Longshore and Harbor Workers' Compensation Act and the Jones Act, and railroad employment under the Federal Employers Liability Act, operate under different remedial schemes, different exclusivity rules, and in the case of FELA and the Jones Act a different standard of causation. The exclusive-remedy analysis in Chapter 19 and the third-party framework that follows from it do not transfer to those settings. A reader with a shipyard, dock, offshore, or rail matter should treat the technical chapters as portable and the liability chapters as not.

Three subject areas are deliberately outside the scope, and saying so is more useful than leaving the reader to infer it from silence. The first is building-related illness — indoor air quality, mold, and Legionella. The reconstruction and sampling-audit methods of Part III transfer to those matters essentially intact, and the litigation they generate (habitability, premises liability, outbreak claims) is substantial. But the exposure standards, the diagnostic literature, the sampling conventions, and the professional guidance that govern building investigations are a distinct body of practice with a distinct evidentiary posture, and treating them in passing would do the reader less good than leaving them out. A hygienist with an IAQ matter should read Chapters 15 through 18 for method and go elsewhere for doctrine.

The second is the psychological and psychosocial claim: workplace stress, post-traumatic stress following a witnessed fatality, and the mental-health dimension of the injury case. These are real, they are litigated, and in a serious incident they frequently travel alongside the physical claims this book does address. They are excluded because the causation analysis, the expert disciplines, and the compensability rules involved are not the ones developed here, and because a safety or industrial hygiene expert who ventures into them is usually testifying outside her qualifications — which is the failure Chapter 1 §II and Chapter 4 §II treat as the most avoidable one in this field.

The third is occupational medicine proper. Chapter 14 sets the boundary between the hygienist's exposure opinion and the physician's diagnosis and holds to it throughout, and nothing in this book equips a non-physician to cross that line. Where a chapter discusses medical evidence, it discusses what the exposure expert must supply to the physician and what the physician's opinion must rest on, not how to form the medical opinion.

PART I

THE EXPERT AND THE COURT

CHAPTER 1

The Expert Is the Case

Selecting and Working with Safety and Industrial Hygiene Experts in Workplace Injury and Occupational Disease Litigation

> *In these cases, the expert does not support the theory of the case, the expert supplies it. For plaintiff and defense counsel alike, the qualities screened for at retention and the practices followed during the engagement determine what the expert can credibly say when it finally matters.*

I. In These Cases, the Expert Carries the Theory

Workplace injury and occupational disease litigation is expert-driven to a degree that most other tort practice is not. Whether the matter is a fall from an unguarded elevation, a machine-guarding amputation, a trench collapse, a confined-space fatality, an arc-flash burn, or a disease claim — silicosis, occupational asthma, noise-induced hearing loss, solvent encephalopathy, an asbestos-related malignancy — the propositions the trier of fact must ultimately accept are technical ones: what hazard existed, what standard of care governed it, what the employer, contractor, or premises owner knew or should have known, what exposure the worker actually sustained, and whether that exposure caused this injury or this disease. No lay witness can carry any of those propositions. The safety professional and the industrial hygienist can, and in a well-tried case they do. Two of those disease claims run as worked examples through Chapters 15 to 18, engineered-stone silicosis and noise-induced hearing loss, chosen because their measurement physics could hardly be less alike and the method that reconstructs them is the same.

The need is symmetric. Plaintiff's counsel needs the standard-of-care breach established against the published benchmarks — the OSHA standards, the ANSI/ASSP consensus series, NFPA codes, the ACGIH Threshold Limit Values — and needs the exposure and causation chain built link by link. Defense counsel needs the same terrain surveyed with equal rigor: the exposure characterized honestly rather than assumed, the plausible alternative causes developed on the record, the workers' compensation and third-party liability boundaries kept straight, and the overreach in the opposing opinion identified while there is still time to move against it. The tools are identical; only the direction of travel differs.

Yet in my experience the expert phase of these cases is largely decided at two moments that counsel controls completely: the day the expert is selected, and the way the engagement is run from that day forward. Everything downstream — the strength of the report, the survivability of the opinion under amended Rule 702, the impression the witness leaves on a jury — is constrained by those two decisions. What follows is a field guide to both, written for counsel on either side of the "v."

II. The Qualities That Separate an Asset from a Liability

Lawyers evaluating a prospective safety or industrial hygiene expert tend to weigh the curriculum vitae. The CV matters, but it is the least discriminating instrument available. The qualities that predict whether an expert will decide your case in your favor, or become the weakest document in your file, sort into four categories.

A. Credentials That Certify Competence, Not Vocabulary

In this field the board certifications that mean something are few and verifiable: the Certified Industrial Hygienist (CIH) and Certified Safety Professional (CSP), awarded and recertified through the Board for Global EHS Credentialing and the Board of Certified Safety Professionals; the Professional Engineer license where the opinions are engineering opinions; the Certified Professional Ergonomist for ergonomic injury claims; board certification in toxicology or occupational medicine where the causation opinion crosses into dose-response and diagnosis. Each requires qualifying education, documented practice experience, a rigorous examination, adherence to an enforceable code of ethics, and continuing recertification. They stand in contrast to the growing catalogue of purchased designations that certify nothing but a registration fee. Counsel should verify certification status directly with the issuing board — active status, recertification currency, and the absence of ethics actions — rather than accepting the letters on the CV. But credentials are a threshold, not a conclusion: they will carry the witness past a qualifications challenge under Rule 702(a) and tell the court nothing about whether the opinion itself is reliable. Every expert who has ever been excluded was credentialed.

B. Experience That Maps to the Facts of This Case

The second quality is fit. A distinguished general safety expert who has never stood on a drill floor, walked a foundry, or rigged a scaffold is a liability in a case that turns on how that work is actually done. The regulatory overlay alone differs by sector — general industry, construction, maritime, mining under MSHA — and an expert who reaches for the wrong volume of the Code of Federal Regulations has handed opposing counsel the first ten minutes of cross-examination. Ask where the prospective expert has actually practiced: which processes she has personally assessed, which exposures she has personally measured, which incident investigations she has personally led. And ask for the ratio of consulting practice to testimonial work. An expert whose professional life consists entirely of litigation gives the jury no independent professional identity to trust and gives opposing counsel the "career witness" theme without effort. The witness you want spends most of her year solving these problems for operating companies and workers, and some of her year explaining them to courts.

C. Methodological Discipline

The third quality is the one the 2023 amendment to Fed. R. Evid. 702 has made decisive: the habits that produce an opinion whose every inferential step a court can independently check. The disciplined expert works from the primary record — sampling data, maintenance histories, training documentation, medical surveillance files — rather than from a narrative received with the retention letter; measures where measurement is possible instead of assuming; ties each analytical step to a published authority, whether an OSHA standard, an ANSI/ASSP or NFPA consensus document, the ACGIH TLVs, a NIOSH criteria document, or the AIHA exposure assessment strategy that governs professional practice in industrial hygiene; states assumptions and tests the conclusion's sensitivity to them; and addresses the obvious alternative explanations before opposing counsel does. These habits are observable in an expert's prior reports and transcripts, and they are worth more than any line on the CV, because after *Daubert*, *Kumho Tire*, and the amended Rule, methodology, not credentials, is where safety and industrial hygiene opinions live or die.

D. Independence and Testimonial Temperament

The final quality is the hardest to read from paper and the most valuable in the courtroom. The expert worth retaining is the one willing to tell you, early and plainly, that the opinion you hoped for is not supportable: a conversation that costs a retention fee and saves a case. Look for a retention history reasonably balanced between plaintiff and defense work; a witness who has only ever testified for one side arrives with a discount already applied to her credibility. Review prior Rule 702 challenges: an exclusion is not disqualifying, but an expert who is not candid about one is. Read prior transcripts and publications for positions that contradict the opinion you need; opposing counsel will. And assess the temperament directly: the ability to teach a technical subject to a jury without condescension, and to remain measured under a cross-examination designed to produce anything but. The simplest test I know is this: the right expert is the one whose report would read the same no matter which party had signed the engagement letter.

III. Best Practices for Running the Engagement

Selection is half the decision. The other half is how counsel runs the engagement, and here the errors are as predictable as they are avoidable. Five practices, followed from the first call, protect the opinion and the case.

Retain before the evidence perishes. The factual basis of a safety or exposure opinion has a shelf life. Incident scenes are remediated, damaged equipment is repaired or scrapped, air that was never sampled can never be sampled, records are destroyed on ordinary retention schedules, and coworker memories converge on the version told at the toolbox talk. The OSHA investigation file, if there is one, arrives by FOIA request on the agency's timetable, not yours. An expert retained in the first weeks can identify what must be preserved, inspect before spoliation letters become spoliation motions, and — just as valuable — shape discovery: the expert knows which documents exist to be requested, from job hazard analyses and industrial hygiene sampling records to audiometric testing, SDS libraries, and maintenance logs, and knows what their absence signifies.

Define the role in writing. Decide at the outset whether the expert is consulting or testifying, and paper the engagement accordingly. Rule 26(b)(4) shields draft reports and most attorney-expert communications from discovery, but the shield has gaps counsel must manage: the facts and data the expert considered, and any assumptions counsel supplied, remain discoverable. Run the file on the assumption that those categories will be produced, because they will be, and keep the consulting and testifying functions separated where the case strategy requires both.

Give the expert the whole record. The most reliable way to manufacture a disastrous deposition is to curate the file: to withhold the unhelpful sampling results, the prior similar incident, the audit finding that cuts the other way. The opinion that survives is the one formed with the bad documents in view, because the expert has either accounted for them in the analysis or calibrated the conclusion to accommodate them. An expert ambushed at deposition with a document she has never seen is not being cross-examined; she is being dismantled, and the record of it will be read to the jury.

Put the expert in the field. Wherever the case allows it, the expert's facts should be the expert's own: a site inspection, examination of the actual or exemplar equipment, measurements taken with calibrated instruments, photographs composed by the person who will authenticate them. In exposure cases this may mean reconstruction or simulation: recreating the task with exemplar materials and sampling the result. Facts the expert gathered personally are the difference between an opinion anchored in the case and an opinion anchored in a stack of paper someone else assembled, and courts applying the amended Rule 702 have shown they can tell the difference.

Budget for the method, not the conclusion. The inexpensive opinion — the records review that skips the inspection, the exposure opinion that skips the dose reconstruction — is the most expensive document in the file the day it is struck or shredded on cross. Scope the engagement around the work the methodology actually requires, and treat an expert's candid estimate of that work as a mark of the discipline you are paying for, not an inflated bid.

IV. Five Cases That Prove the Point

The principles in Sections II and III are not aspirational. They are the documented difference between judgments collected and judgments erased, and the reported decisions in workplace injury and occupational exposure litigation supply the proof. Five cases — four collapses and one survival at the gate — show each principle deciding a real outcome.

Sardis v. Overhead Door Corp. (4th Cir. 2021), the untested opinion. A worker died after the wooden handhold of a shipping container broke, dropping him nine feet onto pavement. The estate's liability case rested entirely on two experts, a packaging-design engineer and a human-factors specialist, and the district court admitted both over objection. The jury returned a $4.84 million verdict. The Fourth Circuit reversed, holding that the trial court had abdicated its gatekeeping role: the design expert had opined that the container's defects caused the death without performing any testing and without citing published literature, and neither opinion was tied to an established industry standard. Because the expert testimony was the only probative evidence of liability, the court did not remand for a new trial, it directed judgment for the defendant. Years of litigation and a verdict in hand produced nothing, because credentialed experts had offered opinions with no method a court could verify.

Tamraz v. Lincoln Electric Co. (6th Cir. 2010), the overstated conclusion. A welder claimed that manganese in welding fumes caused his Parkinson's disease, and a jury awarded $20.5 million. The Sixth Circuit vacated the verdict because the causation opinion of the treating neurologist was a hypothesis dressed as a conclusion: the court traced his reasoning chain link by link and found that he had characterized the supporting literature as theoretical, conceded that the epidemiology had failed to find the correlation, and acknowledged that "speculation" supplied the genetic predisposition his theory required. The court held that reciting the phrase "reasonable degree of medical certainty" does not make a causation opinion admissible: the ipse dixit of the expert is not evidence. The opinion may even have been right; it was simply never built to the standard the rules demand, and the calibration failure cost the verdict.

Pluck v. BP Oil Pipeline Co. (6th Cir. 2011), no dose, no case. In a benzene claim involving non-Hodgkin lymphoma, the plaintiff's causation expert asserted "heavy" exposure without quantifiable exposure data, disregarded recorded levels that were within regulatory limits, and never accounted for the plaintiff's extensive smoking history, a competing benzene source. The Sixth Circuit affirmed exclusion, holding that without reliable exposure data the causation claim was speculation and that the failure to rule out the alternative source independently rendered the opinion unreliable. The case is the template for both sides of the disease docket: for the plaintiff, dose must be reconstructed rather than assumed; for the defense, the unquantified exposure and the unexamined alternative cause are each, standing alone, grounds for exclusion.

Amorgianos v. National Railroad Passenger Corp. (2d Cir. 2002), the method abandoned in application. A bridge painter claimed neurological injury from xylene-based paint. The plaintiff's exposure expert had a legitimate methodology available and departed from it, running the exposure analysis with inputs that did not match the actual work conditions. The Second Circuit affirmed exclusion in language that has governed industrial hygiene testimony ever since: any step that renders the analysis unreliable renders the expert's testimony inadmissible. Summary judgment followed. It is the reliable-application failure in its classic form, a sound method and a real set of facts that never actually met, decided in an occupational exposure case two decades before the 2023 amendment wrote the principle into the rule text.

Milward v. Acuity Specialty Products Group, Inc. (1st Cir. 2011), the survival at the gate. The counterexample matters most. A refrigeration technician's claim that benzene in the products he worked with had caused his acute promyelocytic leukemia — a rare subtype of acute myeloid leukemia — survived because the general-causation expert did the work. No epidemiology linked benzene to that subtype specifically, which is precisely what made the opinion contestable; what carried it was a transparent weight-of-the-evidence analysis that integrated the epidemiology available for the broader disease class, the toxicology, and the experimental evidence of a chromosomal mechanism, with each inference documented and defended. The First Circuit reversed the trial court's exclusion, holding that a rigorously executed and explained methodology was for the jury to weigh. Note the posture, because it is easy to mislabel: the defendants were the manufacturers and suppliers of the benzene-containing products, not the worker's employer, so this is a products case reached around the compensation bar in the manner Chapter 19 describes. The sequel completes the lesson: years later the case ended anyway, when the specific-causation opinion was excluded and that exclusion was affirmed (*Milward v. Rust-Oleum Corp.*, 820 F.3d 469 (1st Cir. 2016)) — a reminder that the expert team is only as strong as its weakest opinion, and that the handoff between the exposure analysis and the medical causation opinion must be built with the same discipline as either opinion alone.

Read together, the five cases share one feature: not one of them turned on credentials. They turned on testing versus no testing, calibration versus overstatement, reconstructed dose versus assumed dose, method applied versus method invoked, and documented rigor rewarded. Every expert in every one of these cases was qualified. The outcomes were decided by the qualities described in Section II and the practices described in Section III, which is precisely the point.

V. Occupational Disease: Where the Expert Matters Most

Everything above applies with double force to occupational disease claims, which add problems that acute-injury cases never face. The first is latency: the exposures at issue may have ended twenty or forty years before diagnosis, at facilities that no longer exist, under employers that have dissolved, with contemporaneous sampling data that was never collected or never kept. Here the industrial hygienist's central contribution is retrospective exposure reconstruction: assembling the dose from work histories, job classifications, the surviving sampling record, the published exposure literature for the industry and era, and, where feasible, exemplar measurements of the recreated task. Done with discipline, the reconstruction converts "he worked around silica" into a defensible quantitative or semi-quantitative exposure estimate; done carelessly, it is the analytical gap on which the entire causation case is excluded, as the plaintiffs in *Pluck* and *Amorgianos* learned at the cost of their cases.

The second problem is the causation structure itself. Disease claims divide causation into the general — is this agent capable of producing this disease, a question answered through the epidemiology and the considerations Bradford Hill articulated — and the specific: did this claimant's exposure produce this claimant's disease, a question that requires the reconstructed dose, the dose-response relationship, and a differential assessment that confronts the competing explanations, from smoking history to non-occupational exposures to the background rate of the disease itself. That structure demands a paired team: the industrial hygienist who builds the exposure, and the occupational physician or toxicologist who carries it across to diagnosis. The handoff between them must be seamless, because any daylight between the hygienist's dose and the physician's assumptions is where the opposing motion will be aimed. Defense counsel should note that the same instruments work in reverse: a rigorous exposure reconstruction is equally capable of showing that the claimed dose could not have occurred, and a disciplined differential assessment is the strongest vehicle for the alternative-cause defense. In disease litigation, whichever side retains the more methodologically serious team usually ends up framing the scientific narrative of the case.

VI. What to Ask in the First Call

The screening described above compresses into a first conversation, and three questions do most of the work.

First, ask how the opinion would be built. Which standards govern this work, which records must exist, what would you need to measure, which alternative explanations will have to be run down? An expert who can sketch the methodology for your fact pattern in the retention interview will be able to defend it in a Rule 702 hearing; an expert who answers with credentials and conclusions will not.

Second, ask for the history. The plaintiff-defense ratio, the prior admissibility challenges and their outcomes, the transcripts and publications on related subjects. The answers matter less than the candor with which they are given: you are auditioning the witness's response to hard questions, and this is the first one.

Third, ask what would change the opinion. What facts, if the record revealed them, would make the opinion you need unsupportable? The expert who answers specifically has just demonstrated independence, intellectual honesty, and command of the subject in a single response, and has told you, before you spend a dollar of the client's money, exactly where the case is vulnerable. That answer is worth more than the CV, the fee schedule, and the references combined.

The symmetry of these cases is the final point. Every quality and every practice described here serves plaintiff and defense identically, because the reliability requirements of the evidence rules are symmetric and the physics, chemistry, and human factors of the workplace do not care who retained the analyst. The side that treats expert selection as a procurement task litigates at a permanent disadvantage against the side that treated it as the case decision it is. In workplace injury and occupational disease litigation, the expert is not part of the case. The expert is the case, choose accordingly, and run the engagement like it.

References

ACGIH. (2026). 2026 TLVs and BEIs: Threshold limit values for chemical substances and physical agents and biological exposure indices. ACGIH.

American Industrial Hygiene Association. (2015). A strategy for assessing and managing occupational exposures (4th ed., S. D. Jahn, W. H. Bullock, & J. S. Ignacio, Eds.). AIHA Press.

Amorgianos v. Nat'l R.R. Passenger Corp., 303 F.3d 256 (2d Cir. 2002).

ANSI/ASSP. (2019). Occupational health and safety management systems (ANSI/ASSP Z10.0-2019). American Society of Safety Professionals.

Daubert v. Merrell Dow Pharms., Inc., 509 U.S. 579 (1993).

Fed. R. Civ. P. 26(b)(4).

Fed. R. Evid. 702 (as amended Dec. 1, 2023).

Hill, A. B. (1965). The environment and disease: Association or causation? Proceedings of the Royal Society of Medicine, 58(5), 295–300.

Kumho Tire Co. v. Carmichael, 526 U.S. 137 (1999).

Milward v. Acuity Specialty Prods. Grp., Inc., 639 F.3d 11 (1st Cir. 2011).

Milward v. Rust-Oleum Corp., 820 F.3d 469 (1st Cir. 2016).

Occupational Safety and Health Standards, 29 C.F.R. pts. 1910, 1926.

Pluck v. BP Oil Pipeline Co., 640 F.3d 671 (6th Cir. 2011).

Sardis v. Overhead Door Corp., 10 F.4th 268 (4th Cir. 2021).

Tamraz v. Lincoln Elec. Co., 620 F.3d 665 (6th Cir. 2010).

CHAPTER 2

Daubert-Proofing Your Safety Expert

What Separates Opinions That Survive from Ones That Get Struck

Since the 2023 amendment to Rule 702, the gatekeeper has teeth again. In workplace injury and OSHA litigation, whether your safety expert reaches the jury now turns less on credentials than on whether the opinion is built to a standard the court can actually check.

I. The Gatekeeper Was Told to Do Its Job

For a generation, the admissibility of expert testimony in federal court was governed by the framework the Supreme Court built in what practitioners call the *Daubert* trilogy: *Daubert v. Merrell Dow Pharmaceuticals, Inc.* (1993), which assigned trial judges a gatekeeping duty to screen expert testimony for reliability; *General Electric Co. v. Joiner* (1997), which held that a court may exclude an opinion when there is too great an analytical gap between the data and the conclusion; and *Kumho Tire Co. v. Carmichael* (1999), which extended the gatekeeping function to all expert testimony, not merely the self-consciously "scientific" kind. That last holding is the one that governs safety experts, accident reconstructionists, and industrial hygienists, whose opinions rest on skill and experience as much as on laboratory science.

The problem, documented for two decades, was that many courts were not actually performing the screening. They recited *Daubert* and then waved the testimony through, treating obvious reliability defects as matters of "weight" for the jury rather than "admissibility" for the court (Bernstein & Lasker, 2015). The Advisory Committee on Evidence Rules concluded that this drift was so widespread it required a textual fix. Effective December 1, 2023, Rule 702 was amended to make two things explicit: the proponent of expert testimony must show, by a preponderance of the evidence, that each admissibility requirement is met; and the expert's opinion must reflect a reliable application of the expert's principles and methods to the facts of the case (Fed. R. Evid. 702; Fed. R. Evid. 702 advisory committee's note to 2023 amendment).

The Committee was careful to say the amendment changed no substantive law: it merely restored the standard the courts were supposed to have been applying all along (Rule 702 advisory committee's note, 2023). But "clarifications" that are backed by new rule text change litigation behavior, and this one has. Federal decisions applying the amended rule now issue weekly, and by mid-2026 the circuit courts have supplied the pattern the district courts initially received unevenly. The Sixth Circuit moved first, within three months of the effective date, affirming the exclusion of an MDL's sole general-causation expert — for relying on a single clinical trial to the exclusion of the other human studies, analyzing animal studies outside his qualifications, and applying the Bradford Hill considerations selectively and inconsistently — and recording in a footnote that the amendment had been drafted to correct decisions incorrectly treating the sufficiency of an expert's basis and the application of the expert's methodology as questions of weight rather than admissibility (*In re Onglyza (Saxagliptin) & Kombiglyze Prods. Liab. Litig.*, 93 F.4th 339, 344, 347, 348 n.7 (6th Cir. 2024)). The Federal Circuit, sitting en banc and dividing eight to two, vacated a twenty-million-dollar damages award and ordered a new trial on damages because the district court had admitted a royalty opinion contradicted by the very license agreements the expert relied on, and had thereby failed in its gatekeeping function (*EcoFactor, Inc. v. Google LLC*, 137 F.4th 1333, 1341–44 (Fed. Cir. May 21, 2025) (en banc), vacating panel op., 104 F.4th 243 (Fed. Cir. 2024), reh'g en banc granted and op. vacated, 115 F.4th 1380 (Fed. Cir. 2024)). The Eighth Circuit, whose case law had for decades treated an expert's factual basis as a question of credibility rather than admissibility, took notice of the amendment and held that opinions lacking an adequate factual basis lack reliability and are properly excluded (*Sprafka v. Med. Device Bus. Servs., Inc.*, 139 F.4th 656 (8th Cir. 2025)). And the Sixth Circuit — in a flood-damage case pleaded as negligence per se against a surface mine, not an exposure case, but decided on grounds that transfer directly — affirmed the exclusion of the plaintiffs' sole causation expert because his report was not based on sufficient facts or data specific to the sites at issue, did not rest on reliable principles and methods, and failed to account for alternative causes of the damage (*Baker v. Blackhawk Mining, LLC*, 141 F.4th 760, 763–66, 770 (6th Cir. 2025)). Borrowing studies of nearby operations did not supply the site-specific basis the rule requires, and without the opinion the plaintiffs could not prove causation. Those same failures — the basis that is not specific to this workplace, the method invoked rather than applied, the alternative cause left unaddressed — are recurring features of safety and exposure opinions, and the reader will meet all of them again in Chapter 17. All four of these decisions are recent enough that their subsequent history is still accruing; each should be confirmed, with pin cites, before it is relied on in a brief.

Honesty about the trend requires noting that it is not uniform, and the exception is an exposure case. In *Sommerville v. Union Carbide Corp.*, No. 24-1491, 149 F.4th 408, 2025 WL 2383496 (4th Cir. Aug. 18, 2025), a divided panel reversed a district court that had excluded a plaintiff's dispersion-modeling expert in an ethylene oxide medical-monitoring action, holding that the district court's concerns about the emissions data the expert had fed into his model went to weight and credibility rather than to admissibility: the very formulation the 2023 amendment was written to retire. The Chief Judge dissented on precisely that ground, rehearing en banc was denied, and the Supreme Court denied certiorari on May 26, 2026, sub nom. *Union Carbide Corp. v. Sommerville*, No. 25-919 — leaving unresolved as between the circuits the question whether a challenge to an expert's factual basis always goes to weight. The same opinion decided a second question that matters at least as much to the readers of this book: the panel also held that the plaintiffs had Article III standing to pursue medical monitoring for exposure without present disease, which enlarges the class of plaintiffs whose exposure reconstructions experts will be asked to build. Chapter 28 §III.D returns to that holding in the context of the ethylene oxide and PFAS dockets. Two practical points follow for the practitioner. The first is that the forum still matters, and counsel should know which line of authority governs before designing the opinion. The second is that the input-validation problem at the heart of *Sommerville* — an expert who adopted a defendant's self-reported emissions figures without independently testing them — is the exposure-reconstruction failure this book returns to in Chapters 16 and 17, and an opinion that survives on appeal because a court called the defect a matter of weight is still an opinion a competent cross-examination will take apart in front of a jury. That is the thesis of this chapter: in the post-2023 environment, a safety expert survives not by being credentialed or persuasive, but by producing an opinion whose every inferential step is anchored in something the court can independently verify, and the work of building that record starts long before the deposition.

II. Where Safety Opinions Actually Get Struck

Having been retained on both sides of exclusion fights, I can report that safety and industrial-hygiene opinions rarely fail because the expert lacks qualifications. They fail at one of four predictable joints, and each maps onto a specific subdivision of Rule 702.

A. Sufficiency of the Basis — Rule 702(b)

Rule 702(b) requires that the testimony be based on sufficient facts or data. The recurring failure here is the expert who reasons from a conclusion outward rather than from the evidence inward. An accident reconstructionist who accepts a client's account of events and then assembles support for it, rather than testing that account against the physical evidence, has inverted the analysis. Courts applying the amended rule have excluded experts whose opinions rested on a characterization of the facts the expert adopted rather than tested — counsel's or the client's account, accepted without independent evaluation and, in the starkest cases, contradicted by the very documents the expert relied on for it (*EcoFactor*, 137 F.4th at 1341–44). For a safety expert, "sufficient facts or data" means the site-specific record — measurements, the actual equipment, the maintenance history, the training documentation — not a general familiarity with how such worksites usually operate.

B. Reliable Methodology — Rule 702(c)

Rule 702(c) requires reliable principles and methods. In the safety context, reliability is usually established by tying the method to published consensus standards: the ANSI/ASSP technical series, NFPA codes, the ACGIH Threshold Limit Values, NIOSH criteria documents, the OSHA standards themselves, and the recognized texts of accident investigation and human-factors analysis. An opinion that traces to those authorities is testing a hypothesis against an external yardstick. An opinion that rests on "in my thirty years of experience" and nothing more is, after *Kumho* and now after the 2023 amendment, exactly the kind of ipse dixit the gatekeeper is instructed to distrust. Experience can inform an opinion, but the expert must be able to explain how that experience leads to the conclusion, why it is a reliable basis, and how it has been reliably applied to the facts (Rule 702 advisory committee's note, 2000, reaffirmed 2023).

C. Reliable Application — the New Emphasis

The most consequential change in the 2023 text is its insistence that the opinion itself, not merely the method in the abstract, reflect a reliable application of the methodology to the facts. This is the *Joiner* analytical-gap problem given rule status. A safety expert may invoke an impeccable standard and still be excluded if the standard was applied to the wrong facts, or applied in a way the standard does not support. The industrial hygienist who cites a permissible exposure limit but never characterizes the actual exposure; the reconstructionist who applies a slip-resistance threshold without measuring the surface; the ergonomist who invokes a lifting equation but omits half its variables — each has a method and a set of facts that never actually meet. This is now the most fertile ground for a motion to exclude, and the place where a well-built opinion most visibly separates itself from a vulnerable one.

D. Fit and Overstatement — Rules 702(a) and 702(d)

Finally, the testimony must help the trier of fact (702(a)) and must not overstate the conclusions that the basis and methodology can reliably support (702(d)). Overstatement is a discipline problem as much as a science problem. The expert who testifies to a "reasonable degree of certainty" about a matter the data can support only as a probability, or who offers a categorical causation opinion where the science supports a contributing-factor opinion, invites exclusion of the overstated portion, and hands opposing counsel a credibility theme for whatever survives.

III. What a Daubert-Resistant Opinion Looks Like

The good news for counsel is that reliability is largely a matter of construction, and construction is controllable. When I am retained early, the file is built to answer the four questions above before they are asked.

Independent factual development. The opinion begins with facts the expert gathered or verified — site measurements, inspection of the actual equipment, review of the primary records — rather than a narrative received from counsel. Where the expert must rely on assumptions, those assumptions are stated, sourced, and tested for sensitivity: does the conclusion hold if the assumed value moves within its plausible range? An opinion that survives a change in its own assumptions is far harder to strike than one that collapses when a single input is questioned.

Method tied to published authority. Each analytical step is linked to a consensus standard or peer-reviewed methodology, and the linkage is documented so the court can follow it without taking the expert's word. The point is not to bury the report in citations; it is to ensure that no inferential step depends solely on the expert's unexplained intuition.

Explicit application. The report shows its work: the specific facts of this incident run through the specific method, arriving at the conclusion by a path the reader can retrace. This is the antidote to the analytical gap. A reviewing judge who can trace the reasoning from data to conclusion has little basis to find the application unreliable.

Calibrated conclusions. The opinion claims exactly what the basis supports, no more. Where the science supports "more likely than not," the opinion says so; where it supports only that a factor contributed, the opinion is confined to contribution. Calibrated conclusions are not weaker conclusions; they are the conclusions that survive cross-examination and, increasingly, the ones that survive the motion in limine.

Consideration of alternatives. A recurring reason courts give for exclusion is the expert's failure to account for obvious alternative explanations. A reconstruction that ignores a plausible competing cause is not a reliable application of any method. Addressing and ruling out alternatives, on the record, is one of the most effective forms of *Daubert* insurance available.

IV. State Courts, *Frye*, and the Multistate Practice Problem

The 2023 amendment governs federal court, but workplace injury and construction cases are litigated heavily in state court, and the states have not moved in unison. A large group of states has adopted *Daubert*-style reliability review by statute or decision; a smaller but significant set, including several with heavy construction dockets, continues to apply some version of the older *Frye* "general acceptance" test or a state-specific hybrid. The practical consequence is that the same safety opinion may face a searching methodological gatekeeping inquiry in one forum and a narrower "general acceptance" inquiry in another.

For counsel, this is a reason to build to the more demanding standard regardless of forum. An opinion constructed to satisfy amended Rule 702 — sufficient basis, reliable method, demonstrated application, calibrated conclusions — will comfortably satisfy *Frye*, because a methodology grounded in ACGIH, NIOSH, ANSI, and NFPA authority is, by definition, generally accepted. The reverse is not true: an opinion built only to clear *Frye*'s general-acceptance bar may still founder on the application requirement if the case is removed to, or filed in, a *Daubert* forum. Building to the ceiling protects the opinion wherever the case lands.

But building to the ceiling does not solve the second half of the state-court problem, which is procedural rather than substantive and which this book's own reliance on New York construction law makes urgent. *Frye* jurisdictions are not simply *Daubert* jurisdictions with a lower bar; several of them operate expert-disclosure regimes that bear almost no resemblance to Rule 26.

New York is the important case, and it is the state supplying much of the falling-object and scaffold law this book relies on. New York applies *Frye* — the inquiry is whether the methodology is generally accepted as reliable in the relevant scientific community, with a separate foundational inquiry into whether the accepted methodology was properly applied in the particular case. And New York requires no expert report at all. CPLR 3101(d)(1)(i) requires a party, upon request, to disclose in reasonable detail the subject matter on which each expert is expected to testify, the substance of the facts and opinions, the qualifications, and a summary of the grounds for each opinion: a disclosure typically two or three pages long, drafted by counsel, and served near the note of issue. There is no Rule 26(a)(2)(B) report, no considered-materials list, no four-year testimonial history, no compensation disclosure, and — critically — no automatic right to depose the expert, since expert depositions in New York state practice occur by stipulation or court order rather than as of right.

Three practical consequences follow, and each is a trap for a practitioner whose instincts were formed in federal court.

First, the disclosure's thinness is a shield and a hazard at once. Counsel receiving a CPLR 3101(d) response learns far less than a Rule 26 report would convey, and the deposition campaign of Chapter 4 may be unavailable to develop it; the *Frye*/foundation motion must therefore be built from the underlying records and the expert's published or prior-testimony history rather than from the expert's own disclosure. Conversely, a disclosure that is vague to the point of non-disclosure can be attacked as inadequate, and courts have precluded testimony that exceeded a disclosure's stated scope, which makes the drafting of a 3101(d) response a strategic act rather than a clerical one.

Second, the same expert may be disclosed under both regimes in related matters — a federal third-party action and a state Labor Law action arising from the same incident are common — and the federal report will be discoverable in, and usable against the witness in, the state case. The expert who wrote a careful Rule 26 report and then adopted a looser 3101(d) summary has created the inconsistency Chapter 4 §III is built to exploit.

Third, the absence of a report shifts the expert's own risk. The federal expert's opinions are fixed on the day the report is served; the New York expert's are fixed at trial, which sounds like freedom and functions as exposure, because the discipline the report imposes — state the assumptions, source every fact, declare the limits — is a discipline no procedural rule will impose in her place. Chapter 3 recommends writing the report even where no rule requires it, and this is why: the document is not primarily a disclosure obligation. It is the instrument by which an opinion is made checkable, first of all by its author.

Counsel handling a multistate practice should therefore resolve four questions at intake, in this order: what reliability standard governs, what the disclosure regime requires and when, whether expert depositions are available as of right, and whether the state applies its rule differently in strict-liability or statutory-liability claims than in ordinary negligence. The answers change the work product, not merely its packaging.

V. Litigating the Gate: The Motion, the Order, and the Appeal

Understanding how a reliable opinion is built is half the discipline; the other half is understanding how the exclusion fight is actually conducted, because the motion practice has its own strategy and its own record. The threshold determinations belong to the court under Rule 104(a), on a preponderance standard the 2023 amendment wrote into the rule text itself, and the proponent carries that burden for each element: a point worth stating plainly in the opposition brief, because the pre-amendment habit of intoning that exclusion is the exception and admission the rule is precisely the drift the Committee acted to end. Timing is set by the scheduling order, and the tactical question is familiar: the early motion educates the court and may shape settlement; the motion filed with summary judgment lets the exclusion argument and the merits argument reinforce each other, since an opinion struck under Rule 702 frequently takes the opposing party's causation or defect case down with it.

The scope of the motion deserves more thought than it usually gets. The undisciplined motion attacks everything and teaches the court to discount all of it; the effective motion concedes the witness's qualifications and the method's pedigree where they are genuinely sound, and drives at the joint that actually fails — the assumption never tested, the standard applied to facts it does not fit, the conclusion pitched a level of certainty above what the analysis can carry. Partial exclusion is a common and often optimal outcome: the overstated causation opinion is struck while the exposure characterization survives, and counsel who anticipated that split has a trial plan for the narrowed opinion while the opponent rebuilds. For the proponent, the mirror discipline is triage in the response: defend the load-bearing opinions with the record, and consider withdrawing the vulnerable peripheral opinion whose defense would spend credibility the central opinion needs.

The record made at the gate outlives the ruling, in both directions. The exclusion order is reviewed only for abuse of discretion under *Joiner*, which makes the trial-court record nearly the whole game: the affidavits, the hearing testimony, and the exhibits must contain everything the appellate court would need, because deference will fill any silence against the appellant. *Sardis* is the instructive counterexample: a judgment reversed because the gate was not meaningfully kept, and a reminder that deference runs out where the district court's reasoning does. And the order follows the expert beyond the case. Chapter 1's screening question —

have you ever been excluded, and why — is answered out of these orders for the rest of the witness's career, which is a reason the expert, and not only the client, has a stake in never offering the overstated opinion the motion was built to catch. The gate is litigated case by case; the record it produces is permanent.

VI. Practice Implications: Hire the Method, Not Just the Name

Three implications follow for counsel selecting and preparing a safety or reconstruction expert in the current environment.

First, engage the expert early, for the reasons Chapter 1 §III sets out at length. The single most common cause of a strikeable opinion is a thin factual basis, and a thin basis is usually a scheduling failure rather than an analytical one.

Second, evaluate the prospective expert's method before the report, not after. Ask how the opinion will be built: which standards, which measurements, which alternatives will be considered and excluded. An expert who cannot answer those questions in the retention interview will not answer them well in a *Daubert* hearing.

Third, use the same lens offensively. The reliability requirements are symmetric: the opposing expert's analytical gaps, unexamined assumptions, and overstated conclusions are now not merely cross-examination fodder but grounds for exclusion under a rule the courts have been told to enforce. The party that understands how a safety opinion is built is best positioned both to build one that survives and to dismantle one that should not. In the post-2023 landscape, that understanding is not a luxury; it is the difference between an expert who testifies and one who never reaches the jury.

References

Baker v. Blackhawk Mining, LLC, 141 F.4th 760 (6th Cir. 2025).

Bernstein, D. E., & Lasker, E. G. (2015). Defending Daubert: It's time to amend Federal Rule of Evidence 702. William & Mary Law Review, 57(1), 1–47.

Daubert v. Merrell Dow Pharms., Inc., 509 U.S. 579 (1993).

EcoFactor, Inc. v. Google LLC, 137 F.4th 1333 (Fed. Cir. May 21, 2025) (en banc), vacating panel op., 104 F.4th 243 (Fed. Cir. 2024), reh'g en banc granted and op. vacated, 115 F.4th 1380 (Fed. Cir. 2024).

Fed. R. Evid. 104(a).

Fed. R. Evid. 702 advisory committee's note to the 2023 amendment.

Fed. R. Evid. 702 (as amended Dec. 1, 2023).

Gen. Elec. Co. v. Joiner, 522 U.S. 136 (1997).

Kumho Tire Co. v. Carmichael, 526 U.S. 137 (1999).

In re Onglyza (Saxagliptin) & Kombiglyze Prods. Liab. Litig., 93 F.4th 339 (6th Cir. 2024).

Sardis v. Overhead Door Corp., 10 F.4th 268 (4th Cir. 2021).

Sommerville v. Union Carbide Corp., No. 24-1491, 149 F.4th 408, 2025 WL 2383496 (4th Cir. Aug. 18, 2025), reh'g en banc denied, cert. denied sub nom. Union Carbide Corp. v. Sommerville, No. 25-919 (U.S. May 26, 2026).

Sprafka v. Med. Device Bus. Servs., Inc., 139 F.4th 656 (8th Cir. 2025).

CHAPTER 3

The Report Is the Deliverable

Writing the Rule 26 Expert Report So the Opinion Survives What Comes Next

The report is the only part of an expert's work most of the case will ever see. It is served before the deposition that will test it and the hearing that will turn on it, and everything the opposing examiner does begins with reading it more carefully than it was written. A report built to the rule's requirements — every opinion stated, every basis disclosed, every method traceable, every assumption declared — is not a formality completed after the analysis; it is the analysis, made discoverable and made to hold.

I. What the Rule Requires, and Why It Is the Defense

A retained expert who may testify must produce a written report, and Fed. R. Civ. P. 26(a)(2)(B) specifies its contents with a precision the safety curriculum never teaches: a complete statement of all opinions the witness will express and the basis and reasons for them; the facts or data considered in forming them; any exhibits that will be used to summarize or support them; the witness's qualifications, including publications from the last ten years; a list of all other cases in which the witness testified at trial or by deposition during the previous four years; and a statement of the compensation to be paid. The list reads like a compliance checklist, and treating it as one is the first mistake. Each requirement is a place the opinion is attacked, and the report is the employer's, or the plaintiff's, first and best opportunity to attack-proof it.

The word that governs the first requirement is complete. An opinion not disclosed in the report is an opinion the witness generally may not offer at trial, because Rule 37(c)(1) makes the exclusion of undisclosed opinions close to automatic absent substantial justification or harmlessness. The discipline this imposes is uncomfortable but non-negotiable: the report must contain the opinions, not gesture at them. "I reserve the right to supplement" is not a reservation of anything; it is a preview of a motion to strike. The expert who realizes, in deposition, that the report omitted the opinion the case actually needs has discovered the omission one stage too late.

II. The Architecture of a Report That Holds

Beneath the rule's enumerated contents sits a structure that the reliability inquiry of Chapter 2 rewards. A report that survives moves in a fixed order: the assignment, stated as the questions the expert was asked and not broader than them; the materials considered, listed completely because the list is discoverable and its omissions are cross-examination; the factual background the expert relied on, with the source of each material fact identified so the opinion does not rest on a narrative received with the retention letter; the methodology, named and tied to a published authority — an OSHA standard, an ANSI/ASSP or NFPA consensus document, the AIHA exposure assessment strategy, a NIOSH method — before it is applied; the analysis, in which each inferential step is shown rather than asserted; the opinions, stated in the language the witness will use on the stand; and the assumptions and limitations, declared by the expert rather than extracted by the opponent.

That last section is the one novices omit and veterans lead with. An opinion that states its own limits — the data it would have preferred and did not have, the assumptions the conclusion depends on, the sensitivity of the result to those assumptions — is an opinion that has already survived the cross-examination designed to expose them, because the expert raised them first and explained why they do not change the conclusion. An opinion that hides its assumptions invites the examiner to find them, and a limitation discovered by the opponent reads as a concealment even when it is merely an oversight. Chapter 4 describes the campaign run against a report from the other chair; this section describes writing the report that the campaign has nothing left to find.

III. Writing for Two Readers

A Rule 26 report is read closely by exactly two people whose reactions decide anything, and they want opposite things from it. The first is a judge, working from the papers, deciding a motion under Fed. R. Evid. 702. That reader is looking for structure: whether each opinion is identified as an opinion, whether the basis for it is stated rather than gestured at, whether the method has a name and a source outside the expert's own practice, and whether the application of that method to these facts can be followed step by step. That reader will never hear the expert speak, is reading under time pressure alongside a stack of other motions, and is looking for reasons to grant or deny without a hearing. What persuades that reader is navigability — headings that announce what is coming, opinions numbered so they can be ruled on individually, a materials list that is complete, and a limitations section that shows the expert already knows where the soft ground is.

The second reader is a jury, and the jury does not read the report at all. The jury hears it, in fragments, read aloud by a lawyer who is either building the witness up or taking the witness apart. Sentences that survive silent reading can fail badly out loud. A sentence with three embedded qualifications is rigorous on the page and evasive in a courtroom. A paragraph that runs 140 words gives a cross-examiner four places to stop and ask whether the witness really meant that. Numbers stated to four significant figures invite the question of how a reconstruction of a 1997 workplace could possibly be that precise. And any sentence that can be read aloud with the emphasis moved — "you say the program was *generally* adequate" — has handed the examiner a theme.

The two demands are reconcilable, and the reconciliation is mostly structural rather than stylistic. Put the rigor in the architecture and the plainness in the sentences. State each opinion in one declarative sentence a juror can hold in memory, then develop the basis beneath it in as much technical detail as the method requires. Let the headings do the qualifying that the sentences should not: an opinion stated flatly under a heading that reads "Opinions Held to a Reasonable Degree of Professional Certainty, Subject to the Assumptions in Section 6" is both readable aloud and fully hedged. Where a qualification is genuinely part of the opinion, give it its own sentence rather than burying it in a subordinate clause, because a qualification the expert volunteers in a short sentence reads as candor, and the same qualification extracted from a long one reads as a concession.

Two habits follow from taking the second reader seriously. Read the opinion section aloud before service; the sentences that are hard to say are the sentences that will be read back. And write every number with the precision the underlying data actually supports, because the report is where the false precision of Chapter 17 §VI enters the case, and it is easier to not write 0.1372 mg/m^3 than to explain it.

IV. Exhibits, Demonstratives, and Disclosure

Rule 26(a)(2)(B)(iii) requires the report to include any exhibits that will be used to summarize or support the opinions. That clause is routinely treated as an afterthought — a line in a checklist, satisfied by attaching the photographs already in the file, and it is the origin of a disproportionate share of the demonstrative problems that Chapter 5 §IV and Chapter 29 §IV describe at trial. The connection is worth making explicitly here, because by the time those problems surface the cure is no longer available.

The practical difficulty is that the rule speaks at the report stage about objects that are usually built much later. The scale diagram, the exposure-over-time chart, the animation of the fall, the overlay showing the guardrail that was removed — these are typically commissioned during trial preparation, months after the report was served, by a graphics vendor working from the expert's conclusions. Courts vary in how strictly they read the clause, and a genuinely illustrative aid created for trial is often admitted over a disclosure objection. But the variance is the problem: a defense worth having should not depend on which side of that split the assigned judge sits, and the sanction when it goes the other way is the loss of the exhibit that was going to carry the opinion.

The disciplined approach costs almost nothing at the report stage. Disclose the exhibits that exist. Then, for the ones that do not yet exist, describe the categories the expert expects to use and the data each will be built from: a time-weighted exposure profile drawn from the sampling data in Appendix 3, a scaled site diagram drawn from the survey of March 14, an annotated enlargement of photograph 22, so that the opposing expert can evaluate the underlying data even before the graphic is made. Anything built later from data already disclosed is then a supplementation under Rule 26(e) rather than a surprise, and the supplementation is served when the exhibit is made rather than when the pretrial order forces it.

Two constraints on content belong in the report stage rather than the trial stage, because both are decided when the underlying file is assembled. The first is that a demonstrative may illustrate the opinion but may not exceed it: a chart showing a modeled concentration is an aid; the same chart with a PEL line drawn across it has made an argument the opinion may not support, and the expert who cannot point to the sentence in the report that the graphic depicts has created an exhibit with no disclosed basis. The second is that any image offered as depiction must be the original file. Upscaling, denoising, and generative fill synthesize detail rather than recovering it, and an enhanced scene photograph offered under the authority of a photograph is an authenticity problem under Fed. R. Evid. 901 before it is anything else. Chapter 29 §IV develops both points; the file discipline that makes them survivable — retain the original, log every processing step and its parameters, disclose the processing — begins when the photographs enter the expert's materials list, which is to say in this chapter.

V. Draft Protection, Communications, and Permissible Bases

The 2010 amendments to Rule 26 changed the discovery landscape for expert work, and both audiences of this book should understand the protection and its edges. Rule 26(b)(4)(B) protects draft reports in any form from discovery, and Rule 26(b)(4)(C) protects communications between the party's attorney and the retained expert, with three exceptions that swallow the incautious: communications that state the compensation for the expert's study or testimony, that identify facts or data the attorney provided and the expert considered, and that identify assumptions the attorney provided and the expert relied on. The practical translation is that an attorney may shape a draft without creating a discoverable trail, but the facts and assumptions the attorney supplies are fair game, which is exactly why the report should identify the source of each material fact, so that the facts the expert gathered independently are visibly distinguished from the facts the expert was handed. The expert who wishes to be seen as the author of the opinion writes a report that shows the opinion was authored from the record.

What the report may rest on is broader than what is admissible. Rule 703 permits an expert to base an opinion on facts or data that need not themselves be admissible, if experts in the field reasonably rely on such facts — the laboratory report, the exposure database, the medical records — but the rule also limits when those inadmissible bases may be disclosed to the jury. The report should therefore make clear that the reliance is of the kind the discipline actually practices, because a Rule 703 objection is, at bottom, the argument that no competent industrial hygienist would have relied on what this one did.

VI. Qualifications, the Testimony List, and Compensation

A model report applying every discipline in this chapter, annotated against the four failure modes of Chapter 2, appears as Appendix F; readers who want the architecture in front of them before reading the rest of this chapter should turn there first.

The three administrative disclosures are treated as clerical and lost as substantive. The qualifications statement is not the full curriculum vitae reproduced; it is the CV curated to the case, foregrounding the experience that maps to these facts — the process personally assessed, the exposure personally measured, the incident personally investigated — because qualifications carry the witness past the Rule 702(a) threshold and, curated well, begin the persuasion the report exists to accomplish. The prior-testimony list is a candor instrument: it must be complete for the four-year window, because opposing counsel will find the omitted case and offer it as evidence that the witness conceals, and because the transcripts on that list will be read for the prior inconsistent statement. The expert should read them first. The compensation statement should state the rate without apology; the vulnerability is never the fee, but the ratio of testimonial to consulting work and the balance of plaintiff and defense retention that Chapters 1 and 27 address, and the report is where a healthy ratio is first put on the record.

VII. Supplementation, Rebuttal, and the Errata Trap

The duty to supplement under Rule 26(e) is a discipline, not an escape hatch. A report is supplemented when the expert learns that a disclosure was incomplete or incorrect — new data, a corrected calculation, a materials list that missed a document — and the supplementation must be timely, well before the pretrial deadline, because a supplement offered late enough to look like a new opinion draws the same exclusion an undisclosed opinion draws. Supplementation is not the vehicle for the opinion the expert wishes had been in the original report; that opinion is either timely disclosed or lost. The rebuttal report, by contrast, responds to the opposing expert's disclosures within the rule's window and is confined to that rebuttal: it is not a second chance at the affirmative case. And the deposition errata sheet, offered under Rule 30(e), should correct transcription, not testimony: a substantive change to sworn deposition answers invites the argument that the witness is rewriting the record, and in many courts the original answer stays in evidence beside the change. The report written completely the first time needs none of these repairs.

VIII. The Non-Retained Expert: Rule 26(a)(2)(C)

The full written report belongs to the retained expert, but workplace injury and OSHA litigation runs on a second class of expert witness the rule treats differently and counsel routinely mishandles: the non-retained expert of Rule 26(a)(2)(C) — the treating physician, the company's own safety director or industrial hygienist, the compliance officer whose opinions ride into the case with their percipient knowledge. For these witnesses the rule requires not a report but a disclosure stating the subject matter of the expected opinion testimony and a summary of the facts and opinions to which the witness will testify. The permissive-sounding standard is a trap in both directions. The disclosure that says only that the treating physician will testify consistent with the medical records discloses nothing, and courts confine such witnesses to percipient facts, stripping the causation opinion the case needed; the disclosure written carefully — subject, opinions, and their factual grounding, summarized — preserves the testimony at a fraction of a report's cost.

The boundary that decides these fights is the source of the opinion. The treating physician who testifies to the diagnosis and causation conclusions formed in the course of treatment is a non-retained expert; the same physician handed the exposure reconstruction and litigation records and asked for an opinion formed for the case has crossed into retained territory, and the opinion formed from materials outside the treatment file generally requires the full report the party never served. The company safety professional presents the same line from the other side, with an added hazard this book's second audience should hear directly: the employer's own safety director, designated to explain the program and opine that it met the standard of care, is offering expert testimony, and the disclosure obligation — and the exposure of everything the witness wrote, audited, and knew — comes with the designation. Chapter 26 describes the day the professional's file is called; the Rule 26(a)(2)(C) designation is one of the doors it is called through, and counsel should make that designation with the file already read.

Two practice points close the loop. First, the disclosure deadline for non-retained experts is the same deadline that governs reports, and the omission is punished the same way, under Rule 37(c)(1): the treating physician never disclosed as an opinion witness is a fact witness at trial, whatever the pretrial memo assumed. Second, the summary disclosure is a commitment, not a placeholder: the opinions it summarizes are the opinions the witness may give, and the vague disclosure that tried to preserve flexibility instead preserves a motion in limine. The report and the summary disclosure are different instruments with the same architecture — the opinion stated, the basis shown, served on time — and the case that treats the second as an afterthought loses testimony it paid nothing to keep.

IX. A Report That Lost an Opinion

Every other major chapter in this book shows a defect and repairs it, and a chapter that asserts an architecture without demonstrating what happens when the architecture is missing has not earned the assertion. The following is composite and simplified, but each of its moves is ordinary.

The report. A safety engineer is retained by the plaintiff in a fall case. The worker stepped backward through an unprotected floor hole while carrying material. The report is competent, twenty-six pages, and states four opinions: that fall protection was required at the opening; that the covering system in use did not meet the marking and securement requirements of the standard; that the employer's program was deficient in three identified respects; and that a compliant cover would have prevented the fall. The fourth opinion is the one that matters, because it is the only one that reaches causation, and it is supported in the report by a single sentence: that a cover meeting the load and securement criteria would have withstood the loading applied and remained in place.

The motion. The defense moves to exclude opinion four under Fed. R. Evid. 702(b) and (d). The motion is short and it does not argue that the engineer is unqualified or that the standard does not apply. It argues that the report discloses a conclusion about load performance without disclosing the load. It asks four questions the report does not answer: what weight was assumed for the worker and the material he carried, what dynamic factor was applied to convert that weight into an applied load, what capacity was assumed for a compliant cover, and where in the record any of those three values came from. It attaches the deposition, in which the engineer explained all three inputs fluently and correctly — and confirmed that the calculation was performed on a legal pad that no longer existed and was not described anywhere in the report.

The order. The court strikes opinion four and leaves the first three standing. The reasoning occupies a page and a half. The engineer is qualified; the standard applies; the program criticisms rest on documents in the record and may be offered. But the causation opinion, the court writes, is a conclusion that the report connects to the facts only by the expert's assurance that the arithmetic works, and the 2023 amendment places the burden of showing reliable application on the proponent. The court declines to treat the deposition as a cure, noting that Rule 26(a)(2)(B)(i) requires a complete statement of the basis in the report and that permitting a first disclosure at deposition would make the requirement optional. The plaintiff's remaining case is a program-deficiency case without a causal bridge, and it settles inside a month for a fraction of what the four-opinion report had supported.

What went wrong, and what the repair costs. Nothing in this sequence involved bad science. The engineer's inputs were defensible, his method was standard, and his conclusion was probably right. What failed was the report's conception of its own job: it treated the opinion as the deliverable and the derivation as backup, when the rule treats the derivation as part of the opinion. The repair is roughly a page. State the assumed worker weight and material weight and cite the source for each — the employment record, the material invoice. Name the dynamic amplification factor and the authority for it. State the capacity criterion from the standard by subsection. Show the comparison. Then add the sentence a good report always adds: the range over which the conclusion holds, and the input that would have to change, and by how much, before it stops holding.

That last sentence is the one the defense motion could not have been written against. An opinion that discloses its own sensitivity has pre-answered the question of whether the result depends on a favorable assumption, and it converts a Rule 702 motion into a cross-examination — which is exactly the conversion this chapter exists to accomplish. The report is the deliverable because it is the only document in the case that can make that conversion happen before anyone is under oath.

The same failure, on the appellate record. The sequence above is composite because it is ordinary, not because it is hypothetical. In *Baker v. Blackhawk Mining, LLC*, 141 F.4th 760 (6th Cir. 2025), the Sixth Circuit affirmed the exclusion of the plaintiffs' sole causation expert on grounds that ran through Rule 702 and Rule 26 at once: the report was not based on sufficient facts or data specific to the sites at issue, it did not rest on reliable principles and methods, it failed to account for alternative causes — and it was independently deficient under Fed. R. Civ. P. 26(a)(2)(B). Without the opinion the plaintiffs could not prove causation, and summary judgment followed. The underlying claim was flood damage pleaded as negligence per se rather than an exposure case, and it is cited here for the procedural holding rather than the substantive one: the report and the opinion are not two documents, and a disclosure defect is not a lesser species of problem that the merits can outrun.

X. Practice Implications

For counsel, the report is the case's spine, and its service date is a deadline behind which the technical work must already be finished: retain on the schedule Chapter 1 §III describes, shape drafts within the protection the rule provides, and read the completed report as the opposing examiner will — hunting the undisclosed opinion, the undocumented assumption, the incomplete list — before serving it. For the expert, the report is the deliverable the entire engagement produces, and it should be written to be read aloud by a hostile examiner without yielding a point that was not already conceded on its face. The opinion that reaches the jury is the opinion that was written to survive the report. Everything in Chapters 2 and 3 is decided, in the first instance, here.

References

American Industrial Hygiene Association. (2015). A strategy for assessing and managing occupational exposures (4th ed.). AIHA Press.

Baker v. Blackhawk Mining, LLC, 141 F.4th 760 (6th Cir. 2025).

Committee Note to the 2010 Amendments to Fed. R. Civ. P. 26 (draft-report and attorney-expert communication protection).

Fed. R. Civ. P. 26(a)(2), (b)(4), (e).

Fed. R. Civ. P. 26(a)(2)(C) (non-retained expert disclosures).

Fed. R. Civ. P. 30(e).

Fed. R. Civ. P. 37(c)(1).

Fed. R. Evid. 702; Fed. R. Evid. 703.

CHAPTER 4

It Dies in a Transcript

Deposing and Cross-Examining the Opposing Expert — Because the Opinion Arrives in a Report but Dies in a Deposition

The opposing safety or industrial hygiene expert is rarely defeated at trial. The expert is defeated months earlier, in a conference room, by an examiner who has read the file more carefully than the witness wrote it. This chapter maps that campaign from both chairs: the file work that precedes the deposition, the architecture that locks and exhausts the opinion, the lines of examination that turn a confident report into an excludable one, and the preparation that armors your own witness against the identical campaign.

I. The Expert Is the Case

In the case types that now dominate the safety and industrial hygiene docket — the third-party construction injury, the engineered-stone silicosis claim, the heat death, the warehouse musculoskeletal case, the chemical-exposure suit — the operative questions are technical before they are legal. What was the dose? What did the standard of care require, drawn from which layer of the regulatory and consensus stack? Which entity controlled the hazard, and what would feasible abatement have cost and prevented? Juries do not answer those questions from the evidence directly; they answer them through the experts, and in most of these cases the verdict is, functionally, a decision about which side's safety professional or industrial hygienist the jury believed and which one it discounted. That makes the opposing expert not a witness in the case but the case itself, and the deposition of that expert the single highest-leverage event in discovery.

The 2023 amendment to Rule 702 and its consequences for safety and industrial hygiene opinions are treated in full in Chapter 2.

One orientation note before the map. Everything that follows is written for both chairs, because the campaign is symmetric: the plaintiff's counsel deposing the defense hygienist and the defense counsel deposing the plaintiff's safety expert are running the same playbook against mirror-image targets, and the safety or IH professional reading this chapter should read every section twice — once as a guide to what will be done to the opposing expert, and once as a warning about what will be done to them. The best protection against the examination described below is not coaching; it is doing the technical work in a way that leaves the examination nothing to find.

II. The Work Before the Deposition

The deposition is won or lost before the witness is sworn. Three bodies of preparation — the file, the qualifications audit, and the methodology map — determine whether the examiner spends the day discovering the expert's opinions or dismantling them, and each is work the retaining lawyer cannot do alone: the consulting expert at counsel's elbow, reading the opposing report with a practitioner's eye, is the difference between an examination that sounds technical and one that is.

A. The File: What Rule 26 Gives and What It Hides

Start with what the rules compel. In federal court, Rule 26(a)(2)(B) requires the retained expert to serve a written report containing a complete statement of all opinions and the basis and reasons for them; the facts or data considered in forming them; any exhibits to be used; the witness's qualifications and a list of all publications from the preceding ten years; a list of all other cases in which the witness testified as an expert at trial or by deposition during the previous four years; and a statement of compensation. Each element is a deposition chapter waiting to be opened. "Considered" is broader than "relied upon" — it reaches everything the expert reviewed, including materials that cut against the opinion and were set aside — and the examiner should demand the complete considered file, not the curated reliance list. The 2010 amendments protect draft reports and most attorney-expert communications as work product, but three categories of communication remain discoverable and are routinely under-pursued: communications about compensation, the facts or data the attorney provided and the expert considered, and the assumptions the attorney supplied that the expert relied upon. The assumptions channel is the richest of the three, because an opinion that rests on counsel's assumed facts rather than the expert's verified ones is an opinion the deposition can detach from the record entirely.

Then demand what the rules do not list but the discipline generates. The industrial hygienist's opinion sits on top of a documentary substrate: field notes, chain-of-custody forms, calibration records, laboratory reports with their analytical methods and detection limits, photographs, instrument logs, modeling inputs and software versions, and the raw spreadsheets behind every summary table in the report. The safety expert's opinion sits on inspection notes, measurement records, photograph metadata, and the marked-up standards and program documents the expert actually consulted. Subpoena or request all of it, and compare it line by line against the report: the gap between what the file shows and what the report claims — the calculation that does not reproduce, the photograph that contradicts the narrative, the note recording a fact the report omits — is the first draft of the cross-examination. And in any matter where an OSHA inspection occurred, obtain the complete inspection file and set it beside the expert's account; the compliance officer's contemporaneous photographs and sampling data are the nearest thing the case will have to a neutral witness, and an expert whose reconstruction quietly disagrees with them has an explaining problem the deposition should make explicit.

B. The Qualifications Audit: Boards and Testimony

The credentials that anchor safety and IH expertise are specific, verifiable, and governed by codes of ethics that the examiner should read before the witness does. The Certified Industrial Hygienist credential, administered by the Board for Global EHS Credentialing, and the Certified Safety Professional credential, administered by the Board of Certified Safety Professionals, each carry eligibility requirements, examination content outlines, continuing-education obligations, and — most usefully — enforceable ethics codes requiring objectivity, disclosure of conflicts, and practice within one's competence. The Certified Professional Ergonomist credential of the Board of Certification in Professional Ergonomics and the state-issued Professional Engineer license carry analogous frameworks. Verify every claimed credential directly with the issuing board, confirm it is current rather than lapsed or retired, and check the disciplinary record. Then audit the résumé the way an investigator would: degrees against registrars, employment against dates, publications against the actual journals, distinguishing peer-reviewed research from trade-magazine columns, and claimed project experience against anything discoverable. A single résumé inflation, however minor, is worth more on cross than a dozen methodological quibbles, because jurors understand embellishment without a tutorial.

The testimonial history deserves equal rigor. The Rule 26 four-year case list is the floor, not the ceiling: run the witness through the verdict reporters, the deposition banks, the *Daubert*-tracking services, and ordinary docket searches, and assemble the full record — every prior report, deposition, and trial transcript obtainable. The harvest is threefold. First, prior inconsistent opinions: the hygienist who testified for a manufacturer that the very modeling approach now offered is unreliable, or the safety expert whose defense-side reports treat OSHA compliance as the ceiling of the duty while the current plaintiff-side report treats it as the floor, has written the cross-examination personally. Second, exclusion history: every prior Rule 702 ruling against the witness, in any court, is both a roadmap and an exhibit. Third, the business profile: the ratio of plaintiff to defense retentions, the fraction of professional income derived from litigation, the repeat relationships with particular firms, and the fee history: all fair game, all quantifiable, and all material the jury weighs in the ancient calculus of the paid witness. The expert with a genuinely mixed practice will say so proudly; the expert without one will resist the arithmetic, and the resistance is itself the answer.

C. The Methodology Map: The Discipline's Own Books

The decisive preparation is a map of what the expert's own discipline says the work should have looked like, because the most damaging cross-examination of a technical witness is conducted not from the examiner's skepticism but from the profession's published expectations. For the industrial hygienist, the canonical framework is the exposure assessment strategy the profession itself teaches: characterize the workplace, workforce, and agents; define similar-exposure groups; make qualitative judgments and then test them with measurement; characterize the exposure profile statistically, respecting the lognormal behavior of occupational exposures rather than averaging it away; and document the uncertainty. Behind that strategy stand the sampling and analytical canons — the NIOSH Manual of Analytical Methods, the OSHA Technical Manual and its sampling procedures, the validated method for each analyte with its flow rates, media, and limits of detection — and, where the case turns on reconstruction rather than measurement, the retrospective exposure-assessment literature with its published hierarchies of data quality, from personal measurements down through area samples, modeling, and analogy. The peer-reviewed studies showing that hygienists' unaided qualitative exposure judgments are frequently inaccurate, and improve only with structured, data-anchored methods, are among the most quietly devastating exhibits available against an expert whose dose opinion rests on adjectives.

For the safety expert, the map is the standards hierarchy itself. The opinion will invoke some blend of OSHA regulations, state-plan variants, consensus standards, industry practice, and the defendant's own program — and each layer has boundaries the examiner should know cold: which consensus standards are incorporated by reference and thereby mandatory, which are voluntary guidance the expert is silently promoting into law; what the General Duty Clause actually requires by way of recognized hazard and feasible abatement, as against the colloquial "OSHA requires safety" the report may imply; what the hierarchy of controls prescribes and in what order; and where threshold limit values sit — professional guidelines with their own documented bases and explicit disclaimers, not bright lines separating safe from harmful, a concession every honest hygienist must make and every overreaching one resists. For the ergonomics opinion, the published assessment tools — the revised NIOSH lifting equation, the observational posture-and-force instruments, the hand-activity and vibration TLVs — each carry documented scopes, input requirements, and validation limits, and an expert who invokes a tool without honoring its inputs has converted the tool into a prop. The consulting expert's task in this phase is to write, for counsel, the examination outline the discipline itself would administer: here is what the method requires; here is what the report shows was done; here is the gap, question by question.

III. The Deposition

The deposition of the opposing technical expert has three simultaneous objectives — lock the opinions so they cannot grow, exhaust the bases so nothing is held in reserve, and build the record the Rule 702 motion and the trial cross will quote — and the architecture that serves all three is patience: open-ended funnels early, surgical confrontation late, and the discipline never to teach the witness the trial cross by rehearsing it. What follows is the sequence, then the substantive lines within it.

A. The Architecture: Lock, Exhaust, and Preserve

Begin by fencing the opinions. Have the witness confirm that the report contains a complete statement of every opinion to be offered, elicit any additions or modifications on the record, and then close each one: is that the entire opinion; are those all the bases; is there anything else, anywhere, that supports it. The exhaustion questions are tedious and indispensable — they are what makes the later supplement, the trial-morning refinement, and the redirect rehabilitation objectionable — and they should be renewed at the end of the day after the substantive examination has jogged the witness's memory. Establish the ground rules of the witness's own process while the mood is still cooperative: what was reviewed and when, what was requested and never received, what was available and never requested, how much time the entire engagement consumed and how it was distributed between reviewing records and writing conclusions. An expert who spent eleven hours on the file and forty on the report has described the engagement's priorities better than any argument could. Then, before any confrontation, walk the witness through the affirmative commitments the discipline compels — the methodological principles, the authoritative texts, the professional obligations of objectivity — because every principle the witness endorses in hour two is a trap the witness cannot escape in hour six. Save the confrontation itself for last, after the concessions are banked, and resist the trial lawyer's itch to win the deposition: the goal is a transcript that wins the motion and the cross, not a witness who leaves the room educated about the weaknesses in time to repair them.

B. The Inputs: Facts, Assumptions, and the Records Not Reviewed

Every technical opinion is a machine that runs on inputs, and the inputs are where safety and IH opinions most often fail. March the witness through the facts assumed, one by one, and source each: measured, observed, taken from a record, supplied by counsel, or assumed. The assumptions supplied by counsel are discoverable notwithstanding work product, and each should be tested against the record, because an exposure reconstruction premised on eight hours of daily dry cutting, when the time records show four and the co-worker depositions show wet methods half the time, is not a weaker opinion but a different case. Then invert the inquiry and catalogue what the expert did not review: the thirty-year exposure records the employer was required to preserve, the OSHA inspection file, the equipment maintenance logs, the training records, the prior incident history, the depositions taken after the report was written. For each omission, close the loop with the two questions that matter: did the expert know the record existed, and would its contents matter to the opinion. A witness who says the unreviewed record would not matter has testified to a closed mind; a witness who concedes it might has testified that the opinion is provisional; there is no third answer, and either serves. Where the opinion rests on the expert's own site visit or testing, reconstruct it forensically: when it occurred relative to the incident and to any changes in the workplace, what instruments were used and how calibrated, what was measured and what was merely observed, and what the field notes actually recorded as against what the report reports. The distance between conditions at inspection and conditions at incident is a standing vulnerability of every after-the-fact evaluation, and the witness should be made to quantify it or concede it cannot be quantified.

C. The Method: Sampling, Modeling, and the Amended Rule 702

The methodological examination should track the reliability factors the gatekeeping cases articulate and the amended rule enforces, translated into the discipline's terms. For the measured exposure case: was the sampling strategy designed to answer the litigation question or borrowed from a compliance context that asked a different one; were similar-exposure groups defined and defensible; do the sample sizes support the statistical claims; were the validated analytical methods followed, and do the laboratory reports match the report's summary; how were non-detects handled, and does the treatment follow the published guidance or flatter the conclusion. For the reconstructed exposure case — the more common and more vulnerable posture — walk the model's every input to its source, obtain the calculations in native form and check that they reproduce, and press on the questions the reconstruction literature itself poses: what data hierarchy does each input occupy, what is the uncertainty band, was any sensitivity analysis run, and would the opinion survive the plausible alternative values of its most fragile input. For the causation opinion, deploy the *Joiner* question in its native habitat: the analytical gap between the data and the opinion is a matter for the court, and the expert's bare assurance does not bridge it, so make the witness display the bridge, publication by publication, and mark every plank that is assertion rather than evidence. And for every opinion of every kind, ask the litigation-genesis questions the courts have long treated as probative: was this method developed for this case or does the witness use it in non-litigation practice; has it been published, tested, or subjected to any review colder than the retaining lawyer's; and can the witness identify anyone outside the courtroom who does the work this way. The examiner is not obligated to believe the answers, only to record them.

Two further lines deserve their own passes. First, the literature: require the witness to identify every published study, standard, and text relied upon; then confront the witness with the principal literature that cuts the other way and record the response to each item — unfamiliar, considered and rejected on stated grounds, or considered and omitted. Cherry-picking is among the most frequently sustained grounds for exclusion under the amended rule, and it is proved not by argument but by this inventory. Second, the boundaries: establish what the witness is not opining — not a physician, so the medical-causation link belongs to someone else; not an economist; not offering a legal conclusion — and nail down each disclaimed territory explicitly, because the boundaries conceded at deposition are the objections sustained at trial when the expert drifts.

D. The Concessions: What Every Honest Expert Must Give

A distinct and underused chapter of the examination consists of propositions so well established in the safety and IH disciplines that the witness must concede them or forfeit credibility, and each concession, banked on the record, becomes a girder in the examiner's own case. The hygienist must concede that occupational exposure limits are not bright lines between safe and harmful, that the TLV documentation itself disclaims use as a legal standard, that exposures vary day to day and the measured or modeled value is an estimate with uncertainty, and that retrospective reconstruction is inherently less certain than contemporaneous measurement. The safety expert must concede that OSHA standards are minimum requirements and compliance does not conclusively establish due care, or, examined from the other chair, that a citation is an allegation resolved under a different burden and does not conclusively establish negligence; that the hierarchy of controls prefers elimination and engineering over administrative controls and PPE; that employers, not third parties, carry certain non-delegable statutory duties; and that hazard recognition is judged by what was known or knowable at the time, not by the incident's outcome. The ergonomist must concede that musculoskeletal disorders are multifactorial, that the assessment tools estimate risk rather than adjudicate causation in an individual, and that personal and non-occupational contributors exist and matter. None of these concessions wins the case; together they shrink the opposing opinion to its honest size, and the expert who refuses them has traded the concessions for something better — a demonstration of advocacy the jury will be shown.

E. Bias, Compensation, and the Career Witness

The bias examination should be conducted with arithmetic rather than adjectives. Establish the hourly rates for review, deposition, and trial; the total billed to date and the estimate through verdict; the annual income from litigation consulting and its share of total professional income; the number of engagements for the retaining firm and for this side of the bar over a defined period; and the last time the witness's opinion, once rendered, disfavored the party who paid for it. Establish the referral channel: who called whom, and what the witness was told about the case before agreeing to serve. Where the audit in Section II.B surfaced prior transcripts, this is where the inconsistencies are marked: the witness is shown the prior testimony, asked whether it was true when given, and asked to reconcile it with today's, and the reconciliation, whatever it is, is locked verbatim. The examiner should conduct this chapter without hostility, because the tone that persuades is bookkeeping, not accusation; the numbers are either a portrait of a professional who sometimes testifies or of a testifier who was once a professional, and the transcript should let them speak in their own voice.

IV. From Transcript to Gate: The Rule 702 Motion

The motion should be outlined before the deposition and written from its page lines. The amended Rule 702 places the preponderance burden on the proponent and directs the court to find, not assume, that the opinion reflects a reliable application of a reliable method to sufficient facts, and the exclusion themes that recur in the safety and IH dockets map directly onto the deposition chapters above. Insufficient facts or data: the inputs chapter, with its assumed-not-verified facts and its catalogue of records never reviewed. Unreliable methodology: the sampling and modeling chapter, with its borrowed compliance strategies, unreproducible calculations, and litigation-born methods. Unreliable application — the ground the amendment most pointedly reinforced: the gap between the method the witness endorsed in hour two and the work the file shows in hour six. The analytical gap and the unexplained assertion: the causation chapter, plank by plank. Cherry-picking: the literature inventory, with the witness's recorded response to each contrary study. The motion that quotes the witness's own concessions — the method the discipline requires, admitted; the step the file omits, admitted; the uncertainty unquantified, admitted — is a different instrument from the motion that argues counsel's disagreement, and in the mass dockets a single exclusion propagates across every case the same expert anchors.

Sequence and scope are strategic decisions, not clerical ones. A partial exclusion that amputates the causation opinion while leaving the standards opinion standing may be worth more than a long-shot total exclusion that educates the opponent; a motion held for trial as a foundation objection may preserve surprise at the cost of the pretrial ruling's settlement value; and in jurisdictions whose state gatekeeping is more permissive than the federal rule, the deposition record serves a different master, the cross-examination itself, and should be built accordingly. What does not vary is the direction of dependence: the motion can be no better than the transcript, and the transcript can be no better than the preparation. The gate is not a separate proceeding from the deposition; it is the deposition, refiled.

V. Cross-Examination at Trial: The Jury's Version

The trial cross is not the deposition performed publicly; it is the deposition distilled to the fraction a jury can absorb and score. The governing constraints are cognitive. Jurors evaluate dueling technical experts less on the merits of the methods — which most cannot independently assess — than on proxies: clarity, consistency, candor, demeanor under pressure, and the tribal arithmetic of who is being paid how much by whom. The cross should therefore be built as a short series of self-contained chapters, each establishing one fact the jury can hold — the record he never read, the calculation that does not reproduce, the sentence in her own prior testimony that says the opposite, the fee total — each chapter closing on a locked deposition answer the witness cannot improve without impeachment, and the whole sequenced to open and close on the strongest material. Leading questions only; one fact per question; no question whose answer is not already in the transcript or the exhibit; and no invitation to explain, because the witness's explanations are redirect's job and the cross-examiner's gift to decline.

Two structural choices shape everything else. The first is the ratio of constructive to destructive cross. Against a competent, likable opposing expert, the higher-value examination is often the constructive one — the sequence of concessions banked at deposition, re-elicited before the jury, that converts the opposing witness into a partial sponsor of the examiner's case: yes, OSHA is a floor; yes, the TLV is not a line between safe and harmful; yes, engineering controls come first; yes, the employer's records would matter. Only after the witness has agreed with the examiner a dozen times does the destructive chapter open, and by then the jury has watched the witness concede that the examiner's framework is the honest one. The second choice is what to do with the mathematics. Every safety and IH case contains a reconstruction — a dose, a wet-bulb temperature, a force-frequency profile — and the cross-examiner must decide whether to fight the arithmetic or the inputs. Fighting the arithmetic in front of a jury is usually a mistake; fighting the inputs rarely is, because a jury that cannot audit a lognormal exceedance calculation can perfectly well understand that the answer assumed eight hours of dry cutting when the timecards say four. Demonstratives serve the same translation: the side-by-side of report claim and file page, the timeline of what was reviewed when, the fee tally — each doing visually what the transcript does textually. And the examiner should end early. The cross that makes three points and sits down beats the cross that makes seven and dissolves them into each other, in every courtroom, every time.

For the jury consultants who will help seat the panel, these contests carry attitudinal structure worth testing rather than assuming. Venire members divide on deference to credentials — some treat board certification as near-conclusive, others as guild self-decoration; they divide on whether compensation discredits testimony or merely prices expertise; and they divide, most consequentially, on quantification itself — whether a reconstruction performed years after the fact reads as science or as advocacy, the same axis that runs through every rising docket in this field. Voir dire should locate each juror on all three axes, and the trial team should test its expert presentation, direct and cross alike, against mock panels drawn across them, because the identical examination that lands as devastating with an audit-minded panel lands as bullying with a deferential one. The cross-examination is a performance scored by the jury the consultants helped choose; the scoring rubric should be known before the performance is designed.

VI. The Other Chair: Defending Your Own Expert

Every principle above has a mirror image, and the retaining side's obligations begin at engagement, not at deposition preparation. The expert should be retained on Chapter 1 §III's schedule; given the complete file, including the documents that hurt, because the opinion that has metabolized the bad facts is armored against them and the opinion that has not is hostage to them; and instructed to build the work for the amended Rule 702 from the first hour — every input sourced, every calculation reproducible, every assumption labeled as such, the contrary literature engaged in the report rather than discovered at deposition, and the opinion's limits stated by its author before the examiner states them for her. The Rule 26 hygiene is procedural armor of the same grade: a report that genuinely contains all opinions and bases, a considered-materials list that is complete on the day it is served, a testimonial list that is accurate, and a communications practice that respects the discoverability of counsel-supplied facts and assumptions, which is to say, a practice of supplying the expert with records rather than narratives.

Deposition preparation for the technical witness is its own discipline. The witness should be rehearsed not on answers but on posture: answer the question asked, completely and then silently; concede the true concession immediately and without garnish, because the fought concession costs triple; distinguish the question that misstates the method from the question that exposes it, and correct the former without quarreling; and treat "I don't know" and "I would have to review that" as professional answers rather than defeats. The witness should re-review the entire file — not the report, the file — because the examiner has, and should be walked through her own prior testimony before the examiner does it for her. At the deposition, defending counsel's work is quiet: preserve privilege where it genuinely exists, object to form where it genuinely fails, and otherwise let the prepared witness be the defense. Afterward, read the transcript the way the opposing motion-writer will, correct genuine errors through the errata process without attempting substantive revision the rules do not permit, and begin the redirect plan for trial: the short, warm examination that lets the expert explain, in her own voice and at last, the answers the cross-examiner would not let her finish. The symmetry deserves stating plainly: the expert who would be unbreakable under Section III's examination is the expert whose work needed no defending, and the professional reading this chapter serves both chairs best by being that expert.

VII. The Program in Either Chair

The practical program compresses to a sequence. Obtain everything: the complete Rule 26 file, the considered materials, the underlying data in native form, the credentials verified at the source, and the full testimonial history — then set the consulting expert to mapping the opposing work against the discipline's own published methods. Outline the Rule 702 motion before the deposition and depose to its elements: lock and exhaust the opinions, source every input, catalogue every omission, walk the method against the canon, bank the discipline's mandatory concessions, and reduce the bias case to arithmetic. Convert the transcript into the gate motion where the forum rewards it and into the trial cross everywhere, and build that cross as a handful of chapters a juror can carry into the deliberation room: constructive before destructive, inputs before arithmetic, three points and sit down. Prepare one's own expert against the identical campaign, beginning at engagement, because the examination described here is coming from the other chair whether or not it is prepared for. And through all of it, hold the orienting fact of the field: in these cases the experts are not witnesses to the dispute — they are the terrain on which it is fought, and the side that has surveyed that terrain earlier, more completely, and more honestly is the side whose expert is still standing when the record closes.

References

Access to employee exposure and medical records, 29 C.F.R. § 1910.1020 (thirty-year retention); General Duty Clause, 29 U.S.C. § 654(a)(1).

American Conference of Governmental Industrial Hygienists, Threshold Limit Values and Biological Exposure Indices (current ed.), including the TLV position statement on use of TLVs as legal standards.

American Industrial Hygiene Association, A Strategy for Assessing and Managing Occupational Exposures (4th ed. 2015).

Board for Global EHS Credentialing (Certified Industrial Hygienist), Board of Certified Safety Professionals (Certified Safety Professional), and Board of Certification in Professional Ergonomics (Certified Professional Ergonomist): eligibility requirements, recertification standards, and codes of ethics.

Committee Note to the 2023 amendment of Fed. R. Evid. 702 (preponderance standard; reliable application; forfeiture of the "goes to weight" default).

Daubert v. Merrell Dow Pharms., Inc., 509 U.S. 579 (1993).

Daubert v. Merrell Dow Pharms., Inc. (on remand), 43 F.3d 1311 (9th Cir. 1995) (litigation-driven methodology).

Fed. R. Civ. P. 26(a)(2) (expert disclosure requirements); Fed. R. Civ. P. 26(b)(4)(B)–(C) (draft-report and attorney-expert communication protections and exceptions, as amended 2010).

Fed. R. Evid. 702 (as amended Dec. 1, 2023); Fed. R. Evid. 703, 705; Fed. R. Evid. 613 (prior statements); Fed. R. Evid. 803(18) (learned treatises).

Gen. Elec. Co. v. Joiner, 522 U.S. 136 (1997).

Kumho Tire Co. v. Carmichael, 526 U.S. 137 (1999).

Logan, P., Ramachandran, G., Mulhausen, J., & Hewett, P., Occupational exposure decisions: can limited data interpretation training help improve accuracy?, 53 Annals of Occupational Hygiene 311 (2009), and the related literature on the accuracy of professional judgment in exposure assessment.

National Institute for Occupational Safety and Health, NIOSH Manual of Analytical Methods (5th ed. 2020); OSHA Technical Manual, TED 01-00-015, and OSHA sampling and analytical methods.

Waters, T.R., Putz-Anderson, V., Garg, A., & Fine, L.J., Revised NIOSH equation for the design and evaluation of manual lifting tasks, 36 Ergonomics 749 (1993); ACGIH TLVs for hand activity and hand-arm vibration; observational ergonomic assessment instruments (RULA, REBA) and their published scopes.

CHAPTER 5

The Opinion Meets the Jury

The Daubert Hearing, Direct and Cross, and Teaching the Technical Case to Twelve Nonspecialists

> *An opinion that survives every earlier stage still has to be delivered to people who have never worn a sampling pump or read a citation. The Daubert hearing decides whether the jury hears it at all; the direct examination decides whether they understand it; the cross-examination decides whether it holds. This chapter is about the last hundred feet: where the technical record, built and defended through every prior chapter, is finally taught to the twelve people who will decide what it means.*

I. The Gate Becomes a Hearing

The gatekeeping obligation examined in Chapter 2 is usually exercised in writing, on a motion in limine, but in a serious case it becomes a live event: a *Daubert* hearing at which the court takes testimony on whether the opinion rests on a reliable application of a reliable method to sufficient facts. The hearing is a trial before the trial, and the expert who has never distinguished the two settings arrives unprepared for the first. Before the judge, the currency is methodology — the published authority, the error rate, the traceable inferential chain — argued to a listener fluent in Rule 702 and unmoved by the jury appeals that will matter later. The expert's task at the hearing is to make the method legible to a lawyer in a robe: to show that each step is the step the discipline prescribes, that the facts were sufficient, and that the application was faithful. Losing the hearing ends the case for that opinion; winning it converts the reliability fight into a weight fight, which is the fight the proponent wants, because weight goes to the jury.

Counsel preparing the expert for the hearing should rehearse the motion the other side will file, not the direct the jury will hear. The most productive preparation is adversarial: the retaining lawyer, or better a colleague, runs the exclusion argument at the expert until the methodology answers are reflexive and the concessions are pre-planned. An expert who can say, at the hearing, precisely which assumptions the opinion depends on and precisely how sensitive the conclusion is to each has taken the exclusion motion's best material and spent it in the forum where it does the least damage.

II. Direct Examination: Teaching, Not Reciting

Direct examination of a technical expert is a teaching exercise disguised as a question-and-answer, and its object is not to prove the expert's brilliance but to leave twelve nonspecialists able to explain the opinion to one another in deliberation. That reframing changes everything about how direct is built. Qualifications come first and briefly, translated into why-this-witness terms rather than recited as a CV — the years on the drill floor, the exposures personally measured, the incidents personally investigated — because a jury credits demonstrated competence over accumulated letters. The assignment is stated plainly, so the jury knows the question before it hears the answer. Then the methodology is taught before the conclusion is delivered, because a jury that understands how the number was built will trust the number, and a jury handed the number first will suspect it.

The vocabulary is the battleground. A permissible exposure limit, a time-weighted average, a respirable fraction, a limit of quantification, an assigned protection factor: each is a term of art the jury does not have, and the expert who uses it without teaching it has spoken to the record and not to the room. The disciplined witness translates once, concretely, and then uses the term with confidence: the time-weighted average is the exposure averaged over the whole shift, the way a day's driving has an average speed even though the car sped and slowed; the respirable fraction is the portion of dust small enough to reach the deep lung, where the larger particles the nose catches never go. Analogy is not condescension; it is the mechanism by which a specialist becomes intelligible to a generalist, and the glossary in Appendix B is a source of exactly these translations.

III. A Direct Examination, Annotated: Building the TWA

This book has shown what a report looks like when it is built to survive (Appendix F). It has not yet shown what the same discipline sounds like out loud, and the omission matters, because the transcript is where the report is either delivered or squandered. What follows is a compressed direct examination of an industrial hygienist in a hypothetical engineered-stone fabrication matter, annotated in the margin notes marked ▶ that Appendix F introduced. The exchange is invented and the witness is a composite; the arithmetic is internally consistent and the method is the one OSHA and NIOSH publish, so the reader can check every step, which is precisely the property the examination is trying to demonstrate to a jury.

The assignment: characterize the respirable crystalline silica exposure of a countertop fabricator during edge-finishing work, and compare it to the permissible exposure limit. The witness has already been qualified and the assignment stated. Direct picks up at the measurement.

Q. Dr. Reyes, before we talk about what you measured, tell the jury what a permissible exposure limit is.

A. It is the maximum concentration of a substance the law allows a worker to breathe, averaged across an eight-hour workday. For respirable crystalline silica the limit is fifty micrograms per cubic meter of air. A microgram is a millionth of a gram. A cubic meter is roughly the volume of air an adult breathes in about an hour of moderate work.

▶ *The term is defined before it is used, once, concretely, with a physical referent for each unit. A jury that has been given a mental picture of a cubic meter will follow the arithmetic; a jury that has not will stop listening at "micrograms per cubic meter" and resume at the conclusion, having taken the number on faith. Faith is what the opposing cross is designed to break.*

Q. How did you measure Mr. Alvarado's exposure?

A. I placed a sampling pump on his belt and ran a tube up to his collar, in what we call the breathing zone. OSHA's sampling guidance describes that zone as a hemisphere in front of the shoulders, six to nine inches out from the nose and mouth; the consensus definition industrial hygienists use draws it a little wider, about thirty centimeters. I sampled to the narrower one, because that is the air he actually inhales and because it is the definition the standard I am measuring against uses. The tube ends in a cyclone — in this case an SKC GS-3, which is the reason the flow rate I will give you in a moment is 2.75 liters per minute rather than some other number; each cyclone model has its own design flow, and running one off its design flow changes what fraction of the dust it lets through. A cyclone is a small device that spins the air and throws out the larger particles, so that what reaches the filter is the respirable fraction: the portion of the dust small enough to travel past the nose and throat and reach the deep lung, which is where silica does its damage. The filter was then sent to an accredited laboratory, which weighed it and measured how much of the dust was crystalline silica.

► *Three terms of art — breathing zone, cyclone, respirable fraction — are each translated once, in the same breath in which they are introduced, and each translation carries a reason: the breathing zone because that is the air he inhales, the respirable fraction because that is the dust that reaches the tissue at issue. Definition plus purpose is what a juror retains; definition alone is vocabulary homework.*

Q. Did you calibrate the pump?

A. Yes. I calibrated it against a primary standard before the shift and again after, using the same sampling train — the same tubing, the same cyclone, the same filter cassette — because the flow rate depends on the resistance of the whole assembly. The pre-shift rate was 2.75 liters per minute and the post-shift rate was within two percent of it. My calibration records are Exhibit 22.

► ***[APPLICATION]*** *Counsel put calibration in the direct, unprompted, before the cross could arrive with it. Chapter 10 sets out why the calibration record is the first place an exposure citation fractures; the mirror of that proposition is that an expert whose own calibration is unimpeachable has taken the first ten minutes of cross off the table and has done it in front of the jury, where the credit accrues.*

Q. What did the sampling produce?

A. Two samples covering the shift. The first ran two hundred forty minutes, from the start of the shift through the morning edge-finishing work. At 2.75 liters per minute for two hundred forty minutes, that is six hundred sixty liters of air, or 0.660 cubic meters. The laboratory reported 72.6 micrograms of crystalline silica on that filter. Seventy-two point six micrograms divided by 0.660 cubic meters is one hundred ten micrograms per cubic meter.

Q. And the second sample?

A. Two hundred minutes in the afternoon, which is 0.550 cubic meters of air. The laboratory reported 20.9 micrograms. That works out to thirty-eight micrograms per cubic meter.

► *The witness performs the division on the record rather than announcing the result. This is the single most under-used technique in technical direct examination. A jury that watches the number get built will defend it in deliberation; a jury handed the number will only be able to report that the expert said so, which is exactly what the opposing expert will also have said.*

Q. You have two numbers, one hundred ten and thirty-eight. The limit is fifty. How does the jury get from those two numbers to one number it can compare to the limit?

A. By weighting each concentration by the time it lasted, which is what "time-weighted average" means. It is the same idea as the average speed of a car that spent part of the trip at highway speed and part in traffic: you do not average the two speeds, you weight each by how long it lasted. One hundred ten micrograms for two hundred forty minutes, plus thirty-eight micrograms for two hundred minutes, is thirty-four thousand microgram-minutes. Divided across the four-hundred-eighty-minute workday, that is seventy-one micrograms per cubic meter.

> ▶ *The analogy carries the concept and then gets out of the way. Note what the witness does not do: extend the metaphor, apologize for it, or ask the jury whether it makes sense. Analogy is a bridge, not a destination.*

Q. Your two samples cover four hundred forty minutes. The workday is four hundred eighty. What did you do about the other forty minutes?

A. I ran the calculation three ways, because that period is an assumption rather than a measurement and I did not want the answer to depend on which assumption I chose. If I assume the exposure during those forty minutes was zero, the most favorable assumption available to the defense, the eight-hour average is seventy-one micrograms per cubic meter. If I assume it equaled the average of the periods I did measure, the answer is seventy-seven. If I assume it equaled the higher, morning task rate, the answer is eighty. All three exceed the limit of fifty.

> ▶ ***[BASIS] [OVERREACH]*** *This is the load-bearing exchange, and counsel elicited the weakness rather than waiting for it. Chapter 3 §II calls the limitations section the one novices omit and veterans lead with; this is that section spoken aloud. The witness has also, in one answer, converted the cross-examination's best question into a demonstration of the opinion's durability: an opinion that survives a change in its own assumptions is far harder to strike, and far harder to shake, than one that collapses when a single input is questioned.*

Q. Is seventy-one micrograms per cubic meter a number you can state to a reasonable degree of scientific certainty?

A. I can state to a reasonable degree of scientific certainty that Mr. Alvarado's eight-hour exposure on the day I sampled exceeded the permissible exposure limit, and that it did so under every assumption I could reasonably make about the unsampled period. I would not tell you that seventy-one is the exact figure;

a measurement carries uncertainty from the pump, the balance, and the analysis. What I can tell you is that the uncertainty is not large enough to bring the result under fifty.

> ► *[OVERREACH] The calibrated answer. The witness declines the number and keeps the conclusion, and, in declining, becomes more believable about the conclusion. Chapter 2 §II.D describes overstatement as a discipline problem before it is a science problem; this answer is the discipline, and a jury reads it as candor rather than weakness because it is candor.*

Q. You sampled one day. Does that tell the jury anything about the other days?

A. By itself, no, and I want to be careful here. What supports the opinion about the other days is not the one day of sampling. It is that the task was the same task, on the same materials, at the same station, with the same absence of local exhaust ventilation, documented in the production records and confirmed by three coworkers and by the employer's own job description. When the task and the controls do not change, the exposure profile does not change, and that is why the profession groups workers who do the same job into what we call a similar exposure group. I would also tell you that if the ventilation had been intermittent, or the material had varied, my confidence would be lower, and I would have said so.

> ► *[APPLICATION] The inference from one day to a career is where exposure opinions are most often struck, and the witness makes the inference visible: not "I extrapolated," but here is the mechanism, here is the corroborating record, and here is the condition under which I would not have made it. Chapter 15 supplies the underlying discipline and Chapter 17 §II shows the same inference being attacked when it is made silently.*

IV. Demonstratives and the Record Made Visible

The technical record is mostly invisible — a number on a laboratory report, a placement described in a compliance officer's notes, a soil classification that was never made — and the direct examination's task is to make it visible without distorting it. A demonstrative that shows the breathing zone, the sampler placement, the geometry of the nip point, or the reconstructed exposure timeline does work no testimony can, because the jury reasons from what it can see. The governing discipline is that the demonstrative must be faithful to the admitted evidence: an exhibit that improves on the record is an exhibit the opponent will use to argue the expert improves on everything. Scale models, annotated photographs, and animation each carry their own foundation requirements, and the expert should be able to testify that the demonstrative accurately represents the facts it depicts. Some cases are decided on what the photographs did not show; the demonstrative is how the expert shows the jury the thing the photograph left out, and shows it honestly.

V. Surviving Cross-Examination

Cross-examination is Chapter 4 run against the witness in front of a jury rather than a court reporter, and the expert who wrote a disciplined report and defended it in deposition has already done most of the surviving. The cross will attack the same four places: the sufficiency of the basis, the reliability of the method, the faithfulness of the application, and the independence of the witness. The prepared expert meets each without the two failures that lose credibility: the reflexive fight over a fair point, and the collapse under an unfair one. Concessions planned in advance are given cleanly, because a witness who cannot concede the small true thing cannot be trusted on the large one; overreach invited by a leading question is declined by answering the question actually asked and stopping. The single most durable protection is the one built long before trial: the opinion the witness would reach whichever side had called her — the principle introduced in Chapter 1 §II.D and worked through the whole of Chapter 6 — cannot be shaken by the demonstration that the witness was retained by one side, because the witness's method did not depend on which side called.

The learned-treatise and prior-testimony materials deserve specific preparation. Under Rule 803(18), a passage from an authority the witness concedes is reliable can be read to the jury against the witness, and the deposition and prior transcripts on the Rule 26 list will be mined for the inconsistent statement. The expert reads all of it before trial, the treatises the field regards as authoritative and every prior transcript, because the examiner has, and the only defense to a genuine inconsistency is to have anticipated and reconciled it before the jury hears it framed as a contradiction.

VI. A Cross-Examination, Annotated: The Four Joints

The cross below is run against the identical opinion, by an examiner who has done the work Chapter 4 describes. It is organized the way an effective cross of a technical witness is always organized, not as an attack on everything, but as a march through the four joints of Chapter 2 §II, conceding the two that are sound in order to be believed on the two that are not. The annotations mark which joint each line is working, and note where the witness's earlier discipline has already closed the door.

Q. You are not a physician.

A. I am not.

Q. You are not offering an opinion that Mr. Alvarado has silicosis.

A. No. That opinion belongs to his physicians. Mine is about the dose.

> ► *Two clean concessions in the first thirty seconds. The examiner is not scoring; he is establishing with the jury that this witness answers questions, so that the later refusals, if any come, will register as evasion. The witness who fights the first two questions has volunteered to make every later concession look extracted.*

Q. Your opinion is that his exposure exceeded the limit. That opinion rests on two air samples.

A. On two samples and on the record establishing that the task was unchanged, yes.

Q. Two samples, on one day, on one worker, in a plant that operated for nine years.

A. That is correct.

> ► ***[BASIS]*** *Rule 702(b) in its purest form, and the strongest line available against any exposure opinion built on limited monitoring. Chapter 15 is the reason it works: workplace exposures are lognormal, a small sample is unlikely to capture the high days, and the profession's own literature says so. Note that the examiner does not argue with the answer. He lets the arithmetic of "two" against "nine years" sit.*

Q. You would agree that occupational exposures vary substantially from day to day.

A. I would.

Q. Several-fold, in some operations.

A. In some operations, yes.

Q. And two samples cannot characterize the shape of that variation.

A. Two samples cannot characterize the distribution. They can establish what the exposure was on the day measured, and that is what I have said they establish.

> ▶ ***[BASIS]*** *The concession is given and then bounded, in the same answer, without argument. This is the technique §V above calls answering the question actually asked and stopping. The witness who says only "I agree" has conceded the case; the witness who says "no, you can absolutely characterize a distribution from two samples" has lost the jury and handed the examiner the profession's own textbooks. The bounded concession is the only answer that survives both.*

Q. Let us talk about the forty minutes you did not sample. You told this jury three different numbers, seventy-one, seventy-seven, eighty.

A. I did.

Q. So you do not actually know what his exposure was.

A. I know it exceeded fifty. Which of the three figures is nearest the truth I do not know, and I said so.

> ▶ ***[APPLICATION]*** *This is the line that would have destroyed an opinion that had hidden the assumption, and it does nothing to an opinion that disclosed it on direct. The sequence is worth studying by both audiences: the value of the direct examination's candor is realized here, forty minutes later, in a question that has no second step. Chapter 3 §II is the reason. A limitation discovered by the opponent reads as concealment; a limitation the witness declared first reads as method.*

Q. You did not sample the ventilation.

A. There was no local exhaust at the edge-finishing station to sample. I photographed the station and measured its dimensions.

Q. You did not test the water-suppression system.

A. The station had none. That is in my report at paragraph forty-one.

> ▶ ***[BASIS]*** *The examiner is probing for the fact the expert did not personally verify: the most productive vein in any cross of a technical witness, and the reason Chapter 1 §III instructs counsel to put the expert in the field. Here the vein is dry, because the expert inspected. An expert who had worked from photographs would be answering "I understand that there was none," and the difference between there was none and I understand there was none is the difference between a witness and a conduit.*

Q. Mr. Alvarado wore a respirator.

A. He was issued one. Whether he wore it, and whether it fit, are disputed in the record.

Q. But if he wore it, his actual intake would have been a fraction of the number you gave this jury.

A. If he wore a properly fitted, properly maintained half-mask in a program that met the respiratory protection standard, the assigned protection factor is ten, and his intake would have been lower. Three things about that. The permissible exposure limit is a limit on the concentration in the air, not on what gets past a respirator, and the standard treats respirators as the last line rather than the control, which is the point of the hierarchy of controls. There is no fit-test record for him in the file I was given. And an assigned protection factor is a program-wide expectation, not a measurement of what happened inside one man's mask on one day.

► ***[APPLICATION] [OVERREACH]*** *The most dangerous question in the sequence, because it invites the witness either to overreach ("a respirator makes no difference") or to collapse ("yes, it would have been a tenth"). The answer does neither: it concedes the physics, names the protection factor, and then puts the burden where the regulation puts it — on the fit-test record the employer did not produce. Note also that the witness has just made the employer's missing document the jury's last impression of the exchange. That is what Chapter 25 means when it says the absent record is itself a finding.*

Q. You have testified for plaintiffs in eleven of the last fourteen matters.

A. In eleven of fourteen, yes. Over my career the split is closer to sixty-forty, and I have written reports for defendants in this same industry, including one that concluded the fabricator's controls were adequate.

► *The independence attack of Chapter 4 §II.B, and the only defense against it is a career, not an answer. The specific counterexample, a defense-side report that reached the defense-side conclusion, is worth more than any characterization of the ratio, which is why Chapter 27 treats the balance of a practice as an asset built over years rather than a talking point assembled before trial.*

Q. Doctor, you were paid four hundred and fifty dollars an hour for the opinion you gave this jury.

A. I was paid four hundred and fifty dollars an hour for the work. The opinion is the same one I would have written for the other side, and the arithmetic on the board is the same arithmetic either way.

► *The fee question is not a joint; it is a closing theme, and the only reply that neutralizes it is the one Chapter 1 §II.D makes the test of the whole engagement — that the report would read the same no matter who signed the letter. A witness who cannot say that sentence and be believed has a problem no preparation will fix, and it was created at the retention stage rather than at trial.*

Read the two examinations together and the shape of the thing is visible. Not one of the cross-examiner's productive questions is about silica. They are about the sufficiency of two samples, the assumption covering forty minutes, the facts the witness took on faith, and the witness's independence — Rule 702(b), Rule 702(d), Rule 702(b) again, and credibility. That is the whole map, and it is the same map in every technical case in this book. The expert who has built the opinion against it, and the counsel who has rehearsed the opponent's campaign against her own witness until it finds nothing the report did not already disclose, arrive at the last hundred feet with nothing left to lose there.

VII. Redirect, Rebuttal, and the Battle of Experts

The direct teaches and the cross tests; the two phases that follow are shorter and worse-prepared in most trials, and they are where a damaged opinion is either repaired or buried. Redirect is a scalpel, not a second direct: it exists to rehabilitate the specific points the cross actually damaged — the answer cut off before its explanation, the exhibit shown out of context, the fair concession the examiner inflated into a surrender — and to do nothing else. The instinct to re-argue the whole opinion signals to the jury that the cross drew blood everywhere; the disciplined redirect asks three questions about the two moments that mattered and sits down. Preparing it is a listening exercise: counsel tracks the cross in real time for the answers that need completing, and the expert, warned in advance, finishes on redirect the explanations the leading questions foreclosed — without the argumentative edge that squanders the recovered ground.

Rebuttal expert testimony is the phase counsel most often fails to plan for, because it is contingent and the pretrial order buries it. The rebuttal witness answers the opposing expert's affirmative case — the alternative-cause theory, the criticized methodology, the competing reconstruction — and its discipline is confinement: rebuttal that restates the case-in-chief invites exclusion, while rebuttal that takes the opponent's opinion on its own terms and shows where its inputs, method, or application fail is often the most persuasive expert testimony in the trial, because it is the only testimony the jury hears after it already knows both stories. The strongest rebuttal is frequently built from the opposing expert's own file — the sensitivity the opinion never tested, run; the alternative the differential never excluded, examined — which is another reason the deposition campaign of Chapter 4 should be conducted with the rebuttal witness already listening.

Underneath all of it is the question counsel should ask before jury selection: how do twelve nonspecialists actually decide between two credentialed witnesses who contradict each other? The honest answer is that they resolve the contest on proxies for trustworthiness, because the merits are, by design of the problem, beyond lay verification. The proxies are learnable: the witness who taught rather than performed; the witness who conceded the small true thing cleanly and explained the large one patiently; the witness whose opinion did not move when its author was pushed; the witness whose account matched the physical exhibits the jurors could inspect themselves. Voir dire and opening should be built to prime those proxies — telling the jury, before any expert testifies, what a reliable method looks like and inviting them to grade both witnesses against it — because the party that supplies the jury's scoring rubric usually wins the category. The battle of the experts is rarely won by the better expert. It is won by the better-prepared jury.

VIII. Practice Implications

For counsel, trial preparation of a technical expert is two distinct exercises that novices merge: the methodology rehearsal for the judge and the translation rehearsal for the jury, run separately because they reward opposite instincts. Build the direct as a lesson, not a resume; foundation every demonstrative to the admitted evidence; and prepare the cross by running the opponent's best campaign at your own witness until it finds nothing the report did not already disclose. For the expert, the courtroom is where every prior discipline is spent or squandered: the report written to survive, the method tied to authority, the assumptions declared, the neutrality maintained — all of it exists to be delivered, intelligibly and unshaken, to the twelve people the entire case was always about. The record decides; the trial is where the jury is finally shown what the record says.

References

Committee Note to the 2023 Amendment of Fed. R. Evid. 702.

Daubert v. Merrell Dow Pharms., Inc., 509 U.S. 579 (1993).

Fed. R. Evid. 702; Fed. R. Evid. 703; Fed. R. Evid. 803(18).

Gen. Elec. Co. v. Joiner, 522 U.S. 136 (1997).

Kumho Tire Co. v. Carmichael, 526 U.S. 137 (1999).

CHAPTER 6

The Record Doesn't Pick a Side

Records Are Only Records Until a CSP or CIH Reads Them With Discipline

> *A permit is a sheet of paper. A near-miss report is a form somebody filled out at the end of a shift, an audit finding is a line in a spreadsheet, a training record is a signature, and a scaffold tag is a punched date. None of it argues, none of it concludes, and none of it knows which party will one day subpoena it. Those artifacts become evidence only when a certified safety professional or certified industrial hygienist reads them against the physics, the governing standard, and the hierarchy of controls, and then states plainly what the reading supports and what it will not carry. This chapter is about that act of reading — the instruments it uses, the discipline it demands, and the reason a disciplined reading returns the same answer from either chair. The dropped-object cases, and the 2025 verdict in Valdivia v. Brock Industrial Services, LLC, No. C-731101, supply the worked example.*

I. Records Are Inert Until Someone Reads Them

A litigation file arrives as a stack of artifacts, not as an account of anything. The permits were issued because the permit system required a permit. The audit was performed because the calendar said the quarter was due. The near-miss report exists because a coworker filled out a form at the end of a shift and dropped it in a box, and the training roster exists because someone had to sign something before the class let out. Not one of those documents was created to prove or disprove a claim. Not one contains a conclusion about duty, breach, foreseeability, or causation — and the document that appears to contain one, the incident report with "employee error" checked on the second page, is asserting a characterization its author had no method for reaching and no authority to state. What the file holds is dates, quantities, signatures, checkboxes, photographs, and gaps. What it does not hold is an opinion.

That is the sense in which the record does not pick a side. It was made before anyone chose sides, by supervisors, auditors, trainers, coworkers, and compliance officers who did not know a lawsuit was coming and had no party to serve when they wrote. It will say the same thing to whichever professional opens the box. But the neutrality of the record is not the neutrality of the reading, and the two are confused constantly — by counsel who believes the documents are the case, and by the expert who believes that having the documents is the same as having an opinion. Two qualified professionals can open the same box and produce opposite reports. When that happens it is almost never because the record was ambiguous about what it recorded. It is because at least one of the two readings was undisciplined: a document measured against a standard that did not govern the task, an absence treated as the absence of information, a number lifted out of the conditions under which it was taken, or a stream of leading indicators skimmed for the three entries that helped and closed before reaching the two that did not.

The expert's contribution, then, is not access to the record. Counsel has that, and by the close of discovery counsel has usually read more of it than the expert has. The contribution is method: a repeatable procedure for converting an inert artifact into a finding, stated explicitly enough that an opposing professional can run the same procedure on the same artifact and check the result. That is what the CSP and the CIH are for, and it is what the December 2023 amendment to Rule 702 now obliges the proponent to demonstrate to the court rather than to the jury on cross. The credential does not make the reading right. The method does. The credential is evidence that the reader was trained in one and can be held to it.

The dropped-object case is the clearest place to watch that work happen, which is why it anchors this chapter. Struck-by injuries from falling objects occupy a peculiar position in workplace litigation: the event is usually over in less than two seconds, there is rarely a dispute that it happened, and yet the legal outcome ranges from a routine workers'-compensation claim to one of the largest single-plaintiff verdicts in American history. Because the event itself is not contested, everything that separates those outcomes is a matter of what somebody read in the paper generated before the object fell. There is no eyewitness dispute to resolve and no competing account of the accident to referee. There is a record, and there are two readings of it.

Consider how quickly a reading can go wrong on a number that nobody disputes. The Bureau of Labor Statistics counted 357 fatal injuries from being struck by a propelled, falling, or suspended object in 2024 — forty-seven percent of the 756 fatal contact incidents that year, and the largest single category within it (CFOI Table 2, released Feb. 19, 2026). That figure is accurate, published, and easy to cite. It is also not comparable to the figures published for 2022 and earlier. BLS implemented version 3 of the Occupational Injury and Illness Classification System for 2023 data forward, and the older category "struck by falling object or equipment"

— 238 deaths in 2022, and in the 217–278 range across the preceding five years — was folded into the broader propelled-falling-suspended category the current tables report. An expert who builds a trend argument across the 2022/2023 boundary has not misread a hard question; he has read a true number without reading the classification scheme that produced it, which is the same unstated-assumption error Chapter 17 catalogues, and opposing counsel who knows the tables will say so. The record was neutral. The reading was not disciplined.

The same lesson runs through the physics. A record in a dropped-object case will often supply two numbers and no analysis: an object weight and a working height. Those numbers mean nothing standing alone. Read against the test protocol behind the head protection the worker was wearing, they become a constraint on every theory in the case. A hard hat meeting ANSI/ISEA Z89.1 is tested against an eight-pound impactor dropped about five feet, on the order of fifty joules of energy. A twenty-pound bar falling thirty feet arrives at roughly thirty miles per hour carrying roughly fifteen times that energy. That arithmetic is a worked illustration of the relationship, not a finding in any particular case — the height, the mass, and the resulting energy must be established from the record of the incident actually at issue, and Section VI's leading case turned on notice rather than on drop distance. The point the illustration makes is the general one: the hard hat is designed for a low-energy crown impact — the dropped fastener, the bumped beam — not for mass at height, which is why every authority from OSHA to NIOSH places dropped-object protection above PPE in the hierarchy of controls: eliminate the overhead exposure, engineer containment through toeboards, screens, and debris netting, and administratively exclude workers from the zone below, and only then rely on the helmet.

Read that way, the PPE issuance log stops being a defense exhibit and becomes something narrower and more useful: proof of what the worker was wearing, and proof of nothing about whether the hazard was controlled. In litigation this means the compliance question is never whether the worker wore a hard hat. It is whether the parties above him controlled the hazard the hard hat was never built to stop — and that question, who was required to control the falling-object hazard, by what method, with what knowledge, and with what result, is a technical question before it is a legal one. It is answered by reading documents, in order, against standards, with instruments. It is where the safety or industrial hygiene opinion becomes the spine of the case for either side.

Which is the premise that governs everything below. The plaintiff's expert and the defense expert are not drawing on two bodies of knowledge. They are running one reading discipline on one record from opposite ends, and the discipline sometimes returns a finding the retaining party did not want. The expert who cannot accept that is not an expert; she is an exhibit, and the amended Rule 702 now treats her accordingly.

II. The Instruments of a Disciplined Reading

Interpretation is not a synonym for narrative. A document is interpreted when it has been asked a question it is competent to answer and the answer has been recorded with its assumptions attached; it has merely been narrated when it has been arranged into a story and the story presented as the finding. What separates the two is a small set of instruments the discipline supplies, each of which takes an artifact and returns a constrained statement. The expert must command all of them before knowing which way the record runs, because until every instrument has been applied there is no reading — only a preference. Opinions built without this foundation are the ones excluded under Rule 702 or dismantled on cross-examination; opinions built on it survive whichever party commissioned them.

A. Physics: What a Recorded Number Will and Will Not Support

Every dropped-object reading begins with mass, height, and the energy equation, because those three numbers frame everything else: whether PPE could have mattered, whether the claimed injuries fit the mechanism, and — in jurisdictions that ask the question — whether the elevation differential was legally significant. The instrument converts recorded quantities into physical limits, and physical limits are the one part of the file that cannot be argued into a different shape. The analysis is not confined to long falls. A heavy object that shifts only a short distance can generate force sufficient to crush a hand or fracture a skull, a point New York's appellate courts have repeatedly recognized in holding that a physically significant elevation differential is measured by the force the object generates, not the distance alone. The expert who quantifies impact energy, states the assumptions, and reconciles the result with the medical record has produced the one part of the opinion neither side can argue with. The expert who skips it has left the record's most objective content unread, and has left the door open for the other side's witness to own the physics.

B. The Standards Matrix: Each Document and Its Duty

A record is read against something or it is not read at all, and the something is rarely a single rule. The governing duties in a dropped-object case are a matrix keyed to the source of the hazard. On scaffolds, OSHA's falling-object provisions require toeboards, screens, guardrail systems, debris nets, catch platforms, canopies, or barricaded exclusion zones below the work. Head protection is separately required where overhead hazards exist, and materials-storage and housekeeping rules govern how tools, stock, and cut debris are staged at height. Above the regulatory floor sits ANSI/ISEA 121, the consensus standard for dropped-object prevention, which addresses tool tethering, attachment points, and containment systems and which, though not directly enforceable, can inform the recognized-hazard analysis under the General Duty Clause and helps establish what a diligent contractor knew or should have known was available. The edition matters, and the expert who cites this standard without dating it has committed the anachronism that Section VI and Chapter 16 §V both identify as a reason opinions get struck. ANSI/ISEA 121-2018 was the first edition and the operative one until the middle of 2024; the revision carrying the designation ANSI/ISEA 121-2023 was approved by ANSI on July 2, 2024 and published that month. For an incident in August 2022 — including the one Section VI examines — the governing consensus document is therefore the 2018 edition, and a state-of-the-art opinion built on the 2023 revision is describing a standard that did not yet exist.

The instrument's discipline is that the mapping runs task first, document second. The expert identifies the specific task being performed at the moment in question, determines which provisions that task triggered, and only then asks what the paper shows about compliance with those provisions. Run the other way — document first, then a search for a standard the document appears to violate — the reading produces citations that do not attach, and a claim framed against the wrong duty, or a defense that ignores an applicable one, collapses the moment a competent opposing expert draws the correct map.

C. The Hierarchy of Controls: Ordering the Reading

The hierarchy of controls is what converts a pile of documents into a sequence of answerable questions, and the sequence is the point: read out of order, the same documents support almost anything. For any overhead exposure, the questions come in order. Could the simultaneous work above and below have been separated in time or space? Were engineering controls — containment, netting, toeboards, tethering — installed and maintained? Were exclusion zones established, barricaded, and enforced beneath the overhead work? Was the permit or job-hazard-analysis system used to identify the exposure before the shift began? Only after those questions are answered does PPE enter the analysis at all.

The documents that answer them — JSAs, permits, toolbox-talk records, audit findings, near-miss reports, disciplinary files — are the same documents in every case. What differs is which way they point once they have been asked the questions in that order. That symmetry is the foundation of everything that follows, because the expert's credibility rests on a demonstrated willingness to run the sequence to the end and report where it stopped, whichever way it ran.

D. The Rest of the Toolbox

Physics, standards, and the hierarchy are the load-bearing instruments, but they are not the whole kit, and the readings that decide cases are the ones that reach for everything the discipline provides. Structured causation methods — fault-tree and bowtie analysis, the incident-investigation frameworks published by CCPS and adopted across industry — force the expert to enumerate every barrier that stood between the elevated object and the worker below, and to document which barriers were absent, which failed, and which were never designed in. That enumeration is what keeps a reading from stopping at the first failure it finds and calling it the cause. Leading-indicator analysis reads the employer's own near-miss, audit, and observation data as the predictive stream it was designed to be, and asks the disciplined question: what did this system know, and when? Human-factors analysis addresses the hand that actually released the object — workload, fatigue, task design, the normalization of shortcuts — and distinguishes the error a system should have anticipated from the violation no system can. Scene reconstruction, from photographs, measurements, and the object's documented resting position, anchors the trajectory to physical fact rather than to recollection, and it is the instrument that tells a court what a lost object or a reconfigured scaffold would have shown. And exposure-and-probability reasoning — how many workers, how many hours, beneath how much unsecured overhead work — is what converts an anecdote into a rate and a rate into a statement about foreseeability.

None of these instruments is proprietary to either side. Each of them, applied honestly, will sometimes build the claim and sometimes break it, which is exactly why the expert must apply all of them before knowing which. An opinion that reports the output of the three instruments that favored the client, and is silent about the two that did not, is not a reading of the record. It is a selection from it, and Section VI describes what courts have begun doing about that.

III. What the Record Holds, and How Fast It Disappears

A reading is only as good as the record it is run on, and in dropped-object litigation the record is perishable on a scale of days. The object is returned to service or scrapped; the scaffold is reconfigured by the next shift; the exclusion-zone tape comes down before the ambulance is out of the gate; and the staging of materials at height, the single most probative physical fact in the case, exists afterward only in whatever photographs someone thought to take. The expert should therefore be on the file before discovery is served, not after it closes — Chapter 1 §III makes the general case — because the expert is the only participant who knows which records the safety-management system generates, what each should contain, and which absences are themselves findings.

That last clause is the part most often skipped. In a functioning safety-management system, a document's absence is not an information gap; it is a recorded fact about the system, because the system was designed to produce that document. A scaffold that was erected but has no inspection entry, an overhead task with no permit, a unit with a near-miss reporting program and no near-miss reports for eleven months, a corrective action opened and never closed — each of those silences is readable, and each reads differently depending on what the program itself said should have been there. Reading a silence requires the governing document first: the written program, the contract scope, the site safety plan. Absent that, the expert who announces that a missing record proves indifference has read nothing and asserted much, and the expert who declines to draw any inference from a silence has abandoned half the record.

From the employer and site defendants, the demands follow the instruments. The object itself, preserved with its documented resting position, and the photographs and measurements of the elevated work area as it stood. The permits, job-hazard analyses, and pre-task plans for the overhead work, together with the site plan showing any exclusion zone and the records of its enforcement. The scaffold erection, inspection, and modification records: the paper that shows whether toeboards, screens, and netting were present, and when. The tethering and containment equipment issued for the task, against the inventory the program said should be issued. The near-miss, audit, observation, and stop-work records for the unit — the whole file and not the single form for the day in question, because a near-miss stream is read as a stream and a single report read alone can be made to mean anything. The training, competency, and disciplinary files of the workers above, including, where the record raises it, the file of the specific employee whose conduct released the object. And the complete OSHA inspection file, obtained in discovery or by records request, rather than the citation summary: the compliance officer's photographs and measurements are frequently the only professional documentation of the scene that will ever exist, which is to say the only part of the record made by someone trained to record it.

From the plaintiff, the defense demands the mirror record and reads it with the same instruments. The complete medical file, pre- and post-incident, tested against the impact-energy calculation and the claimed mechanism. The plaintiff's training acknowledgments, prior discipline, and assigned position at the moment of the strike, inside or outside any designated zone. The PPE issued and worn — not because the hard hat could have stopped the bar, but because the use record speaks to the worker's and the crew's relationship to the program. Prior claims and subsequent activity bearing on causation and damages. The symmetry is the point: the defense expert who audits the plaintiff's record with the same instruments the plaintiff's expert turned on the employer is practicing the same discipline, and the jury can tell.

Preservation is where the record's readability is won or forfeited before anyone opines. Both sides should move on the first day: the plaintiff with a specific preservation letter identifying the object, the elevated work area, the equipment, and the records; the defendant by preserving them, documenting the preservation, and photographing the scene before it changes. The stakes are not theoretical. New York's appellate courts, in recent construction-injury decisions, apply a settled framework — a party seeking sanctions must show an obligation to preserve at the time of destruction, a culpable state of mind that can be satisfied by ordinary negligence, and the relevance of what was lost — and the available sanctions run from an adverse-inference charge to the striking of a pleading. But the same courts have declined to sanction where the lost evidence was not central to the case or its loss caused no real prejudice. Which is to say: whether spoliation decides the case turns on a technical question — what would a competent reading of the lost object or record have established? — and that question is answered by an expert affidavit, not by counsel's assertion. The expert who can explain precisely what the scrapped bar or the discarded scaffold plank would have revealed, and by what method, is the difference between a spoliation motion that wins the case and one that draws a shrug.

IV. Reading the Record: Hazard, Notice, Certainty

When the reading runs in the plaintiff's direction, it climbs. Three layers rest on one another — that a falling-object hazard existed and was controllable; that the responsible parties knew or should have known of it; and, where the theory demands it, that the injury was not merely possible but the predictable product of continuing the operation as it stood — and each layer is a different question put to the same documents. The higher the legal theory reaches, the more the reading must deliver, and the discipline is knowing where the record stops supporting the climb.

The base layer is the uncontrolled hazard, read out of the defendant's own system. The expert establishes what the task above the worker was, what falling-object controls the standards matrix required for that task, and which of them were absent or unenforced. This is documentary work: the scaffold inspection records that do or do not show toeboards and screens, the permit that does or does not identify overhead work, the site plan that does or does not establish an exclusion zone, the tethering equipment that was or was not issued. The strongest plaintiff readings are drawn almost entirely out of the defendant's own paper, because a hazard the defendant's own program was designed to control is a hazard the defendant cannot call unforeseeable — and because a finding assembled from the opponent's records requires the expert to have interpreted, rather than supplied, the content.

The second layer is notice, and it is read out of the leading-indicator stream. Safety management generates a continuous flow of predictive data — near-miss reports, audit findings, prior incidents, stop-work records, disciplinary files — and the discipline here is to read the whole flow for the specific hazard at issue rather than to search it for the entries that help. A prior dropped-tool near miss in the same unit, an audit noting missing toeboards, a supervisor's email about debris falling from a deck: each converts the general risk of overhead work into specific, dated, employer-held knowledge, and the date is the finding. Where the record shows a particular employee repeatedly involved in unsafe acts, the reading sharpens further, because continued assignment of a known-unsafe worker to overhead tasks above unprotected colleagues is a management decision, made with knowledge, repeated over time: precisely the material from which the highest-culpability theories are built.

The top layer, substantial certainty, is where the reading must do its hardest work and where an undisciplined one is most easily exposed. In most states the exclusive-remedy bar of workers' compensation forecloses a tort suit against the injured worker's own employer unless the employer's conduct was intentional or the injury was substantially certain to follow from it. That is a far higher bar than negligence, and the expert cannot meet it with adjectives. What meets it is a probability argument grounded in the record: a specifically identified hazard, dated management knowledge of it, feasible controls identified by the employer's own program and by consensus standards, and a decision, renewed shift after shift, to continue the exposure unchanged. The function of the reading is to show that on this record the incident was not a bolt from the blue but the arrival of an outcome the system had been announcing, and that the only genuinely uncertain variable was which worker would be standing in the zone when it arrived. Where the record will not carry that, the disciplined expert says so, and says so to retaining counsel before it is said on cross.

In statutory-liability jurisdictions the reading maps mechanism to statute, and a word of orientation is owed here, because the decisions this chapter draws on come disproportionately from one state. New York's Labor Law § 240(1) imposes absolute liability on owners and contractors for elevation-related injuries, admits no comparative-negligence defense, and has generated more falling-object case law than the rest of the country combined. It is used here because it produced the decided cases, not because it describes the law where most readers practice. Most states have no analogue; a worker injured by a falling object in an ordinary negligence jurisdiction must prove duty, breach, causation, and damages in the conventional way, and the physics that establishes a "physically significant elevation differential" under § 240(1) instead goes to foreseeability and the standard of care. Read the New York material, then, as a worked example of what statutory strict-liability regimes ask an expert to establish — other examples include the scaffold and structural work acts several states retain, and the safe-place statutes of a handful more — and localize it before relying on it. Where the applicable scheme is a statutory one, the assignment changes shape: the expert establishes that the object was one that required securing for the purposes of the undertaking, that the securing was absent or failed, and that the elevation differential was physically significant given the object's weight and the force it generated. Recent appellate decisions have made the technical content of that mapping decisive, holding that a heavy object falling a short distance still triggers the statute, that recently installed components are not exempt as permanent parts of the structure, and that the worker's own conduct in placing a ladder or standing in the area sounds only in comparative negligence, which is no defense to the statute at all. Each of those holdings is, at bottom, a ruling about physics and construction practice — which is to say a ruling that the side whose expert read the record more precisely was positioned to win.

V. Reading the Record for What It Will Not Support

Run from the defense chair, the discipline is the same and the direction reverses. The assignment is not to deny the injury or to minimize the hazard: a defense reading that argues for gravity's innocence loses the jury in the first hour. The assignment is to test each layer of the plaintiff's construction against the documents and the standards, and to hold the case at the level of culpability the record actually supports — because in the employer-defendant case the entire financial exposure turns on whether the conduct stays within negligence, where the exclusive remedy holds, or crosses into intent, where it does not.

First, audit the duty mapping. The defense reading begins where the plaintiff's should have, with the standards matrix, and asks whether the duties invoked actually attached. Was the object one the applicable rule required to be secured for that task, or material being deliberately moved in the ordinary course? Was the injured worker inside a designated exclusion zone or outside it? Did the falling-object provision cited govern the surface the object fell from? A meaningful fraction of dropped-object claims are framed against duties that did not apply to the task at hand, and the defense expert who demonstrates the mismatch with the text of the standard, rather than with rhetoric, has removed the foundation from everything built above it.

Second, read the program as it actually operated. A written plan is not a defense, but an enforced one is evidence, and the defense expert assembles the affirmative record: training completed before the incident, JSAs that identified overhead work, permits issued, toolbox talks documented, audits performed and findings closed, discipline imposed for prior violations. The purpose is double. It answers the claim of indifference directly, and it re-reads the notice evidence: a near-miss report inside a functioning reporting system, followed by a dated corrective action and a closed finding, is proof the system worked, not proof the employer ignored danger. The distinction between an employer that generated safety data and acted on it and one that generated the data and filed it is frequently the whole case, and nothing but the underlying records — read in sequence, with the closure dates — can make it.

Third, confront hindsight with base rates and intervening conduct. The substantial-certainty theory depends on the proposition that this incident was the predictable culmination of the record rather than a low-probability event that happened to occur. The defense reading tests that proposition with the industry's own numbers — tens of thousands of struck-by injuries annually across every sector, the large majority occurring at employers no jury would call intentional actors — and with a rigorous sequence analysis of the incident itself. If the object fell because of a single unforeseeable act, a coworker's momentary mishandling or a failure invisible to inspection, the causal chain runs through conduct the employer could not have scheduled around, and the case belongs in negligence, where the statutory bargain places it. The expert must make this argument with care, because hindsight bias runs the other way in the jury box: after a catastrophic outcome, every prior warning looks like prophecy. The job of the defense reading is to put the warnings back into the context in which they were actually received — what the entry said, what it was ranked, what the reporting volume was that month, and what the system did with it.

Fourth, test the fit between mechanism and injury. The physics that anchors the plaintiff's case is equally available to the defense. Impact energy calculated from the actual mass and the actual height either does or does not account for the claimed injury pattern; the object's documented resting position either is or is not consistent with the described trajectory; the medical record either does or does not match the mechanism. Where the fit is clean, the honest defense expert says so and the case is about liability, not causation. Where it is not, the mismatch is often the most powerful evidence the defense possesses, and it is evidence only a technically disciplined reading can develop.

Finally, know what the record will not be made to say. The defense expert's most valuable contribution is sometimes delivered before the answer is filed: a candid assessment that the file will not hold the position counsel wants to take. Where the record shows a specifically identified hazard, feasible controls, and continued exposure over documented internal objection, the reading that pretends otherwise does not survive cross-examination — it merely raises the verdict. The recent history of nine-figure dropped-object awards is in large part a history of defendants who tried their cases against records their own experts could not honestly read in their favor.

VI. What the Recent Cases Teach

A note on posture before the cases. The verdict that anchors this chapter is a trial-court judgment, not an appellate holding. As of this writing it was on the public record as headed for appeal, with related insurance-coverage litigation pending, and nothing in the discussion below should be read as a prediction of what survives review. That is not a caveat about this case only. A jury verdict is evidence of what a record did to twelve people; it is not law, it establishes no rule, and a substantial-certainty finding is precisely the kind of finding appellate courts examine closely. The chapter uses the verdict as a teaching artifact — an unusually complete illustration of what a notice record does once it has been read properly, and of what happens when the other side cannot answer the reading — and an expert who cites it in a report should describe it as what it is. Chapter 1 §IV collects the decisions in which verdicts were erased on appeal because an opinion should never have been admitted, and Chapter 2 §II sets out the four joints at which such opinions fail.

Valdivia v. Brock Industrial Services, LLC, No. C-731101 (La. 19th Jud. Dist. Ct., jury verdict, Feb. 2025) — the record read all the way up. A twenty-five-year-old scaffold builder at a Lake Charles refinery suffered catastrophic brain and spinal injuries when a twenty-pound metal scaffold bar struck his hard hat: dropped, on the sworn accounts, by a coworker handling it overhead after the injured worker had himself just passed it up by hand, so that it slipped from the coworker's grip and came back down on him. The jury returned $411 million, reported as the largest single-plaintiff personal-injury award in Louisiana history, after finding the employer's conduct intentional or the incident substantially certain to occur, the finding required to escape the exclusive-remedy bar. The twelve-person jury deliberated, on the contemporaneous trial reporting, about two hours before returning a unanimous verdict. Note carefully what the verdict did not turn on: this was a short drop at close quarters, not the fall from height that the energy arithmetic of Section II.A illustrates, and drop distance was never the plaintiff's case. What carried it was notice — which is to say, what carried it was a reading of the internal record. Evidence that the employer knew a specific coworker was a recurring danger, described in the litigation as a walking hazard, converted a generic overhead-work risk into dated, specific, management-level knowledge, and the continuation of the operation unchanged supplied the substantial-certainty inference. Nothing in that record was created to make the point. Every document in it was made for an operating reason by someone with no side to serve. It is equally the cautionary illustration for the defense: once the internal record establishes specific notice, the defense reading must engage those documents on their own terms, and a two-hour deliberation is what it looks like when it cannot.

Cevallos v. WBB Construction, Inc. and *Colon v. New York City Department of Education* (App. Div. 2024) — the mechanism mapped to the statute. New York's recent falling-object decisions under Labor Law § 240(1) show the statutory version of the same contest between readings. In *Cevallos*, a cut piece of wood fell, struck the plaintiff on the leg and then the ladder he was standing on, and brought worker and ladder down together; the argument that the worker should not have positioned himself beneath the cutting area sounded only in comparative negligence, which is no defense to the statute. In *Colon*, a heavy stone that shifted a mere two feet still presented a physically significant elevation differential because its weight generated substantial force in even a short descent. Both rulings reward the side whose expert quantified force rather than arguing distance, and both illustrate what the statutory jurisdictions actually reward: a reading that maps mechanism onto statutory element — securing required, securing absent, force significant — has done the work the verdict will rest on.

Molina v. 114 Fifth Avenue Associates (App. Div. 2024) — a defense theory and its technical limit. Defendants in falling-object cases frequently argue that the statute does not reach objects that are permanent parts of the structure. *Molina* marks the boundary: a just-installed sprinkler pipe that fell was held not yet permanent for the statute's purposes, because material the workers themselves had recently placed remained part of the undertaking that required securing. The lesson for the defense expert is about precision in reading the construction sequence: the permanence theory is viable, but only where the installation record genuinely supports it, and an opinion that stretches the theory past what the record shows hands the plaintiff a credibility gift.

Marina v. Bama Reinforcing, LLC (Ala. Nov. 2024) — the third-party path and the shared jobsite. The Alabama Supreme Court's reversal of a directed verdict for a rebar subcontractor — in a suit by a concrete finisher injured by a cut piece of metal rebar support left on the ground at a federal construction project — is a reminder that most dropped-object and debris litigation proceeds not against the plaintiff's employer but against the other contractors on a shared site, where ordinary negligence suffices and the exclusive-remedy bar does not apply. For the expert in either chair, the multi-employer worksite adds a document layer before any of the questions in this chapter can be answered: whose task created the hazard, whose contract allocated control of the area, whose program governed housekeeping and overhead protection in the zone, and what each employer knew. The subcontract, the site-specific plan, and the coordination minutes are read first, because until control is mapped the safety records cannot be assigned to anyone.

Teodoro v. C.W. Brown, Inc. and *Yi Jiang Pai v. Nelson Senior Housing Development Fund Corp.* (App. Div. 2021, 2024) — the record preserved, the record lost, and the expert who defines which mattered. New York's recent construction-injury decisions on spoliation frame the discovery stakes with unusual clarity. The courts apply a three-part showing — a duty to preserve at the time of the loss, a culpable state of mind that ordinary negligence can satisfy, and the relevance of what was destroyed — and hold broad discretion over sanctions, from an adverse-inference charge down to nothing at all. The dispositive variable in practice is centrality: where the lost evidence is not central to the case, or its loss works no real prejudice, the sanction shrinks or disappears. Centrality is a technical question in legal dress. Whether the discarded object, the reconfigured scaffold, or the overwritten inspection log was central depends on what a competent reading of it could have established — the trajectory, the securing method, the presence of the toeboard — and only a qualified expert can tell the court, concretely, what the lost record would have answered. Read together, the two decisions instruct both sides identically: preserve everything, document the preservation, and when the other side fails to, put an expert affidavit — not an adjective — behind the spoliation motion.

Koger v. Costco Wholesale Corp. (N.D. Cal. 2023) and the amended Rule 702 — the reading policed at the gate. In a falling-injury premises case, the court's rulings on cross-motions to exclude both sides' experts illustrate the boundaries every safety and industrial hygiene opinion now lives within: opinions phrased as legal conclusions are not expert testimony, and opinions on either side are tested for whether the expert considered the full evidentiary record and applied a reliable method to it. The December 2023 amendment to Fed. R. Evid. 702 sharpened the gate, requiring the proponent to demonstrate — by a preponderance, to the court, not to the jury on cross — that the opinion reflects a reliable application of a reliable method to the facts. The early appellate returns show the amendment has teeth precisely where selective readings live: the Sixth Circuit has recognized that the amendment was drafted to correct decisions treating the sufficiency of an expert's basis as a question of weight rather than admissibility, and that an opinion built on cherry-picked data warrants exclusion; the Acetaminophen MDL court excluded a slate of experts for engaging with only the favorable data and dismissing the rest, holding that reliability requires meaningful engagement with the evidence that cuts both ways. That is the doctrinal statement of this chapter's thesis. The record is not the opinion; the reading is, and a reading that skipped the unhelpful documents is not a reliable application of anything. For the expert, the instruction is concrete: quantify the physics, cite the specific standard provision, tie every inference to a document or a measurement, address the records that point the other way, state the limits of the reading, and leave the words negligent, intentional, and substantially certain to the lawyers. The one-sided reading is no longer merely unpersuasive; increasingly, it is inadmissible.

VII. The Discipline That Makes the Reading Believed

The two constructions described above — the ascent from hazard to notice to certainty, and the audit of duty, program, sequence, and fit — are not opposing methodologies. They are one reading discipline run with equal rigor from opposite ends of the same file, and the expert who has genuinely done both is stronger for it. The plaintiff's expert who has spent a career auditing safety programs knows exactly which enforcement records would have saved this defendant, and their absence is the finding. The defense expert who has built substantial-certainty cases knows exactly what the leading-indicator record must contain to support one, and its thinness is the finding. Retaining counsel on either side should want the expert the other side would have wanted.

The practical program follows directly. Early retention matters more in dropped-object cases than in almost any other injury mechanism, because the most readable evidence — the object itself, the scaffold configuration, the staging of materials at height, the exclusion-zone markings — is altered or returned to service within days, and an unreadable record cannot be rescued by any amount of later diligence. The expert should drive the preservation and discovery demands: the object and its resting position, photographs and measurements of the elevated work area, the permits and JSAs for the overhead task, the tethering and containment equipment issued, the near-miss and audit record for the unit in full, the training and disciplinary files of the workers above, and the complete OSHA inspection file rather than the citation summary. And the report itself should be written for the amended Rule 702 — method stated, application shown, contrary documents addressed, limits acknowledged — because the reading that survives the gate is the reading built to be checked by someone hoping to find it wrong.

This chapter's title is its argument, and the argument reduces to one observation. The record is what it is. It was written by supervisors, auditors, trainers, coworkers, and compliance officers who did not know a lawsuit was coming and had no side to serve when they wrote, and it will sit in its boxes saying nothing at all until a professional trained to read it does so. The certification on the report is a claim about how that reading was performed: that the physics was calculated rather than asserted, that each document was measured against the duty that governed the task, that the hierarchy was run in order, that the silences were interpreted against the program that should have filled them, and that the documents cutting against the client were read as carefully as the ones supporting it. An expert who would perform the same reading in the opposite chair — and who tells retaining counsel plainly when the record runs against them — is the witness a jury believes, the witness the amended Rule 702 admits, and the witness whose opinion decides the case rather than merely accompanying it. The $411 million verdict that anchors this chapter was not produced by advocacy, and it was not produced by the documents either; the documents had been sitting in that employer's own files for years. It was produced when one side's expert read them into an answerable account and the other side could not answer it. Your record doesn't pick a side. Your expert shouldn't either, because the expert who would reach the same conclusion in the opposite chair is the only one whose conclusion is worth anything in yours.

References

In re Acetaminophen — ASD-ADHD Prods. Liab. Litig., 707 F. Supp. 3d 309 (S.D.N.Y. 2023).

American National Standards Institute / International Safety Equipment Association. (2014, as revised). ANSI/ISEA Z89.1, Industrial head protection.

American National Standards Institute / International Safety Equipment Association. (2024). ANSI/ISEA 121-2023, Dropped object prevention solutions (approved July 2, 2024; revising ANSI/ISEA 121-2018, the edition in force until mid-2024).

Cevallos v. WBB Constr., Inc., 227 A.D.3d 657 (2d Dep't 2024).

Colon v. N.Y.C. Dep't of Educ., 228 A.D.3d 422 (1st Dep't 2024).

Head protection (construction), 29 C.F.R. § 1926.100; head protection (general industry), § 1910.135.

Koger v. Costco Wholesale Corp., No. 3:20-cv-08759 (N.D. Cal. 2023) (orders on motions to exclude expert testimony).

Marina v. Bama Reinforcing, LLC, No. SC-2023-0651 (Ala. Nov. 8, 2024).

Molina v. 114 Fifth Ave. Assocs., LLC, 231 A.D.3d 543 (1st Dep't 2024).

National Institute for Occupational Safety and Health. Hierarchy of controls. Centers for Disease Control and Prevention.

N.Y. Labor Law § 240(1); Occupational Safety and Health Act § 5(a)(1) (General Duty Clause); Fed. R. Evid. 702 (as amended Dec. 1, 2023).

In re Onglyza (Saxagliptin) & Kombiglyze Prods. Liab. Litig., 93 F.4th 339 (6th Cir. 2024).

Scaffolds — falling object protection, 29 C.F.R. § 1926.451(h); materials handling and storage, § 1926.250; general housekeeping, § 1926.25.

Teodoro v. C.W. Brown, Inc., 200 A.D.3d 999 (2d Dep't 2021).

U.S. Bureau of Labor Statistics. (2023, December 19). National census of fatal occupational injuries in 2022 (USDL-23-2615), Table 2. U.S. Department of Labor. (Reporting the pre-OIICS v3 "struck by falling object or equipment" series.)

U.S. Bureau of Labor Statistics. (2026, February 19). National census of fatal occupational injuries in 2024 (USDL-26-0230), Table 2: Fatal occupational injuries for selected events or exposures, 2023–24. U.S. Department of Labor.

U.S. Bureau of Labor Statistics. Occupational Injury and Illness Classification System (OIICS) version 3.0, implemented for 2023 data forward.

Valdivia v. Brock Indus. Servs., LLC, No. C-731101 (La. 19th Jud. Dist. Ct. Feb. 28, 2025) (jury verdict of $411,687,387.36 on a finding of intentional conduct or substantial certainty; contemporaneous trial reporting records a week-long trial and approximately two hours of deliberation by the twelve-person jury).

Yi Jiang Pai v. Nelson Senior Hous. Dev. Fund Corp., 2024 N.Y. Slip Op. 05816 (2d Dep't Nov. 20, 2024).

PART II

THE ANATOMY OF AN OSHA CASE

CHAPTER 7

Best Practices During an OSHA Inspection

What the Compliance Officer Must Do, Where Counsel Belongs, and How to Prepare Your Workforce

An OSHA inspection is one of the few encounters with the federal government that is scripted in advance. The script is the Field Operations Manual, the compliance officer is directed to follow it, and the employer is free to read it. This chapter walks the inspection as the manual directs it, explains why counsel belongs there from the first knock rather than the first citation, and closes with the decisions in which evidence was suppressed because an inspection exceeded its bounds.

I. The Inspection Has a Published Script

Every federal OSHA inspection is conducted by a compliance safety and health officer working from the same document: the Field Operations Manual, OSHA Instruction CPL 02-00-164, the agency's enforcement policies and procedures manual, posted in full on the agency's public website and updated chapter by chapter — most recently the penalty chapter, revised July 14, 2025 (Occupational Safety and Health Administration [OSHA], 2025). The manual governs how inspections are scheduled, how they are opened, how the walkaround is conducted, how employees are interviewed, how evidence is documented, and how observations become citations and citations become penalties. The inspection priorities it establishes are themselves a preview of who gets inspected and why: imminent danger situations first, then fatalities and catastrophes — which employers must report within eight hours for a fatality and twenty-four hours for an inpatient hospitalization, amputation, or loss of an eye — then employee complaints and referrals, and finally programmed inspections under site-specific targeting and national or local emphasis programs (OSHA, 2025). One thing the manual does not permit is a courtesy call. Section 17(f) of the Occupational Safety and Health Act makes the unauthorized giving of advance notice of an inspection a criminal offense, and the implementing regulation confines authorized advance notice to a handful of narrow circumstances approved by the area director (Occupational Safety and Health

Act of 1970 [OSH Act], 29 U.S.C. § 666(f); 29 C.F.R. § 1903.6). The compliance officer arrives unannounced because the law requires it.

The stakes of the encounter are set annually. For citations issued in 2026, the maximum penalty is $16,550 for each serious or other-than-serious violation, $16,550 per day for each failure to abate beyond the abatement date, generally capped at thirty days, and $165,514 for each willful or repeated violation; the 2025 amounts were carried forward into 2026 without an inflation adjustment (OSHA, 2026). Because penalties are assessed per violation rather than per inspection, the difference between an inspection that stays within its intended scope and one that expands into a wall-to-wall examination is not a matter of inconvenience; it is frequently the difference between a five-figure event and a six-figure one, before any willful or repeat classification enters the arithmetic, and before the employer's citation history begins attracting the attention of the Severe Violator Enforcement Program.

A candid word about what the manual is and is not. The Field Operations Manual binds the agency's practice as a matter of internal policy; it is not a statute, and the courts have consistently held that it confers no independent procedural rights on employers — a technical departure from the manual, standing alone and without prejudice to the employer, will not invalidate a citation (*Accu-Namics*, Inc. v. Occupational Safety and Health Review Commission, 1975). But two consequences follow from the manual's existence, and both favor the prepared employer. The first is predictability: because the government has published the choreography of its own examination, an employer can rehearse it — can know, before the credentials are presented, exactly what will be asked for, in what order, and by what right. The second is boundary: where the inspection exceeds not merely the manual but the constitutional limits the manual is written to respect — consent, warrant, probable cause, and scope — the remedy is no longer academic, and Section V collects the decisions in which citations did not survive the excess. The pages between here and there describe the inspection as it is supposed to happen, because the employer who knows the script is the one who notices the departure.

II. What the Compliance Officer Must Do

A. Arrival, Credentials, and the Opening Conference

The inspection begins at the gate, and the manual's requirements begin there too. The compliance officer must locate the owner, operator, or agent in charge and present credentials: a photo identification card with serial number that the employer may, and should, verify by calling the area office (29 C.F.R. § 1903.7(a);

OSHA, 2025). At that moment the employer faces the first decision of the inspection: consent or compulsory process. Under *Marshall v. Barlow's, Inc.* (1978), the employer has a Fourth Amendment right to refuse entry and require the agency to obtain an administrative inspection warrant, and the exercise of that right cannot itself be penalized. Whether to exercise it is a judgment call that belongs in a protocol written with counsel long before the knock — most employers consent, because a warrant is ordinarily obtainable and the refusal purchases a day or two at the cost of an inspector's goodwill — but the decision, either way, should be deliberate rather than improvised, and the scope of any consent should be stated and documented at the threshold.

The opening conference that follows is not a courtesy; it is a required stage of the inspection with required content. The compliance officer must explain the purpose and nature of the inspection, its intended scope, and the records that will be requested; where the inspection is complaint-based, the officer must furnish a copy of the complaint, with the complainant's identity withheld if confidentiality was requested, a protection the statute itself guarantees (29 U.S.C. § 657(f)(1); OSHA, 2025). The officer must determine whether an OSHA-funded consultation visit is in progress, whether the employer participates in a voluntary program that exempts it from programmed inspection, and whether appropriations act limitations apply. And the officer must advise both the employer and the employees of their right to have representatives accompany the walkaround: the statute gives an employer representative and an employee-authorized representative each the right to participate (29 U.S.C. § 657(e)), and since 2024 the regulation has permitted employees to designate a third party who is not an employee — including, controversially, a union representative at a non-union site — where the compliance officer determines the third party is reasonably necessary to the inspection (Worker Walkaround Representative Designation Process, 2024). The manual permits an abbreviated opening conference where site conditions argue for prompt entry, but abbreviation is not elimination; the required elements are delivered as the walkaround proceeds. Table 7-1 summarizes the obligations the manual imposes at each stage.

Table 7-1. What the Field Operations Manual Requires of the Compliance Officer at Each Stage

Inspection stage	What the compliance officer must do	Primary authority
Arrival	Present credentials to the owner, operator, or agent in charge; no advance notice absent narrow, approved exceptions; honor a refusal of entry by seeking a warrant rather than forcing access	29 C.F.R. §§ 1903.6, 1903.7; *Barlow's* (1978)
Opening conference	Explain purpose, nature, and intended scope; furnish the complaint (complainant confidential); identify records to be requested; advise employer and employees of walkaround representation rights; check consultation-in-progress and exemptions	29 U.S.C. §§ 657(e), 657(f)(1); FOM Ch. 3
Records review	Request OSHA 300/300A/301 records and required written programs; review injury and illness data relevant to the inspection's scope	29 C.F.R. pt. 1904; FOM Ch. 3
Walkaround	Confine the inspection to its stated scope, subject to plain-view observations; permit employer and employee representatives to accompany; comply with the employer's safety rules and PPE requirements; protect trade secrets observed en route	29 U.S.C. §§ 657(a), 664; 29 C.F.R. §§ 1903.8, 1903.9
Interviews	Question employees privately; honor the voluntary character of non-managerial interviews; permit management witnesses their chosen representative, including counsel	29 U.S.C. § 657(a)(2); FOM Ch. 3
Closing conference	Describe apparent violations and abatement expectations; furnish employer rights and responsibilities information; explain the informal conference and the fifteen-working-day contest period	29 C.F.R. § 1903.7(e); FOM Ch. 3

Inspection stage	What the compliance officer must do	Primary authority
Citation	Issue citations with reasonable promptness and never more than six months after the occurrence of the violation; compute penalties by gravity, then apply size, good-faith, and history adjustments	29 U.S.C. §§ 658(a), 658(c); FOM Ch. 6

Note. FOM refers to the Field Operations Manual, OSHA Instruction CPL 02-00-164, as amended through the July 14, 2025 revision of Chapter 6 (OSHA, 2025). The table summarizes the principal duties discussed in the text; it is not an exhaustive restatement of the manual.

B. The Walkaround, the Sampling, and the Interviews

The statute authorizes the compliance officer to enter without delay and at reasonable times, and to inspect within reasonable limits and in a reasonable manner (29 U.S.C. § 657(a)): language that does real work, because it is the textual home of the scope discipline the courts enforce. An inspection opened on a complaint about a specific condition is an inspection of that condition, the areas and records reasonably related to it, and whatever violations sit in plain view along the reasonable route to it; it is not, without more, a license to open every door in the building. The manual directs the officer to walk the route with the employer's representative and the employees' representative, to comply with the employer's own safety rules and personal protective equipment requirements while on site, and to safeguard any trade secrets observed: a duty backed by criminal penalties for disclosure (29 U.S.C. § 664; 29 C.F.R. § 1903.9). The officer documents as the officer goes: photographs, video, instrument readings, and, in health inspections, personal and area sampling. The employer's representative should be doing exactly the same thing at the officer's shoulder — matching photograph for photograph, noting what was measured and where — because the employer's contemporaneous record of what the inspection actually covered is the raw material of every scope argument Section V describes, and because an employer conducting side-by-side monitoring is entitled to know what the agency's instruments are collecting.

The interviews are where the inspection's center of gravity actually sits. Section 8(a)(2) of the Act authorizes the compliance officer to question any employee privately, and the manual treats private interviews of non-managerial employees as a standard element of every inspection: the employer has no right to be present, the interview is confidential, and a statement given in confidence is protected from disclosure to the employer (29 U.S.C. § 657(a)(2); OSHA, 2025). What the manual does not make the interview is compulsory: a non-managerial employee may decline to be interviewed, may end the interview at any time, and may request a representative of the employee's own choosing, and absent a subpoena the conversation happens only with the employee's cooperation. Interviews of management personnel stand on entirely different footing: a supervisor speaks as the company, the supervisor's statements are admissions attributable to the employer, and the employer is entitled to have its representative — including counsel — present. The distinction between those two categories of conversation, and the fact that most employees have never been told which category they occupy, is the single strongest argument for the workforce preparation Section IV describes.

C. The Closing Conference, the Citation, and the Penalty Arithmetic

The inspection ends as formally as it began. The compliance officer must conduct a closing conference with the employer and with the employees' representative, describe the apparent violations observed, discuss abatement and the timeframes likely to be set, and furnish the employer with information on its rights and responsibilities following an inspection, including the informal conference procedure and the right to contest (29 C.F.R. § 1903.7(e); OSHA, 2025). What follows the conference is governed by two clocks that run against the government and one that runs against the employer. Citations must be issued with reasonable promptness, and section 9(c) of the Act flatly bars any citation issued more than six months after the occurrence of the violation (29 U.S.C. §§ 658(a), 658(c)). The employer's clock is shorter and less forgiving: fifteen working days from receipt of the citation to file a notice of contest, after which an uncontested citation becomes a final order of the Review Commission — unreviewable, and permanently available as the predicate for a repeat classification at ten times the serious penalty.

The penalty the citation proposes is not an arbitrary number; it is the output of a calculation the manual prescribes in detail, and the calculation is where preparation is literally priced. The officer assigns a gravity-based penalty from the severity of the harm threatened and the probability of its occurrence, and then applies reductions that the July 14, 2025 revision of Chapter 6 makes materially more generous than before, in the sequence that revision fixes — size, then good faith, then history, then quick fix. The size reduction still tops out at 70 percent, but the threshold moved, and the movement is the part a reader with a small-employer client needs: the maximum now reaches employers with twenty-five or fewer employees rather than ten or fewer, and on serious-willful items an 80 percent reduction reaches employers of twenty or fewer. Good faith is worth up to 25 percent for an effective, documented safety and health management system in operation before the officer arrived. A clean recent citation history is worth up to 20 percent, doubled from the previous 10 percent and now available to employers never inspected as well as to those inspected within five years without a serious, willful, repeat, or failure-to-abate violation. And a 15 percent quick-fix reduction rewards prompt correction: the revision expanded it beyond abatement performed in the officer's presence to abatement completed as soon as possible afterward, allowing several days where correction requires purchasing materials, fabrication, or training, and it applies to other-than-serious and low-to-moderate-gravity items (OSHA, 2025). Figure 7-1 displays the structure. The good-faith reduction deserves particular attention, because it is the line item the employer controls completely and earns entirely in advance: it is awarded or withheld on the strength of the written programs, training records, and self-inspection documentation the employer can produce at the opening conference, and no amount of charm during the walkaround substitutes for it.

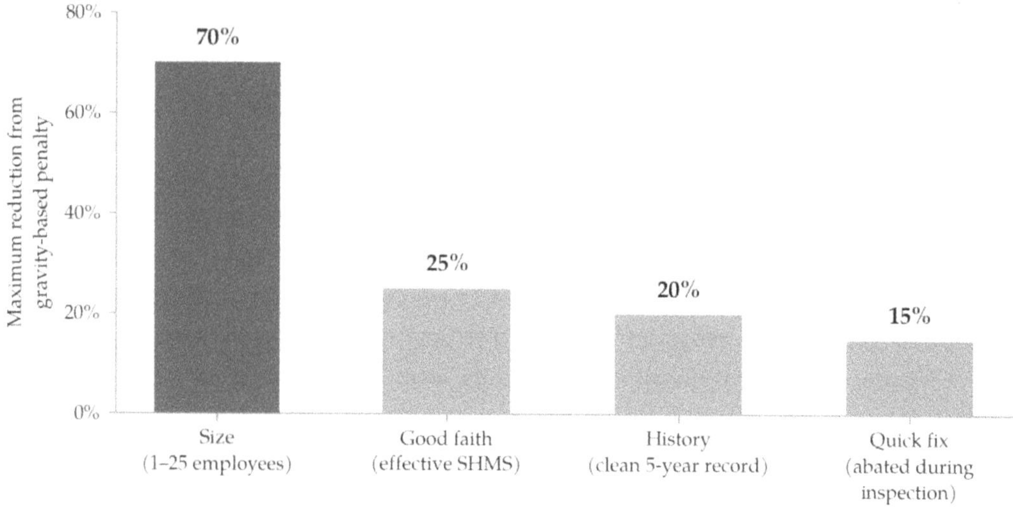

Figure 7-1 Maximum Penalty Adjustment Factors Under the Field Operations Manual Note. Adjustment factors are applied to the gravity-based penalty calculated by the compliance officer, as set out in Chapter 6 of the Field Operations Manual, revised July 14, 2025 (OSHA, 2025). Maximum size and history reductions depend on employer size and citation history criteria detailed in the manual; good faith requires a documented, operating safety and health management system; the quick-fix reduction applies to other-than-serious and low-to-moderate-gravity items abated during the inspection or as soon as possible afterward. *Good-faith and history reductions are not applied to willful, repeat, or failure-to-abate violations;* size reductions remain available for serious willful violations on the separate schedule the manual sets out, which reaches 80 percent for employers of twenty or fewer.

III. Where Counsel Belongs in the Process

The instinct that calling a lawyer will antagonize the inspector has cost employers far more than it has ever saved them, and it misunderstands what counsel actually does during an inspection. Counsel's first and most valuable contribution is made before any inspection begins: a written inspection protocol that settles, in advance and in calm, the questions that cannot be competently answered at the gate — who greets the officer and who is called, whether and on what terms the company consents to entry, how the scope of a consented inspection is stated and confirmed in writing, which records are produced on request and which requests are referred to counsel, and who accompanies the walkaround. The consent-or-warrant decision in particular is a legal judgment with facility-specific stakes: consent given without a stated scope is the cheapest thing an employer will ever give away, because an inspection whose boundaries were never articulated is an inspection whose boundaries can never be exceeded. Counsel's role is not to slam the gate: in the ordinary case it is to open it, deliberately, with the scope stated, the consent documented, and the ground rules established while everyone is still being polite.

During the inspection itself, counsel manages the two channels through which the employer's own conduct creates evidence: documents and statements. On documents, the discipline is to produce promptly and completely what the law requires — the injury and illness records the employer must keep and make available, the written programs the standards mandate — while ensuring that anything beyond the legally required production is a considered decision rather than a reflex, and that a log is kept of every page that changes hands. On statements, counsel attends management interviews as of right, prepares supervisors for the difference between answering a question and narrating a theory, and ensures that any post-incident internal analysis — the root-cause investigation, the expert's preliminary assessment — is structured from the outset under counsel's direction if privilege is to be claimed for it, because privilege attaches to how a document was created, not to how badly the employer later wishes it were protected. Counsel is also the natural custodian of the parallel record: the side-by-side photographs, the duplicate sampling data, the contemporaneous notes of where the officer went and what the officer said, all of which exist for exactly one future reader — the judge who will someday be asked whether the inspection stayed within its bounds.

The compliance officer arrives and you are not the host. On a multi-employer site the opening conference has a participant problem the single-employer script does not anticipate, and the wrong instinct is common in both directions. A subcontractor whose crew is the subject of the inspection is a cited-employer candidate in its own right and should not delegate its participation to the general contractor, whose interests diverge the moment the citation-policy analysis of Chapter 11 begins: the controlling employer's best position is frequently that the creating employer's conduct was unforeseeable, which is a position adverse to the creating employer. Nor should a subcontractor decline to participate on the theory that the site is someone else's problem; the walkaround, the interviews, and the measurements are producing the record that will be cited against it. The workable practice is that each employer designates its own representative, each accompanies the walkaround as to its own work, each takes its own parallel documentation, and no employer's representative answers questions about another employer's operations, personnel, or programs, which are questions that employer's representative is present to answer. Joint-defense arrangements can be sensible on a multi-employer site, but they are agreements to be made deliberately by counsel, not postures to be assumed in a trailer at eight in the morning. The host's obligations to grant access and the subcontractor's obligations to its own workforce are separate, and an employer that lets the general contractor speak for it has adopted statements it did not make.

After the closing conference, the process becomes procedural, and the deadlines are unforgiving of self-help. The informal conference with the area director — available only within the fifteen-working-day contest window — is routinely the highest-yield hour in the entire enforcement process: classifications are amended, penalties reduced, and abatement dates extended there, and an employer who attends without counsel negotiates against the Field Operations Manual without having read it.

Because the chapter calls it the highest-yield hour, it should describe it. The conference is informal in the sense that it is not adjudicative — no rules of evidence, no transcript, no sworn testimony — and formal in the sense that it is a negotiation with a federal official who has settlement authority and a directive telling him how to use it. It is typically attended by the area director or an assistant, the compliance officer who conducted the inspection, the employer's representative, counsel, and, where the workforce is represented, the union, which has a right to participate. Requesting it does not extend the contest deadline; an employer that lets the fifteen working days lapse while a conference is being scheduled has converted its citation into a final order, and the standard practice is therefore to file the notice of contest and confer in parallel, withdrawing the contest if the conference resolves the matter.

What to bring is the abatement evidence, and it should be assembled as if for a hearing. Photographs of the corrected condition with dates. The purchase orders, work orders, or invoices proving the correction was bought and performed. The revised written program with the change dated. The training roster showing the affected workforce was retrained, signed. The inspection record showing the correction was verified in place after installation. And — where the employer intends to argue the size, good-faith, or history reductions — the employee count, the safety-management-system documentation, and the citation history that support each. An employer that arrives with an explanation gets an explanation's worth of credit. An employer that arrives with a closed corrective-action package frequently gets the classification amended, which is worth far more than the penalty.

What a settlement looks like on paper is worth previewing, because employers are routinely surprised by it. The informal settlement agreement will typically recite the amended classification and penalty item by item, fix new abatement dates, require abatement certification and, for higher-gravity items, documentation and tagging under 29 C.F.R. § 1903.19, and include a withdrawal of the notice of contest. Two provisions deserve counsel's attention every time. Modern agreements frequently include enhanced terms — third-party audits, corporate-wide abatement across establishments, interim monitoring, periodic reporting to the area office — and every one of those is independently enforceable, with per-day penalties for failure to abate, so each should be treated as an engineering specification with a named owner and a verification record rather than as boilerplate. And agreements commonly include a clause stating that the settlement is not an admission for purposes other than the Act; that clause is worth negotiating for, and it is worth knowing that its effect in a later civil case is a question of state evidence law rather than a matter the area director can guarantee. The contest decision itself is a litigation judgment with consequences beyond the penalty on the page, because an uncontested serious citation is a permanent fixture in the employer's history: the predicate for repeat classifications across a five-year, potentially corporate-wide lookback, and the plaintiff's first exhibit in any injury lawsuit the underlying facts later generate. And throughout the process, counsel polices the line the company must never approach: statements to a federal compliance officer are statements to the federal government, false statements are a felony under both the Act and the general false-statements statute (29 U.S.C. § 666(g); 18 U.S.C. § 1001), and the surest way for a routine civil inspection to become a criminal referral is a manager who improvises, a record that was improved after the fact, or an interview that was interfered with. Counsel's presence is not obstruction; it is the mechanism by which the employer participates fully, truthfully, and no further than the law requires.

IV. Preparing the Workforce: What You Can and Cannot Say

An OSHA inspection is conducted through conversations, and nearly everyone the compliance officer will speak with is someone the employer never prepared. The receptionist decides how long the officer stands unescorted in the lobby watching the forklift traffic. The supervisor decides whether the answer to a question is what happened or a theory of why it was not the company's fault. The third-shift operator decides whether the written lockout procedure describes anything that actually occurs at his machine. None of these people can be scripted — and, as discussed below, must not be — but all of them can be trained, and the training must happen in calm times, as a standing element of the safety program, because training conducted in the parking lot while the officer waits in the conference room is both worthless and conspicuous.

For the management team, the affirmative practices are concrete. Designate an inspection coordinator and an alternate for every shift, and train reception to verify credentials, notify the coordinator immediately, and escort the officer to a conference room: promptly and courteously, because delay reads as concealment. Accompany the officer everywhere; the walkaround right exists to be used. Mirror the record: photograph what the officer photographs, note every instrument reading, log every document produced. Answer questions truthfully, responsively, and narrowly — the truthful answer to a question no one asked is the most common self-inflicted wound in enforcement history — and treat every answer as testimony, because a supervisor's words are the company's admissions. Never guess, estimate, or speculate; the correct answer to an unknown is that you will find out. And fix what can be fixed on the spot: immediate abatement in the officer's presence is rewarded in the penalty calculation and remembered in the report, and the manual's quick-fix reduction exists precisely to encourage it (OSHA, 2025). Table 7-2 collects the disciplines.

For employees, preparation means information, not instruction. Employees should understand, before any inspection, what the law already provides: that a compliance officer may ask to interview them privately; that for non-managerial employees the interview is voluntary and confidential — the employer will not be present and is not entitled to a report of what was said; that they may decline, may stop the interview at any time, and may ask for a representative of their own choosing; and that if they do speak, the only rule is the truth. That is the entire message, and its delivery must be scrupulously neutral, because the line between preparing employees and tampering with them is bright and federally enforced. An employer may not coach employees toward preferred answers, may not instruct or pressure them to decline interviews, may not ask them afterward what was said, may not surveil the interviews, and may not retaliate in any form against an employee for speaking with the government or filing the complaint that brought the officer to the gate: section 11(c) of the Act prohibits retaliation, the complaint's confidentiality is statutory, and an interference or retaliation case converts a citation dispute into something categorically worse (29 U.S.C. §§ 657(f)(1), 660(c)). The same bright line governs the physical evidence: nothing is altered, created, backdated, or destroyed in anticipation of or during an inspection, not the training roster with the missing signatures, not the inspection tag on the crane, not the 300 log. The false record is the one document that can never be explained at the informal conference, because it is not a citation problem; it is a felony (18 U.S.C. § 1001). Table 7-2 sets out, role by role, what each part of the workforce should and should not do while the inspection is under way.

Table 7-2. Workforce Conduct During an Inspection

Role	What you can and should do	What you cannot do
Reception / gate	Verify credentials; notify the inspection coordinator immediately; escort the officer to a conference room without delay	Leave the officer waiting or wandering unescorted; deny entry without instruction from the coordinator or counsel
Inspection coordinator	Confirm and document the stated scope; accompany the entire walkaround; mirror all photographs and readings; log every document produced; correct hazards on the spot where possible	Consent to expansions of scope without deliberation; volunteer records or areas beyond the request; obstruct or interfere with the inspection
Supervisors and managers	Answer truthfully, responsively, and narrowly; say "I don't know, I will find out" when true; request that counsel attend their interviews	Guess, estimate, or speculate; offer theories, excuses, or blame; forget that their statements are the company's admissions
Employees	Know that interviews are private, confidential, and voluntary; decline or end an interview at will; request their own representative; tell the truth if they speak	Nothing is required of them; the prohibitions run against the employer
Everyone	Preserve every record and physical condition as it exists; refer legal questions to counsel	Alter, create, backdate, or destroy any record; coach, question, surveil, or retaliate regarding employee interviews or complaints

Note. Interview rights of non-managerial employees derive from 29 U.S.C. § 657(a)(2) and the Field Operations Manual's interview procedures; the anti-retaliation prohibition is 29 U.S.C. § 660(c); the false-statement and false-records exposure arises under 29 U.S.C. § 666(g) and 18 U.S.C. § 1001. Preparation of the workforce must inform employees of their rights without directing the exercise of those rights in the employer's favor.

V. When the Government Exceeds Its Bounds

The boundaries described above are not decorative, and the case law enforcing them begins at the Supreme Court. In *Barlow's* (1978), the Court held that OSHA inspections are searches within the meaning of the Fourth Amendment and that an employer may refuse warrantless entry: the agency must either obtain consent or secure an administrative warrant supported by probable cause, which may rest on specific evidence of an existing violation or on the employer's selection under a neutral administrative plan. Everything that follows in this section is an application of that holding, and every one of these cases was won or lost on the record of what the inspection was authorized to cover and what the compliance officer actually did.

The foundational scope case is *Sarasota Concrete*. An employee complaint directed at specific conditions became the basis for a warrant authorizing, and a compliance officer conducting, a comprehensive, wall-to-wall inspection of the entire facility. The Review Commission held that the warrant was not supported by probable cause for an inspection of that breadth, suppressed all of the evidence obtained, and vacated the citations in their entirety, and the Eleventh Circuit affirmed, announcing the rule that governs complaint-based inspections to this day: the scope of the inspection must bear an appropriate relationship to the violation alleged in the complaint (*Sarasota Concrete Co.*, 1981; *Donovan v. Sarasota Concrete Co.*, 1982). The court was notably unmoved by the government's plea of good faith, observing that officials who knowingly risk a questionable search cannot then escape its consequences by claiming they meant well. The Tenth Circuit enforced the same boundary one step earlier in the process in *Marshall v. Horn Seed Co.* (1981), holding an inspection warrant invalid where the agency had done essentially nothing to establish that the employee complaint behind it was credible: a reminder that probable cause is a fact to be demonstrated, not a box to be checked.

The most instructive modern decision is *Trinity Industries*, Inc. v. Occupational Safety and Health Review Commission (1994), because it shows both the boundary and the limits of the remedy in a single opinion. OSHA had received a three-item employee complaint about a Sharonville, Ohio plant and, invoking an internal directive — CPL 2.45A — that allowed a complaint to trigger a full-scope inspection when combined with injury-rate criteria, obtained a single warrant authorizing a wall-to-wall examination. The Sixth Circuit held the directive unconstitutional: an administrative plan whose trigger is an employee complaint is not the neutral plan *Barlow's* requires, and a complaint cannot bootstrap a limited special inspection into a comprehensive one. The proper procedure, the court explained, is a warrant limited to the complaint, followed — if the limited inspection and records review justify it — by a second warrant for anything broader. Trinity nonetheless kept its citations, because the officers had relied in objective good faith on a facially valid warrant that two courts had already upheld (*United States v. Leon,* 1984), but the concurrence made the forward-looking consequence explicit: with the directive now invalidated, evidence from any future inspection conducted the same way would be suppressed. The suppression remedy itself had been settled eight years earlier in *Smith Steel Casting Co. v. Brock* (1986), where the Fifth Circuit held that the exclusionary rule applies in Commission proceedings whenever the government seeks to assess penalties for past violations on the strength of illegally obtained evidence: the rule the Sixth Circuit adopted wholesale in Trinity.

Honesty about the limits of these cases is part of using them well. Citations are rarely vacated for a bare departure from the Field Operations Manual: the manual creates no employer rights, and under *Accu-Namics* and its progeny a procedural misstep under section 8 of the Act — an imperfect opening conference, a walkaround technicality — will not disturb a citation unless the employer can show it was actually prejudiced (*Accu-Namics,* 1975). The good-faith exception can preserve evidence gathered under a warrant later held defective, as it did for the government in Trinity. What the cases do reliably punish is the inspection that exceeds its constitutional authorization — the search broader than its warrant, the warrant broader than its probable cause, the consent broader in the citation than it ever was at the gate — together with the citation that misses the statute's own six-month bar (29 U.S.C. § 658(c)). And each of those challenges is won or lost on a contemporaneous record: what scope was stated at the opening conference, what was consented to and in what words, where the officer actually went, and what was photographed and sampled along the way. That record does not assemble itself in litigation; it is created in real time by a trained coordinator walking at the officer's shoulder and a counsel who insisted, at the threshold, that the boundaries be spoken aloud.

One caution, because the chapter can be misread. Nothing in it counsels an adversarial inspection, because the adversarial inspection is almost always the expensive one; compliance officers are professionals executing a published procedure, and the employer's interests are served by an encounter that is prompt, courteous, truthful, and precisely as broad as the law requires — no narrower, and not one door wider. The employer who knows what the compliance officer must do cannot be surprised by any of it. The employer whose counsel framed the entry cannot be cited for what was never inspected. The employer whose workforce knows its rights cannot have those rights used against it. The inspection will be conducted from the government's script either way. Best practice is simply to have read it first.

References

Accu-Namics, Inc. v. OSHRC, 515 F.2d 828 (5th Cir. 1975).

Donovan v. Sarasota Concrete Co., 693 F.2d 1061 (11th Cir. 1982).

False statements or entries generally, 18 U.S.C. § 1001.

Inspections, citations and proposed penalties, 29 C.F.R. pt. 1903 (2025).

Marshall v. Barlow's, Inc., 436 U.S. 307 (1978).

Marshall v. Horn Seed Co., 647 F.2d 96 (10th Cir. 1981).

Occupational Safety and Health Act of 1970, 29 U.S.C. §§ 651–678 (1970).

Occupational Safety and Health Administration. (2025). Field operations manual (CPL 02-00-164; Chapter 6 rev. July 14, 2025). U.S. Department of Labor. https://www.osha.gov/enforcement/directives/cpl-02-00-164 Occupational Safety and Health Administration. (2026). 2026 annual adjustments to OSHA civil penalties [Memorandum]. U.S. Department of Labor.

Occupational Safety and Health Administration. (n.d.). Employer rights and responsibilities following a federal OSHA inspection (OSHA 3000). U.S. Department of Labor.

Sarasota Concrete Co., 9 BNA OSHC 1608 (O.S.H.R.C. 1981), aff'd, 693 F.2d 1061 (11th Cir. 1982).

Smith Steel Casting Co. v. Brock, 800 F.2d 1329 (5th Cir. 1986).

Trinity Indus., Inc. v. OSHRC, 16 F.3d 1455 (6th Cir. 1994).

United States v. Leon, 468 U.S. 897 (1984).

Worker Walkaround Representative Designation Process, 89 Fed. Reg. 22558 (Apr. 1, 2024) (codified at 29 C.F.R. § 1903.8(c)).

CHAPTER 8

The Citation Is Not the Verdict

Affirmative Defenses, the Contest Before the Commission, and the Settlement Calculus

A citation is the government's opening position, not its judgment. Between the closing conference and a final order lies a sequence of decision points — the informal conference, the notice of contest, the answer and its pleaded defenses, the settlement table, the hearing — and at each one the employer can strike a citation, downgrade its classification, group its items, or cut its penalty, provided the technical record supports the move. This chapter maps the defenses that actually succeed and the ones that merely consume fees, walks the contest through the Review Commission, and explains why a consent order, once signed, must be treated as a specification to be executed rather than a document to be filed.

I. The Fifteen Working Days That Decide Everything

Every option this chapter describes expires on the same clock. Under Section 10(a) of the OSH Act, an employer has fifteen working days from receipt of a citation to file a written notice of contest, and a citation that is not contested within that window becomes a final order of the Occupational Safety and Health Review Commission "not subject to review by any court or agency" (29 U.S.C. § 659(a)). The finality is genuine. Courts and the Commission have excused a blown deadline only in the narrow circumstances that excuse a default judgment, and an employer who let the period lapse because the citation was routed to the wrong desk, or because negotiations with the Area Office felt promising, has converted every defense in this chapter into an academic exercise. The first act of citation defense is therefore clerical: the date of receipt is documented, the deadline is computed in working days rather than calendar days, and the contest decision is calendared for a date early enough to leave room for the informal conference.

The informal conference is the most underused bargain in the enforcement system, and also the most misunderstood. The regulation permits an employer, an employee, or their representatives to meet with the Area Director to discuss any issue raised by an inspection, citation, notice of proposed penalty, or abatement date (29 C.F.R. § 1903.20), and the Field Operations Manual authorizes the Area Director to resolve the matter there — by reclassifying violations, deleting items the file will not support, adjusting penalties, and extending abatement dates — through an informal settlement agreement (OSHA, 2025a). Two features of the conference determine how to use it. First, it does not stop the clock: the fifteen working days run through it, so the conference belongs in the first week, not the last. Second, it is a preview of the government's file conducted before positions harden. An employer who arrives with a technical audit already in hand — the sampling record analyzed as Chapter 10 describes, the elements mapped as Chapter 9 describes, the abatement already begun and photographed — negotiates from evidence. An employer who arrives to express disappointment negotiates from hope. In my experience the difference between those two conferences is routinely the difference between the standard penalty reduction and a reclassification that changes the employer's citation history for the next five years.

What the informal settlement agreement buys the government is the waiver of contest rights, and that trade should be priced honestly. For a citation that is factually sound, fairly classified, and cheaply abated, the informal settlement is usually the correct outcome: it caps the expense, avoids the docket, and — negotiated well — can convert a serious item into an other-than-serious one, which matters enormously because classification, not penalty, is what a repeat characterization or a plaintiff's lawyer will be built on later. For a citation whose measurement will not survive audit, whose classification is inflated, or whose abatement demand would re-engineer a process the standard does not actually reach, the fifteen-day window exists precisely so the employer can decline the bargain. Contest is not an act of war against the Area Office; it is the statutory mechanism for making the Secretary prove the case.

II. Rebutting the Case in Chief Before Pleading Anything

The defenses with the best win rate are not affirmative defenses at all. They are failures of the Secretary's own burden — the elements of the prima facie case examined at length in Chapter 9 and the measurement chain audited in Chapter 10 — and they deserve first position in any answer because the employer proves nothing to win on them; the government simply fails. Four of them recur in reported decisions with particular force.

Applicability and preemption. The cited standard must actually govern the cited condition, and where a specific standard applies to a condition, it preempts the general one (29 C.F.R. § 1910.5(c)). Citations drafted in the field are regularly written to the standard the compliance officer knows rather than the standard the operation is governed by — general industry paper for construction work, a general housekeeping provision for a condition a substance-specific standard addresses — and the mismatch is a legal defect the employer can establish from the face of the documents.

Employee exposure. The Secretary must show that employees were exposed to the violative condition or had access to the zone of danger, a test the Commission framed in Gilles & Cotting, Inc., 3 BNA OSHC 2002 (1976), as reasonable predictability of access rather than certainty of contact. The proof is operational: where employees actually work, walk, and reach, established by task analysis rather than by the geometry of a photograph, which is to say, it is the safety professional's evidence, not the lawyer's.

Knowledge. The Secretary must prove the employer knew, or with reasonable diligence could have known, of the violative condition, and the Commission's knowledge decisions — Dover Elevator Co., 16 BNA OSHC 1281 (1993), among them — make the question turn on what the employer's inspection and supervision systems would have revealed. The knowledge element also carries the most consequential circuit split in citation defense: whether a supervisor's own misconduct supplies the employer's knowledge of that misconduct. The Fifth and Eleventh Circuits require the Secretary to prove the supervisor's violation was foreseeable before imputing it (W.G. Yates & Sons Construction Co. v. Occupational Safety and Health Review Commission, 459 F.3d 604 (5th Cir. 2006); *ComTran Group, Inc. v. U.S. Department of Labor*, 722 F.3d 1304 (11th Cir. 2013)), while a supervisor's knowledge of a subordinate's violation is imputed in the ordinary way (*Quinlan v. Secretary, U.S. Department of Labor*, 812 F.3d 832 (11th Cir. 2016)). Because Commission cases are appealable to the circuit where the violation occurred, where the employer is headquartered, or the D.C. Circuit (29 U.S.C. § 660(a)), counsel should identify the governing imputation rule at intake: the same facts can be a defense verdict in one circuit and an affirmed citation in another.

Time and form. Citations must issue with reasonable promptness and describe the violation with particularity (29 U.S.C. § 658(a)), and no citation may issue more than six months after the occurrence of the violation (29 U.S.C. § 658(c)). The six-month limit has teeth: in *AKM LLC d/b/a Volks Constructors v. Secretary of Labor*, 675 F.3d 752 (D.C. Cir. 2012), the D.C. Circuit rejected the Secretary's theory that unrecorded injuries were continuing violations that kept the limitations period open for the five-year retention life of the records, and vacated recordkeeping citations issued years after the entries should have been made. Counsel auditing a citation should date every alleged violative condition against the six-month line: particularly in recordkeeping, exposure, and program-deficiency citations, where the government's theory of when the violation "occurred" is often the weakest sentence in the file.

III. The Affirmative Defenses and Their Records

An affirmative defense concedes the elements and excuses the violation, which means the employer carries the burden of proof, and carries a pleading obligation as well: under the Commission's rules, affirmative defenses must be stated in the answer or they are subject to waiver (29 C.F.R. § 2200.34(b)(3)–(4)). The safe practice is to plead every defense the facts could plausibly support and abandon the surplus after discovery; the fatal practice is to discover the defense at the hearing.

A. Unpreventable Employee Misconduct

The unpreventable employee misconduct defense is the workhorse, and its four elements have been stable for decades: the employer must show that it (1) established work rules designed to prevent the violation, (2) adequately communicated those rules to its employees, (3) took steps to discover violations of the rules, and (4) effectively enforced the rules when violations were discovered (Jensen Construction Co., 7 BNA OSHC 1477 (1979)). Every element is a records question, and the fourth is where the defense lives or dies. The Commission's decision in Valdak Corp., 17 BNA OSHC 1135 (1995), aff'd, 73 F.3d 1466 (8th Cir. 1996), states the recurring failure: a disciplinary program that exists on paper but produces no discipline — or discipline that begins only after the accident — does not establish effective enforcement; it establishes the opposite.

For the safety professional, the defense is a specification written years before the citation. The work rule must map to the standard that will someday be cited, which argues for rules written against the regulatory text rather than in aspirational generalities, the drafting discipline Chapter 25 develops in full. Communication must be provable per employee: signed training rosters, competency verifications, toolbox-talk records tied to the rule at issue. Discovery of violations must be documented as a program: scheduled and unscheduled inspections, supervisor audits, near-miss reporting that actually captures rule-breaking. And enforcement must show a graduated, consistently applied response with dates, names, and escalation. There is a symmetry here that both audiences of this book should sit with: the same disciplinary file that proves the fourth element of the defense also proves, entry by entry, that the employer knew this rule was being broken, which is why a file showing five warnings for missing fall protection and no further action is not a misconduct defense but a knowledge exhibit for the Secretary. The defense rewards employers who enforce, not employers who document leniency.

When the misconduct is a supervisor's own, the defense converges with the imputation split described above, and the practical rule is the one the Fifth and Eleventh Circuits have written: the employer's proof must show the supervisor's defiance of the rules was not foreseeable — that supervisors were themselves trained, audited, and disciplined. A misconduct program that monitors the crew but exempts the foreman fails precisely where these cases are decided.

B. Infeasibility of Compliance

The infeasibility defense requires the employer to prove that literal compliance with the standard was technologically or economically infeasible under the circumstances, and that it either used alternative protective measures or that none were available. Both halves matter. The Commission has little patience for infeasibility claims that end at "it couldn't be done": the employer that could not install the guard must show what it did instead, and an employer that did nothing in the alternative has effectively pleaded that it identified a hazard and left employees exposed to it. The proof is engineering proof — equipment geometry, process constraints, the vendor correspondence and trial installations that show compliance was actually attempted — and it is the rare defense that is stronger when the safety professional, rather than counsel, is its principal witness. Economic infeasibility is harder still: the cost must threaten the viability of the operation, not merely its margin, and the financial record that proves it will be discoverable.

C. Greater Hazard

The greater hazard defense asks the Commission to accept that compliance would have endangered employees more than the violative condition did, and its three-part structure — the hazards of compliance exceed the hazards of noncompliance; alternative means of protection were unavailable; and a variance application was unavailable or inappropriate — makes it among the hardest defenses to win. The third element is the trap: the Act provides a variance procedure precisely for conditions where a standard fits badly (29 U.S.C. § 655(d)), and an employer who never sought one must explain why self-help was chosen over the lawful route. The defense succeeds, when it succeeds, on quantified comparative risk — an exposure or energy analysis of both conditions performed to the same rigor the rest of this book demands — not on a supervisor's testimony that the tie-off point looked more dangerous than the fall.

D. Vagueness, Fair Notice, and Interpretive Overreach

A standard — or the Secretary's reading of one — can be too indefinite to enforce against the employer's conduct. The fair-notice principle holds that a regulation cannot be applied in a way a reasonable employer in the industry could not have anticipated from its text; the D.C. Circuit's decision in Gates & Fox Co. v. Occupational Safety and Health Review Commission, 790 F.2d 154 (D.C. Cir. 1986), vacating a citation built on an interpretation the regulatory language did not fairly disclose, remains the canonical statement. The defense has gained force since 2024, but the doctrinal ground must be described precisely, because the two deference regimes at work here are routinely conflated and a brief that conflates them will be corrected from the bench. *Loper Bright Enterprises v. Raimondo*, 603 U.S. 369 (2024), ended Chevron deference to an agency's construction of the statute it administers. That holding reaches the Secretary's reading of the OSH Act itself — the multi-employer citation doctrine, the scope of Section 5(a)(1), the meaning of "recognized hazard" — and those readings now receive independent judicial construction. What *Loper Bright* did not disturb is the separate deference an agency receives for interpreting its own regulation, which is the deference at issue whenever the fight is over what a standard means rather than what the Act permits. *Martin v. Occupational Safety and Health Review Commission* (CF&I Steel Corp.), 499 U.S. 144 (1991), allocated that deference to the Secretary rather than to the Commission; it did not enlarge it, and it survives *Loper Bright* intact. The case to brief is therefore *Kisor v. Wilkie*, 588 U.S. 558 (2019), which withholds deference to an agency's reading of its own rule unless the regulation is genuinely ambiguous after the ordinary tools of construction have been exhausted, the proffered reading is reasonable, and the reading is the agency's authoritative and considered position, implicates its substantive expertise, and works no unfair surprise on regulated parties. Applied to a standard adopted in 1971 and never revisited, every one of those conditions is contestable, and the last of them merges with the fair-notice principle of Gates & Fox, which is why the strongest vagueness briefs now run *Kisor* and Gates & Fox together rather than reaching for *Loper Bright*, which does not govern the question. The constitutional pressure that *SEC v. Jarkesy*, 603 U.S. 109 (2024), places on penalty adjudication is examined in Chapter 28. Counsel defending a citation built on a creative reading of an old standard is litigating in the most favorable interpretive climate in a generation: provided the brief argues *Kisor* for the regulation and *Loper Bright* for the statute, and does not mistake one for the other.

IV. Reclassification, Grouping, and the Penalty Arithmetic

Most contested citations are not vacated; they are reduced, and the reductions follow the structure of the government's own arithmetic. A proposed penalty is a gravity-based figure — severity times probability — adjusted for the statutory factors of size, good faith, and history (29 U.S.C. § 666(j); OSHA, 2025a), at the maximums set out in Table 8-1. Each input is contestable, and the Commission assesses penalties de novo: it owes the Secretary's calculation no deference and may reweigh every factor on the hearing record.

Table 8-1. Maximum Federal OSHA Civil Penalties for Citations Issued in 2026

Classification	Maximum	Minimum (statutory or policy)
Serious	$16,550	$1,085 (policy)
Other-than-serious	$16,550	None
Failure to abate (per day beyond the abatement date; generally capped at 30 days)	$16,550	None
Willful	$165,514	$11,823 (statutory)
Repeat	$165,514	$4,256 (policy)

Note. No inflation adjustment was made for 2026; the 2025 amounts carry forward because the Bureau of Labor Statistics did not publish the October 2025 CPI-U figure the adjustment statute requires. The failure-to-abate figure is a daily rate, but not an unbounded one: agency practice generally limits it to thirty days, so the practical ceiling on a single unabated item is roughly $496,500 rather than an open-ended accrual — a number worth having before the reserve conversation of Section VII. The reduction factors these maximums are discounted against — size, good faith, history, and quick fix — were revised by the July 14, 2025 revision of Chapter 6 of the Field Operations Manual, set out at Chapter 7 §II.C and discussed in Chapter 24 §III. The May 21, 2026 memorandum restated those revisions and updated certain minimum proposed penalties to reflect them; the $11,823 willful minimum is statutory and unchanged, while the minima shown for serious and repeat violations are policy figures derived from the lowest gravity-based penalty after the maximum permissible reductions.

Classification is the highest-value target. A willful violation requires proof of intentional disregard of the standard or plain indifference to employee safety, a state of mind, and evidence that the employer held a good-faith belief that its conduct complied, or was actively working the problem, defeats willfulness even where it does not defeat the violation. The documented self-audit program described in Chapter 24 is the classic willfulness antidote, and an uncorrected audit finding the classic willfulness exhibit. A repeat characterization requires a substantially similar prior violation that became a final order before the conduct at issue, under the test of Potlatch Corp., 7 BNA OSHC 1061 (1979), and it is attacked on both prongs: dissimilarity of the hazard and standard, and the finality timeline, which is one more reason the classification bought at an informal conference years earlier turns out to matter. Serious versus other-than-serious turns on whether death or serious physical harm could result, a medical and physical question on which the technical expert, not the compliance officer, is the qualified witness.

Grouping and duplication are the quiet second target. The Field Operations Manual itself authorizes combining related violations into a single item for penalty purposes, and the Commission vacates duplicative items where two citations demand the same abatement for the same condition: the same missing guard cited once as a machine-guarding failure and again as a training failure, the same unrecorded injury multiplied across paperwork provisions. Counsel should map every cited item to its abatement: where two items collapse into one fix, there is an argument that they should collapse into one item, and penalties assessed per violation make the arithmetic of that argument immediate.

V. The Contest: From Notice to the Commission and Beyond

The notice of contest itself is a short document with long consequences: it must be filed with the Area Director within the fifteen working days, and it should state whether the employer contests the citation, the proposed penalty, the abatement date, or all of them, item by item, because what is not contested becomes final. Affected employees and their representatives hold their own, narrower right to contest the reasonableness of abatement periods and to elect party status in the employer's contest (29 U.S.C. § 659(c)), and counsel should expect a union local to exercise it in any serious case.

Filing transfers the case from the agency to the Commission: the independent adjudicatory body whose Article III legitimacy the Supreme Court sustained in

Atlas Roofing Co. v. Occupational Safety and Health Review Commission, 430 U.S. 442 (1977), and ordinary civil litigation begins in miniature: the Secretary files a complaint, the employer answers within the Commission's deadlines, and the affirmative defenses are pleaded or imperiled (29 C.F.R. § 2200.34). Smaller cases may proceed under the Commission's simplified proceedings, which dispense with formal pleadings and most discovery (29 C.F.R. §§ 2200.200–.211); conventional cases proceed through discovery to a hearing before an administrative law judge. Discovery is where the technical audit this book keeps describing becomes admissible: the complete inspection file, the sampling and calibration records, the compliance officer's notes and photographs, and the deposition of the compliance officer against the Field Operations Manual's own script (Chapter 7). Two structural facts favor a well-prepared contest. First, a good-faith contest stays the abatement obligation for the contested items until the order becomes final (29 U.S.C. § 659(b)), which is leverage, but leverage that must never be mistaken for permission to leave employees exposed while the docket runs; interim protection is both an ethical obligation and, at the settlement table, evidence. Second, the Secretary controls the citation, not the Commission: under Cuyahoga Valley Railway Co. v. United Transportation Union, 474 U.S. 3 (1985), the Secretary may withdraw or amend a citation, and a technical audit that persuades the Solicitor's office a citation cannot be proven ends cases the Commission never sees.

The judge's decision becomes a final order of the Commission unless a commissioner directs review on a petition for discretionary review, and final orders are appealable to the courts of appeals under the venue choices already described (29 U.S.C. § 660(a)). Most contested cases end far earlier: the Commission's rules provide for settlement at any stage and for settlement judges in appropriate cases (29 C.F.R. §§ 2200.100, 2200.120), and the realistic purpose of most contests is not a hearing but a better agreement than the fifteen-day window offered — items withdrawn on the strength of the audit, classifications corrected, penalties re-figured, abatement reengineered to what the process actually needs.

VI. Consent Orders, Settlements, and Compliance

Whether it arrives as an informal settlement agreement inside the fifteen-day window or a stipulated settlement entered as a consent order on the Commission's docket, the concluded agreement changes the employer's legal position in ways that outlast the penalty check. Three features deserve negotiation and, afterward, obedience.

First, the agreement's terms are the new standard. Modern settlements routinely go beyond item-by-item abatement to enhanced terms — third-party audits, corporate-wide abatement across establishments, training commitments, interim monitoring, reporting to the Area Office — and every enhanced term should be treated as an engineering specification with an owner, a deadline, and a verification record, because each becomes independently enforceable. A final order, consented to or litigated, is enforceable in the court of appeals (29 U.S.C. § 660(b)), and the failure to abate a violation covered by a final order draws per-day penalties under 29 U.S.C. § 666(d) at the maximums in Table 8-1, subject to the thirty-day limit agency practice generally observes: arithmetic that turns a neglected settlement term into a six-figure line item, and turns several of them into the most expensive one in the company's history. Abatement itself must be certified and, for higher-gravity items, documented and tagged under the abatement-verification rule (29 C.F.R. § 1903.19), and the certification is a representation to a federal agency that should be executed with the care that description implies.

Second, the agreement writes the employer's enforcement future. Settled citations — at whatever classification survives negotiation — populate the establishment's history for the repeat analysis and the history adjustment on the next inspection's penalties, and a settlement that leaves willful or multiple high-gravity serious items on the record can land the employer in the Severe Violator Enforcement Program, with the mandatory follow-up inspections, corporate-wide reach, and public designation that program carries (OSHA, 2022; CPL 02-00-169). Follow-up inspections should be assumed, not feared: the employer that treats the settlement as a project plan greets the follow-up with a verification file; the employer that treated it as a closing document greets it with a failure-to-abate exposure.

Third, the agreement echoes in civil litigation. Standard practice is to negotiate non-admission language, the settlement resolves contested claims without admitting the violations, and to state that the agreement is entered to avoid the expense of litigation. Counsel should assume the citation history will nonetheless be requested in discovery in any later injury case, and should read Chapters 11 and 19 on how OSHA outcomes travel into negligence litigation before deciding which classifications are worth paying to change. In my experience, the reclassification an employer buys at settlement is rarely about this year's penalty; it is about what the file will look like when it is Exhibit 14.

VII. The Reserve, Mediation, and Coverage

The Introduction named four readers in the room after a serious incident, and one of them, the claims professional, has appeared only glancingly since. This section is for that reader, and for the counsel and experts whose work reaches a file through her. The settlement calculus of the preceding sections is conducted in the enforcement forum, but it is priced somewhere else: in a reserve set within weeks of the loss, revised on a schedule, and defended internally against a number that moves for reasons the technical file explains better than anything else in the claim.

A. The Technical Finding as a Valuation Input

A reserve is a prediction, and like any prediction it is only as good as the earliest reliable information available to it. In workplace injury and exposure claims the earliest reliable information is almost never the pleading, the demand, or the citation. It is the answer to four technical questions, and an expert can usually answer them within days of retention: Did the standard the Secretary cited actually apply to this equipment in this configuration? Is the measurement chain beneath the exposure allegation intact, calibration, placement, duration, chain of custody, laboratory method? Which entity controlled the hazard, and what does the contemporaneous record show about notice? And what does the employer's own file say when it is read by someone who is looking for what is missing rather than for what is there?

Each of those answers moves a reserve, and each moves it in a direction the pleading cannot predict. A willful citation with a $165,514 proposed penalty and a defective calibration record beneath it is a smaller exposure than its face value; a serious citation with a five-year run of unclosed audit findings behind it is a much larger one, because the citation is not the claim — the citation is a discoverable admission attached to a wrongful death case that has not been filed yet. The claims professional who prices the citation is pricing the wrong instrument. Section III above describes what a technical audit finds; the point here is that the same audit, done at intake rather than at expert-disclosure, is the cheapest reserve-accuracy purchase available on the file.

The corollary is a scheduling one. The audit is worth its cost only while the evidence still exists, and the perishability described in Chapter 6 §III and Chapter 21 runs on the calendar of the loss, not the calendar of the litigation. A consulting expert retained in month one costs a fraction of the reserve movement she will produce; the same expert retained in month twenty is reconstructing from whatever the file happens to contain.

B. What the Record Does at Mediation

Mediation in these cases is usually described as a negotiation about value and is more accurately a contest about which documents the neutral believes will reach a jury. That reframing changes what belongs in a mediation submission.

The most powerful exhibit at a workplace-injury mediation is rarely the expert report, because both sides have one and each discounts the other's. It is the document the defendant created before anyone chose sides and cannot now explain: the audit finding that identified the hazard and was closed out with no corrective action, the near-miss report describing the same mechanism eight months earlier, the training roster with signatures and no comprehension check, the 300 Log entry that records the third laceration on the same unguarded machine. Those documents have a property no expert opinion has, they are unrebuttable as to their own existence and date, and a neutral evaluating settlement posture prices them accordingly.

The same logic works in reverse and is underused by defendants. A defense mediation submission built on the technical record — the calibration gap in the government's sampling, the reconstruction that the plaintiff's expert never ran a sensitivity analysis on, the alternative exposure source developed rather than merely named — moves a plaintiff's number in a way that a recitation of legal defenses does not, because the plaintiff's counsel is also pricing what survives a Rule 702 motion. Chapter 17 is written as an offensive playbook; at mediation it is a valuation memorandum.

C. The Coverage Conversation

Coverage is the third destination of the technical finding and the one most likely to be reached last, which is usually a mistake. The general observation, subject entirely to the policy language and the governing state's law, is that the findings that most improve an employer's position in the enforcement forum sometimes worsen its position under its own policies, and the reverse. A finding of employer intent, or of conduct substantially certain to cause injury, is what defeats workers'-compensation exclusivity and opens the third-party or direct action, and it is also the finding most likely to trigger an expected-or-intended exclusion or take the loss outside employers' liability coverage. The anchor case of Chapter 6 is the illustration: a nine-figure verdict resting on a substantial-certainty finding, followed by coverage litigation and a public dispute about whether the insurers should have resolved the claim before trial.

The practical instruction is that the same technical audit that prices the claim should be read, early, by whoever is responsible for the coverage position, because the theory that minimizes the citation and the theory that preserves the coverage are not always the same theory, and choosing between them is a decision that should be made deliberately rather than discovered at trial. That choice is a legal and underwriting judgment on which this book takes no position and on which the policy, the jurisdiction, and coverage counsel govern. What the expert supplies is the factual predicate underneath it, stated the same way it would be stated to a court, which is the only form in which it is worth anything to a reserve, a mediator, or a carrier.

VIII. Practice Implications

For counsel, the sequence is fixed and the calendar is sovereign: document receipt, compute the fifteen working days, get the technical expert into the file before the informal conference, contest item by item, plead every supportable defense in the answer, and litigate toward the settlement the audit justifies. For the safety professional, this chapter is a design brief written in reverse: the misconduct defense is built from work rules, training records, inspection programs, and enforced discipline created years before any citation; the willfulness antidote is the audit-and-correct record of Chapter 24; the infeasibility and greater-hazard defenses are engineering analyses that exist only if someone performed them at the time; and the consent order, once signed, is a safety project to be executed and verified, not a legal artifact to be filed. The citation is the government's theory of what the record shows. The defense, in every form this chapter has described, is the disciplined demonstration of what the record actually contains, which is to say, it is the argument of this entire book, made under deadline.

References

29 C.F.R. § 1903.19 (abatement verification); 29 C.F.R. § 1903.20 (informal conferences); 29 C.F.R. § 1910.5(c) (application of standards).

AKM LLC d/b/a Volks Constructors v. Sec'y of Labor, 675 F.3d 752 (D.C. Cir. 2012).

Atlas Roofing Co. v. OSHRC, 430 U.S. 442 (1977).

ComTran Grp., Inc. v. U.S. Dep't of Labor, 722 F.3d 1304 (11th Cir. 2013).

Cuyahoga Valley Ry. Co. v. United Transp. Union, 474 U.S. 3 (1985).

Dover Elevator Co., 16 BNA OSHC 1281 (Rev. Comm'n 1993).

Gates & Fox Co. v. OSHRC, 790 F.2d 154 (D.C. Cir. 1986).

Gilles & Cotting, Inc., 3 BNA OSHC 2002 (Rev. Comm'n 1976).

Jensen Construction Co., 7 BNA OSHC 1477 (Rev. Comm'n 1979).

Kisor v. Wilkie, 588 U.S. 558 (2019).

Loper Bright Enters. v. Raimondo, 603 U.S. 369 (2024).

Martin v. OSHRC (CF&I Steel Corp.), 499 U.S. 144 (1991).

Occupational Safety and Health Act of 1970, 29 U.S.C. §§ 655(d), 658(a), 658(c), 659(a)–(c), 660(a)–(b), 666(d), 666(j).

Occupational Safety and Health Administration. (2022). Severe Violator Enforcement Program (CPL 02-00-169). U.S. Department of Labor.

Occupational Safety and Health Administration. (2025a). Field Operations Manual (CPL 02-00-164, as amended). U.S. Department of Labor.

Potlatch Corp., 7 BNA OSHC 1061 (Rev. Comm'n 1979).

Quinlan v. Sec'y, U.S. Dep't of Labor, 812 F.3d 832 (11th Cir. 2016).

Rules of Procedure of the Occupational Safety and Health Review Commission, 29 C.F.R. §§ 2200.34, 2200.100, 2200.120, 2200.200–.211.

SEC v. Jarkesy, 603 U.S. 109 (2024).

Valdak Corp., 17 BNA OSHC 1135 (Rev. Comm'n 1995), aff'd, 73 F.3d 1466 (8th Cir. 1996).

W.G. Yates & Sons Constr. Co. v. OSHRC, 459 F.3d 604 (5th Cir. 2006).

CHAPTER 9

Five Fatal Flaws

Where OSHA Citations Break Down — Three Fractures in the Secretary's Own Case, and the Two Affirmative Defenses Treated at Chapter 8

> *Every element of the Secretary's prima facie case rests, in the end, on a technical proposition that a compliance officer asserted and no one has yet tested. Chapter 8 supplies the procedural architecture: the burdens, the affirmative defenses, the contest, and the settlement calculus. This chapter does something narrower and, for the technical reader, more useful: it shows where a safety professional or industrial hygienist actually finds the fracture in the government's file, and what evidence closes it.*

I. The Citation as an Engineering Document

A citation arrives. It names a standard, describes a condition, proposes a penalty, and sets an abatement date. Most defense strategy from that point forward is conducted as a legal exercise: read the standard, read the case law, argue the elements.

That approach leaves money on the table, because every one of the Secretary's four elements, and every affirmative defense the employer can raise, ultimately turns on a technical proposition that a compliance officer asserted and that no one has yet tested. Did the standard actually apply to this equipment, configured this way? Was the employee actually within the zone of danger, or does the geometry say otherwise? Was compliance actually feasible on this jobsite, or is that an assumption the compliance officer made from the tailgate of a truck? These are not legal questions. They are engineering and industrial hygiene questions that arrive dressed as legal questions, and they are where contested citations are won.

The stakes justify the work. Maximum penalties currently stand at $16,550 per serious or other-than-serious violation and $165,514 per willful or repeat violation, with a statutory floor of $11,823 for a willful (Occupational Safety and Health Administration [OSHA], 2026). Notably, there was no inflation adjustment for 2026: the 2025 amounts carry forward because the Bureau of Labor Statistics could not produce the October 2025 CPI-U data required by statute (OSHA, 2026). But the penalty is rarely the real exposure. A serious citation that becomes a final order is a predicate for a repeat citation for five years, an admission in the wrongful death case that follows, and a discoverable document that plaintiff's counsel will hand to a jury. What gets contested, and how well, matters long after the check clears.

Five things fracture these cases, and that is where this chapter's title comes from. Three of them are failures in the Secretary's own case, and they are this chapter's subject, because each is an engineering or industrial hygiene question wearing a legal costume. The other two, infeasibility and unpreventable employee misconduct, are affirmative defenses on which the employer carries the burden, and they belong to Chapter 8, which treats them with the controlling Commission authority they require. Setting them aside here is not a demotion. It is the recognition that they are proved differently, pleaded differently, and lost differently, and that a chapter written about where a technical expert finds the fracture in the government's file should be about the government's file.

What follows, then, are the three places I am asked to look when defense counsel retains me, and the analytical soft spots I find there.

II. What This Chapter Assumes

The procedural framework is Chapter 8's, and this chapter does not restate it. In brief, and by way of orientation only: to prove a violation of a specific standard under Section 5(a)(2) the Secretary must establish that the cited standard applied to the condition, that its terms were not met, that employees were exposed or had access to the violative condition, and that the employer knew or with reasonable diligence could have known of it (*Atlantic Battery Co.*, 16 BNA OSHC 2131, 2138 (No. 90-1747, 1994); *Astra Pharmaceutical Products, Inc.*, 9 BNA OSHC 2126, 2129 (No. 78-6247, 1981), aff'd in relevant part, 681 F.2d 69 (1st Cir. 1982)). Only if the Secretary carries all four does the burden shift to the employer, and the affirmative defenses that then become available — unpreventable employee misconduct, infeasibility, and greater hazard — must be pleaded in the answer on pain of waiver (29 C.F.R. § 2200.34(b)(3)–(4)). Chapter 8 §III treats each of those defenses, with the controlling Commission authority and the records that prove them; Chapter 8 §II treats the applicability, exposure, knowledge, and limitations attacks as procedural defenses. Nothing in this chapter should be read as a substitute for that treatment.

What this chapter adds is the part no procedural treatment supplies. Three of the Secretary's four elements are, underneath the doctrine, engineering and industrial hygiene propositions: applicability is a reading of a specification, exposure is a question of geometry and operating envelopes, and constructive knowledge is a safety-program audit. Those three are where a technical expert earns the retention, because the evidence that resolves them is evidence a compliance officer either did not gather or did not know how to read. The fourth element, whether the terms of the standard were met, is usually the least contested and, where the standard is an exposure standard, is the subject of Chapter 10. The affirmative defenses are Chapter 8's; the flaws below are the government's.

III. Flaw One: The Standard Does Not Actually Apply

The first element is also the most frequently conceded and the most frequently wrong. Compliance officers work from a finite mental catalogue of standards, and when a condition looks like a hazard, the nearest plausible standard gets cited. Whether that standard, read on its own terms, actually reaches the cited equipment or activity is a separate question, and it is a question of technical definitions, not legal interpretation.

The analysis is textual and mechanical. Standards are built from defined terms, and those definitions have engineering content. Does this device meet the standard's definition of the equipment class it regulates? Was the employee engaged in the activity the standard governs, or an adjacent one? Does an exception, exemption, or scope limitation apply? Is the condition covered by a more specifically applicable standard, which would displace the general one the Secretary chose?

This is the least glamorous of the three and often the most decisive, because it is dispositive rather than merely mitigating: if the standard does not apply, the item is vacated regardless of how hazardous the condition was. The technical expert's function here is to read the standard the way an engineer reads a specification — term by term, against the actual equipment — rather than the way an inspector reads it, which is by pattern recognition.

IV. Flaw Two: Exposure Was Assumed, Not Proven

The Secretary must prove that employees were exposed to the violative condition — and the governing test, settled since *Gilles & Cotting, Inc.*, 3 BNA OSHC 2002, 2003 (No. 504, 1976), is not whether a hazard existed in the abstract, but whether it was reasonably predictable, by operational necessity or otherwise, that employees would be, are, or have been in the zone of danger created by it. The Commission has applied that formulation consistently since (*Kokosing Construction Co.*, 17 BNA OSHC 1869, 1870 (No. 92-2596, 1996); *Capform, Inc.*, 19 BNA OSHC 1374, 1376 (No. 99-0322, 2001), aff'd, 34 F. App'x 152 (5th Cir. 2002)), and exposure of a single employee satisfies the element. There is also a jurisdictional limit that is easy to overlook: no citation may issue more than six months after the occurrence of the violation (29 U.S.C. § 658(c)), a bar the D.C. Circuit enforced against a continuing-violation theory in *AKM LLC d/b/a Volks Constructors v. Secretary of Labor*, 675 F.3d 752 (D.C. Cir. 2012).

"Zone of danger" is a geometric and physical concept, and that is precisely why it is contestable. It has an actual boundary, determined by the mechanics of the hazard: the reach radius of the swinging counterweight, the fall trajectory from the unguarded edge, the collapse zone of the unshored excavation wall, the dispersion pattern of the contaminant plume. Compliance officers routinely assert exposure from a photograph and a visual impression of proximity. A technical expert can do considerably better: reconstructing the site geometry, the equipment's actual operating envelope, and the employee's documented position and task, and then asking whether the worker was in fact within the boundary the physics defines.

The same rigor cuts the other way and is worth knowing about in advance: where the hazard is mobile — an excavator tracking along a trench line, for instance — the zone of danger moves with it, and a defense premised on a single snapshot of separation will fail. The point is not that exposure is always defeasible. The point is that it is an evidentiary proposition with a physical answer, and it is frequently asserted rather than proven.

V. Flaw Three: Knowledge Manufactured From the Violation

Knowledge is the element most likely to be litigated and the one where the law has moved most in employers' favor. The Secretary may prove that the employer actually knew of the hazardous condition, or that it could have known through the exercise of reasonable diligence — and reasonable diligence, the Commission has held, comprises the obligation to inspect the work area, to anticipate hazards to which employees may be exposed, and to take measures to prevent the occurrence (*Frank Swidzinski Co.*, 9 BNA OSHC 1230, 1233 (No. 76-4627, 1981); *Precision Concrete Construction*, 19 BNA OSHC 1404, 1406 (No. 99-0403, 2001)). Because a corporation acts through its agents, a supervisor's actual or constructive knowledge of a violative condition is ordinarily imputed to the employer, and that imputation alone can satisfy the Secretary's burden without any independent showing that the safety program was deficient (*Dover Elevator Co.*, 16 BNA OSHC 1281, 1286 (No. 91-862, 1993)).

The critical exception arises when the supervisor is himself the wrongdoer. In *ComTran* Group, Inc. v. U.S. Department of Labor (2013), a project manager dug a six-foot excavation and piled the spoil at its edge, violating the standards himself, and the Commission imputed his knowledge of his own misconduct to the company. The Eleventh Circuit reversed. It held that where the Secretary seeks to establish employer knowledge through a supervisor's own misconduct, she must offer evidence independent of the misconduct itself, such as evidence of lax safety practices; otherwise the violation (element two) would automatically establish knowledge (element four), collapsing the Secretary's burden into a tautology. In so holding, the court aligned itself with the Third, Fourth, Fifth, and Tenth Circuits (*ComTran*, 2013; see also W.G. Yates & Sons Construction Co. v. Occupational Safety and Health Review Commission, 2006).

The exception has a limit worth knowing. In *Quinlan v. Secretary, U.S. Department of Labor* (2016), the same court declined to extend *ComTran* where the supervisor was simultaneously involved in violative conduct alongside a subordinate: there, the supervisor's knowledge of the subordinate's violation is imputed, because the Secretary proves the violation independently of the knowledge.

Here is why this doctrine is a technical fight rather than a purely legal one. When the Secretary cannot rely on imputation, she must prove constructive knowledge — which means proving that the employer's safety program and supervision were inadequate enough that the misconduct was reasonably foreseeable and that reasonable diligence would have revealed the condition (*ComTran*, 722 F.3d at 1316; *Precision Concrete*, 19 BNA OSHC at 1406). That is a safety-program audit, and it is exactly the analysis a qualified safety professional performs: Were the hazards identified? Were work rules written to address them? Was training delivered and documented? Were inspections conducted at a frequency proportionate to the risk? A defensible program, documented and produced, does not merely support an affirmative defense later: it defeats the Secretary's prima facie case at the knowledge element, before the burden ever shifts.

VI. What Each Flaw Leaves Behind: The Evidence File

Read together, the three flaws describe a single evidentiary program, and it is worth setting out as one because counsel who ask for it in these terms get it, and counsel who ask for "the safety records" do not.

For applicability, the file is a specification file: the equipment's nameplate, model documentation, and manufacturer's classification; the as-built or as-configured drawings; the process description that establishes which activity was underway; and the text of every standard in the candidate set, including the more specifically applicable one that displaces the Secretary's choice under 29 C.F.R. § 1910.5(c). The finding is textual and it is either there or it is not.

For exposure, the file is a geometry file: dimensioned site measurements rather than photographs, the equipment's documented operating envelope and reach radius, the task analysis establishing where the employee actually worked and traveled, the badge, time, and assignment records fixing who was present, and — where the hazard is a contaminant rather than a mechanism — the dispersion or ventilation analysis that establishes the plume's actual boundary. Where the hazard was mobile, the file must include the machine's travel path, because a snapshot of separation proves nothing about an excavator tracking a trench line.

For constructive knowledge, the file is the program itself, and it is the same audit an insurer, a corporate parent, or a plaintiff's expert would run: written programs mapped to the hazards actually present; hazard assessments dated before the inspection; training records that name the specific hazard and bear a date and a signature; an inspection cadence with documentation of what was looked for and what was found; corrective actions with closure evidence rather than closure dates; and the disciplinary file, which cuts both ways and should be read honestly by whichever side holds it. Chapter 24 describes how to build that file before it is needed; Chapter 25 describes how its wording will read when it is read aloud.

The affirmative defenses draw on the same materials but are proved to a different standard and are treated where they belong, in Chapter 8 §III: unpreventable employee misconduct at §III.A, infeasibility at §III.B, greater hazard at §III.C. One point of overlap deserves emphasis here rather than there, because it is technical rather than procedural. The four elements of the misconduct defense — a work rule addressed to the hazard, communicated through training, monitored by inspection, and enforced by documented discipline — are the elements of a functioning safety management system, which means the defense cannot be constructed after the inspection. It is a factual showing about a program that either existed or did not, and the constructive-knowledge audit above and the misconduct defense below are the same audit graded against two different burdens. An employer who can produce, for this specific hazard, the rule, the training record, the inspection record, and the disciplinary record has both defeated the Secretary's knowledge element and built the affirmative defense. An employer who cannot has neither.

VII. Practice Implications: Earning the Fee

Engage before the informal conference, not after. Employers have fifteen working days from receipt of a citation to contest it or to seek an informal conference with the Area Director, and penalty reductions and reclassifications are routinely negotiated there. That window is also when the technical record is most malleable and the evidence most recoverable: the equipment is still configured as it was, the witnesses still remember, the site still exists. An expert brought in after the pleadings are closed is reconstructing; an expert brought in during the contest window is preserving.

Attack the elements in order of dispositiveness, not order of appeal. Applicability, if it wins, vacates. Exposure, if it wins, vacates. Knowledge, if it wins, vacates. The affirmative defenses of Chapter 8 §III are the fallback, not the opening, and they carry a burden the employer does not otherwise bear. Counsel who reflexively lead with unpreventable employee misconduct have conceded three elements they may not have needed to concede, and have volunteered for a burden of proof the Secretary was going to have to carry alone.

Understand what the classification is really worth. A willful violation requires the Secretary to show heightened awareness of the violation's illegality — an intentional, knowing, or voluntary disregard for the requirements of the Act, or plain indifference to employee safety (*Valdak Corp.*, 17 BNA OSHC 1135, 1136 (No. 93-0239, 1995), aff'd, 73 F.3d 1466 (8th Cir. 1996); *Williams Enterprises, Inc.*, 13 BNA OSHC 1249, 1256–57 (No. 85-355, 1987)). That is a state-of-mind showing that a well-documented, actively managed safety program directly undercuts. Reclassifying a willful to a serious does more than cut the penalty by an order of magnitude; it removes the single most damaging document plaintiff's counsel could put in front of a jury in the civil case that follows.

And build the record before you need it. Every one of these three flaws is easier to exploit — and every one of Chapter 8's affirmative defenses easier to prove — when the underlying technical work was done in advance: the hazard assessments, the feasibility evaluations, the inspection logs, the disciplinary records. The employer who commissions that work after the citation arrives is buying a defense at retail. The employer who has it already is simply producing it.

OSHA's case is a technical assertion dressed in legal language. Undress it, and you find the fracture.

References

AKM LLC d/b/a Volks Constructors v. Sec'y of Labor, 675 F.3d 752 (D.C. Cir. 2012).

Applicability of standards, 29 C.F.R. § 1910.5(c); Rules of Procedure of the Occupational Safety and Health Review Commission, 29 C.F.R. § 2200.34 (2025).

Astra Pharm. Prods., Inc., 9 BNA OSHC 2126 (No. 78-6247, 1981), aff'd in relevant part, 681 F.2d 69 (1st Cir. 1982).

Atlantic Battery Co., 16 BNA OSHC 2131 (No. 90-1747, 1994).

Capform, Inc., 19 BNA OSHC 1374 (No. 99-0322, 2001), aff'd, 34 F. App'x 152 (5th Cir. 2002).

ComTran Grp., Inc. v. U.S. Dep't of Labor, 722 F.3d 1304 (11th Cir. 2013).

Dover Elevator Co., 16 BNA OSHC 1281 (No. 91-862, 1993).

Frank Swidzinski Co., 9 BNA OSHC 1230 (No. 76-4627, 1981).

Gilles & Cotting, Inc., 3 BNA OSHC 2002 (No. 504, 1976).

Kokosing Constr. Co., 17 BNA OSHC 1869 (No. 92-2596, 1996).

Occupational Safety and Health Act of 1970, 29 U.S.C. §§ 651–678, particularly §§ 654(a)(2), 658(a), 658(c), and 666.

Occupational Safety and Health Administration. (2024). Field operations manual, CPL 02-00-164, chs. 4 and 6. U.S. Department of Labor.

Occupational Safety and Health Administration. (2026, May 21). 2026 annual adjustments to OSHA civil penalties [Memorandum]. U.S. Department of Labor.

Precision Concrete Constr., 19 BNA OSHC 1404 (No. 99-0403, 2001).

Quinlan v. Sec'y, U.S. Dep't of Labor, 812 F.3d 832 (11th Cir. 2016).

Valdak Corp., 17 BNA OSHC 1135 (No. 93-0239, 1995), aff'd, 73 F.3d 1466 (8th Cir. 1996).

W.G. Yates & Sons Constr. Co. v. OSHRC, 459 F.3d 604 (5th Cir. 2006).

Williams Enters., Inc., 13 BNA OSHC 1249 (No. 85-355, 1987).

CHAPTER 10

Attacking OSHA's Sampling Methodology

When the Industrial Hygiene Data Does Not Support the Citation

In an overexposure case, the citation is only as strong as the measurement beneath it. Every exposure-based citation OSHA issues — silica, lead, benzene, asbestos, noise — rests on a chain of technical steps that the Secretary must prove and that a qualified industrial hygiene expert can independently audit. When any link fails, the citation fails with it. This chapter maps the vulnerabilities and the documented cases where they decided the outcome.

I. The Exposure Citation Rests Entirely on the Measurement

Most OSHA citations turn on what a compliance officer saw. An overexposure citation turns on what a compliance officer measured, and that difference is the entire defense opportunity. When the Secretary alleges that employees were exposed above a permissible exposure limit under 29 C.F.R. § 1910.1000 or a substance-specific standard, or above the noise limits of Table G-16 of § 1910.95, she carries the burden of proving the violation by a preponderance of the evidence, and the proof is almost always a number: a time-weighted average constructed from air samples collected on a handful of employees during one or two shifts, or a noise dose accumulated on a dosimeter clipped to a shirt collar. That number is not an observation. It is the output of a long technical chain — sampling strategy, instrument selection, calibration, media handling, placement in the breathing zone, timing, shipment, laboratory analysis, and a final calculation with an error adjustment — and the Occupational Safety and Health Review Commission has made clear, in decisions spanning five decades, that a break anywhere in that chain can be fatal to the citation.

Defense counsel are accustomed to litigating the legal elements of a citation: applicability of the standard, employer knowledge, employee exposure, feasibility of abatement. In an exposure case there is an element underneath all of these that is purely technical, whether the measurement is valid, and it cannot be evaluated from the citation narrative or the compliance officer's testimony alone. It requires someone who has personally run sampling pumps, worn out dosimeters, argued with laboratories, and applied the same OSHA procedures the compliance officer was supposed to follow. This chapter describes where the sampling case breaks, how a technical expert finds the breaks in the government's own file, and what the reported decisions teach about which attacks succeed and which fail.

II. Where the Sampling Case Breaks

OSHA has published, in remarkable detail, the procedures its own personnel are expected to follow. The OSHA Technical Manual (TED 01-00-015), Section II, Chapter 1, governs personal sampling for air contaminants; the mandatory and non-mandatory appendices to the substance-specific standards prescribe sampling and analytical protocols; and OSHA's enforcement directives instruct compliance officers on how exposure must be documented before a citation issues. These documents are the yardstick. Every deviation between what the compliance officer did and what the agency's own procedures require is a fact the defense is entitled to develop, and the deviations cluster in five places.

A. Calibration and Pump Performance

A personal sampling result is a mass of contaminant divided by a volume of air, and the volume is only as good as the pump's flow rate. OSHA's asbestos sampling protocol is explicit: each sampling pump is to be calibrated before and after sampling, against a primary standard, and preferably at the sampling site, because environmental conditions affect flow. Comparable expectations run through the Technical Manual's treatment of air sampling generally. The case file should therefore contain pre- and post-sampling calibration records for every pump, traceable to the specific instrument by serial number. When post-sampling calibration is missing, when the records cannot be matched to the pumps actually used, or when the flow drifted during the shift — filter loading in dusty environments is a recognized cause — the computed air volume is unreliable, and every concentration calculated from it inherits the defect. The same logic governs noise instrumentation: dosimeters and sound level meters must be properly calibrated and used, and the Commission's decisions treat calibration as a threshold question in weighing the readings.

B. Placement and the Breathing Zone

Exposure standards regulate the air an employee breathes, so the sample must be collected from the breathing zone — OSHA's asbestos protocol, for example, specifies the cassette be positioned about ten centimeters from the employee's nose and mouth, secured at the collar or lapel. Placement disputes are among the most heavily litigated sampling issues, and they cut in both directions. Where a compliance officer cannot attach the pump to the employee and instead stations it on nearby equipment, the defense will ask whether an area sample is being passed off as a personal one. Where the employee wears a welding helmet or face shield, the fight is over whether the inlet belongs inside or outside the protection: a question the Commission answered for lead and cadmium in *Southern Scrap Materials Co.*, discussed in Section IV, by looking to how the standard itself defines the exposure limit. The lesson of that line of cases is precise: a placement attack succeeds or fails on the governing protocol for the specific contaminant and equipment, not on generalized intuitions about where the air is, which is exactly the analysis a qualified industrial hygienist is trained to perform.

C. Duration, Timing, and the Construction of the TWA

A permissible exposure limit expressed as an eight-hour time-weighted average must be proven as one. The Technical Manual instructs compliance officers on minimum sampling times and volumes needed to reach the analytical method's limit of quantification, and it instructs them not to exceed a method's recommended sampling time, because analyte breakthrough on sorbent tubes and filter overloading corrupt the sample. Between those bounds lies the arithmetic: partial-shift samples extrapolated to a full shift, unsampled intervals assumed rather than measured, zero-exposure or worst-case assumptions applied to the gaps, and short-term samples repackaged as full-shift evidence. Each of these choices is discoverable, reconstructable, and testable. An expert who re-runs the TWA computation from the raw sampling times and laboratory masses, rather than accepting the summary figure on the violation worksheet, regularly finds that the number the citation depends on is an artifact of assumptions the Secretary never proved. In noise cases the analogous issue is what the instrument was allowed to hear: under § 1910.95(b)(2), noise variations with maxima at intervals of one second or less are treated as continuous, and the treatment of impulse noise — power press strokes, impacts, drops — within a dosimeter's accumulated dose was the precise issue on which the citations in *Collier-Keyworth Co.* were vacated.

D. Media Handling, Blanks, and the Laboratory

Between the workplace and the laboratory bench, the sample must remain what it was. OSHA's own protocols build in the controls: field blanks — clean media opened at the sampling site, handled, stored, and shipped with the samples but never used to collect air — exist to detect contamination introduced in handling, and the asbestos protocol specifies how many blanks accompany each sample set. The laboratory record contains its own vulnerabilities: chain-of-custody gaps between collection and receipt; analysis by a method other than the validated one; and quality-control failures within the analysis itself — the asbestos counting protocol, for instance, requires statistical homogeneity testing of fiber counts and directs the laboratory to flag samples whose true concentration may differ significantly from the reported value. The substance-specific standards also impose accuracy requirements on the monitoring method itself; the benzene standard requires accuracy, at a 95 percent confidence level, of plus or minus 25 percent for concentrations at or above 0.5 ppm. The laboratory file — bench sheets, calibration curves, QC recoveries, blank results — is discoverable, and in a case where the margin over the PEL is thin, it is often where the citation quietly dies. The discovery history of Southern Scrap is instructive: analytical data omitted from the Salt Lake Technical Center's final report surfaced only on the eve of the hearing, after repeated document requests.

E. Sampling and Analytical Error: OSHA's Own 95 Percent Rule

The final vulnerability is the one many citations never survive: the error band. OSHA acknowledges that a measured exposure will rarely equal the true exposure, because error is inherent in sampling, analysis, and pump flow: the agency's term is sampling and analytical error, or SAE. Under the procedures in the Technical Manual, and as restated in enforcement directives such as the respirable crystalline silica inspection directive, a violation of a PEL is established only if the measured exposure exceeds the limit after the SAE correction is applied at a 95 percent confidence level. The consequence is arithmetic and unforgiving: a result modestly above the PEL, minus the applicable error, may prove nothing at all. Every exposure citation file should be audited for this calculation: whether the SAE was applied, whether the correct error factor was used for the method and analyte, and whether the adjusted value still clears the limit. The same principle appears in the noise cases as a margin-of-error requirement on dosimeter doses, and, as Collier-Keyworth shows, a dose comfortably above 100 percent can still be legally insufficient once the required allowances are made.

III. Auditing the Government's File, Link by Link

None of these vulnerabilities is visible from the citation. They live in the inspection case file — the sampling worksheets, calibration logs, dosimeter printouts, photographs, laboratory reports and bench data, and the compliance officer's notes — which counsel obtains in discovery or, earlier, by Freedom of Information Act request. The expert's first task is a document-level audit: reconcile every sample number against a pump, a calibration record, a location on the employee, a start and stop time, a shipment record, and a laboratory result, and re-derive every reported concentration and TWA from the raw entries. The audit is conducted against OSHA's own published procedures, which is what gives it force at hearing: the defense is not asking the Commission to prefer the employer's methodology over the agency's, but to hold the agency to the methodology it published.

The second task is context. A single day of compliance sampling is a snapshot, and the employer's own exposure record — historical industrial hygiene surveys, insurer or consultant sampling, audiometric data, process and production records for the sampled shift — frequently shows that the day OSHA measured was unrepresentative: an upset condition, an atypical task, a ventilation system down for maintenance. The professional framework for judging exposures from limited data, reflected in the American Industrial Hygiene Association's exposure assessment strategy, gives the expert a disciplined vocabulary for explaining to a judge why one number is not an exposure profile. And the same expert can design and execute post-citation sampling of the operation, properly done, to show what the exposures actually are: evidence that matters both to the merits and to the feasibility and abatement conversation that follows.

The third task is restraint, and it is the mark of the expert worth retaining. Not every file supports an attack, and an overreached sampling challenge damages the credible ones — Section IV's final case shows the Commission dismantling a sampling defense that ignored the magnitude of the readings. The expert's honest early read — these results survive their error band; this calibration record is intact — is worth as much to case strategy as any exclusion motion, because it redirects the defense to the elements the Secretary actually failed to prove.

IV. Two Cases and a Warning from the Commission

Collier-Keyworth Co., 13 BNA OSHC 1208 (No. 80-2848, 1987), later vacated per stipulation on appeal, the instrument heard what the standard excludes. OSHA cited the company after measuring plant noise with sound level meters and personal dosimeters, and an administrative law judge and then the full Commission confronted a technical question with case-dispositive stakes: the plant noise included impulse noise — sharp energy peaks lasting less than a second, spaced more than a second apart, of the kind produced by power press strokes — and the instruments had accumulated it into the measured doses. The Commission held that if impulse noise of that character must be excluded from dose calculations under Table G-16, then dosimeter readings that include it read too high; OSHA's own inspecting industrial hygienist agreed under cross-examination that impulse noise was not supposed to be included in her dosimeter or sound level meter measurements. The Commission also confronted the accuracy limits of the instrument itself, concluding that a dosimeter reading of 158.5 percent of the permissible dose was insufficient to prove overexposure where the employee's noise was predominantly high frequency and the required error allowance was applied. Every citation item was vacated. The case remains the template for the instrumentation attack: identify what the standard regulates, identify what the instrument actually recorded, and put the daylight between them on the record through the compliance officer herself.

Southern Scrap Materials Co., OSHRC Docket No. 94-3393 (2011): the placement war, won and lost on the protocol. Following a 1994 inspection of a Baton Rouge scrap recycling facility, OSHA issued citations alleging numerous violations of the lead and cadmium standards and proposed penalties totaling $1,937,000 for the contested items. The employees at the center of the exposure case were torch cutters who wore face shields, and the compliance officers had clipped the sampling cassette inlets outside the shields. The administrative law judge, relying on the Commission's earlier welding-fume decision in *Equitable Shipyards* — which involved sampling inside welding helmets — held the placement improper, invalidated OSHA's exposure monitoring, and vacated numerous lead and cadmium items on that basis. On review, the Commission reversed the placement holding: because the lead and cadmium standards define the action level and PEL without regard to personal protective equipment, exposure is measured in the breathing zone outside a face shield, and all but one of OSHA's monitoring results were held valid and reliable. The employer nonetheless prevailed on dozens of items on other grounds, and the judge had assessed $266,500 against the nearly $2 million proposed. The double lesson is the point: sampling placement is genuinely litigable — it carried the case at the trial level — but the winning argument is the one anchored in the specific standard's own definition of exposure, which is a technical reading an industrial hygienist, not a generalist, is equipped to brief. The case also rewards discovery persistence: analytical data omitted from the OSHA laboratory's final report was produced only after repeated requests, the day before the hearing began.

Bonney Forge Div., Gulf & Western Energy Products Group, 14 BNA OSHC 1968, 1972–73 (No. 79-4053, 1991), the warning. The administrative law judge vacated a noise citation in language every defense lawyer would like to frame, finding the industrial hygienist's data collection "imprecise, confused, fraught with inconsistencies" and unreliable. The Commission reversed, and its reasoning is the map of a failed sampling attack. The record showed the dosimeters were properly calibrated and used; the one calibration concession — that a couple of instruments were "reading a little bit low" — meant any error understated the exposure and thus favored the employer; the microphones were attached at the shirt collar in conformance with accepted OSHA and industrial hygiene practice; and, decisively, the readings were of a magnitude no error theory could absorb: even after allowing a very large margin of error, the employer's own noise-control expert accepted ten of thirteen readings as establishing overexposure, conceding that doses over 250 percent revealed overexposure. Sampling attacks are attacks on the margin. Where the measurements tower over the limit, the margin does not matter, and pressing the challenge anyway spends credibility the defense will need for the arguments that can still win.

V. The First Thirty Days After an Exposure Citation

The sampling defense is perishable, and the work belongs at the front of the case. Within the contest period, counsel should obtain the complete inspection file, including the sampling worksheets, calibration records, dosimeter data, and the full laboratory package with bench-level data rather than the summary report, and should treat gaps in that production as facts in themselves. The employer's side of the record should be preserved simultaneously: the production and process records for the sampled shifts, ventilation maintenance logs, the identity and recollections of the sampled employees, and any historical exposure monitoring, before ordinary retention schedules erase it. The retained industrial hygienist should rederive every number the Secretary intends to rely on — flow rates, air volumes, TWAs, SAE adjustments, noise doses — and map each against the OSHA Technical Manual, the applicable standard's sampling appendix, and the governing enforcement directive. The output of that audit is the informed settlement posture: which items are exposed to a methodology challenge, which will survive it, and where the technical record supports reclassification or withdrawal. In an exposure case, that analysis is not support for the defense theory. On the items that matter most, it is the defense theory.

The credibility of all of this rests on a single condition. The disciplines described here are not tricks for making valid measurements disappear; they are the profession's own quality requirements, applied evenhandedly. An expert who audits OSHA's sampling against OSHA's procedures, and who says so when the sampling passes, is the witness a judge believes when the sampling fails. In the exposure docket, the citation is only as good as the sample. Counsel who tests that proposition early, rigorously, and honestly litigates from strength; counsel who accepts the number on the violation worksheet has conceded the government's entire case in its opening line.

References

Air contaminants, 29 C.F.R. § 1910.1000.

American Industrial Hygiene Association. (2015). A strategy for assessing and managing occupational exposures (4th ed., S. D. Jahn, W. H. Bullock, & J. S. Ignacio, Eds.). AIHA Press.

Asbestos, 29 C.F.R. § 1910.1001, App. B (detailed procedures for asbestos sampling and analysis).

Benzene, 29 C.F.R. § 1910.1028, App. D (sampling and analytical methods).

Bonney Forge Div., Gulf & Western Energy Products Group, 14 BNA OSHC 1968 (No. 79-4053, 1991).

Collier-Keyworth Co., 13 BNA OSHC 1208 (No. 80-2848, 1987), vacated per stipulation, No. 88-2140 (1st Cir. Mar. 22, 1989).

Equitable Shipyards, Inc., 13 BNA OSHC 1177 (No. 81-1685, 1987).

Lead, 29 C.F.R. § 1910.1025; Occupational Exposure to Lead, 43 Fed. Reg. 52,952 (Nov. 14, 1978).

Occupational noise exposure, 29 C.F.R. § 1910.95, Table G-16.

Occupational Safety and Health Administration. (1999, as updated). OSHA Technical Manual (TED 01-00-015), Section II, Chapter 1: Personal sampling for air contaminants. U.S. Department of Labor.

Occupational Safety and Health Administration. (2020). Inspection procedures for the respirable crystalline silica standards (CPL 02-02-080). U.S. Department of Labor.

Occupational Safety and Health Administration. (n.d.). Sampling and analysis: Data validation and interpretation. U.S. Department of Labor. https://www.osha.gov/sampling-analysis/validation

Respirable crystalline silica, 29 C.F.R. § 1910.1053.

Southern Scrap Materials Co., OSHRC Docket No. 94-3393 (Rev. Comm'n Sept. 28, 2011).

CHAPTER 11

Multi-Employer Worksites

Who Really "Controlled" the Hazard?

> *On a construction site with four contractors and one general, the citation, and the civil complaint, follow control. And control, as OSHA and the courts define it, is a question of what actually happened on the site, not what the subcontract says. But control alone is not liability: since 2019, the Review Commission has insisted that a controlling employer's duty is one of reasonable care in a secondary safety role, not strict liability, and has vacated citations against general contractors that could prove they met it.*

I. One Hazard, Four Possible Defendants

A single unguarded excavation, a single missing guardrail, a single de-energized panel that was never locked out: on a multi-employer construction site, one hazardous condition can generate liability for several employers at once. OSHA has said so, formally, since 1999, when it issued its Multi-Employer Citation Policy as Directive CPL 02-00-124. Under that policy, more than one employer may be cited for a hazardous condition that violates an OSHA standard, and the agency works through a two-step analysis: first, classify each employer on the site as a creating, exposing, correcting, or controlling employer; second, determine whether that employer met the obligations its role carries (OSHA, 1999).

The four roles are defined by function, not job title. The creating employer caused the hazardous condition. The exposing employer's own employees are exposed to it. The correcting employer is the one responsible, in the common undertaking, for fixing that class of hazard: typically the trade that installs or maintains the safety system. And the controlling employer has general supervisory authority over the site, including the power to correct violations itself or to require others to correct them (OSHA, 1999). An employer can occupy more than one role at once; a general contractor is frequently both a controlling and an exposing employer, and a subcontractor that damages another trade's guardrail becomes a creating employer as to a hazard its own workers never touch.

For litigators, the policy is more than an enforcement mechanism. Because an OSHA citation documents a specific standard violation and the employer's notice of the hazard, the multi-employer analysis frequently sets the map for the civil case that follows: the third-party negligence suit against the general contractor or the controlling employer whose supervisory authority, plaintiffs will argue, carried a duty of reasonable care. That is the thesis of this chapter: on a multi-employer site, the decisive question is rarely whether a hazard existed — it is who controlled it, and "control" is established by the exercise of authority in practice, which contracts can describe but cannot contract away. But classification is only the first half of the fight. Because the controlling employer's duty is one of reasonable care in a secondary safety role, a general contractor that can document its diligence can defeat the citation even where the hazard, the exposure, and its own controlling status are all conceded.

II. The Doctrine Survived Its Court Challenges

For years, employers argued that the multi-employer doctrine exceeded OSHA's statutory authority, reasoning that the Act obligates an employer to protect its own employees, not someone else's. Two decisions largely closed that argument.

A. Solis v. Summit Contractors, Inc. (8th Cir. 2009)

In *Summit Contractors*, the Eighth Circuit upheld OSHA's authority to cite a controlling employer even when only another employer's employees were exposed to the hazard. The court grounded the holding in the text of the Act, reasoning that an employer's duty to comply with occupational safety and health standards under 29 U.S.C. § 654(a)(2) is not limited to the protection of its own employees, and that reading the statute otherwise would render its reference to "places of employment" superfluous (*Summit Contractors*, 2009). *Summit Contractors* reinstated the multi-employer policy within the Eighth Circuit and gave the doctrine its most-cited modern textual defense.

B. Acosta v. Hensel Phelps Construction Co. (5th Cir. 2018)

The more dramatic decision came nine years later. In *Hensel Phelps*, a general contractor building a public library in Austin was cited as a controlling employer after an OSHA compliance officer found a sub-subcontractor's employees working beside an unprotected, nearly vertical wall of Type C soil roughly twelve feet high, in plain view of the general contractor's on-site management, in violation of the excavation standard at 29 C.F.R. § 1926.652(a)(1) (*Hensel Phelps*, 2018). An administrative law judge found that *Hensel Phelps* met the definition of a controlling employer but felt bound by older Fifth Circuit precedent, *Melerine v. Avondale Shipyards* — holding that OSHA regulations protect only an employer's own employees, and so vacated the citation. The Fifth Circuit reversed, overruling its own precedent and holding that the Secretary has authority to cite a controlling general contractor for a hazardous condition affecting another employer's employees (*Hensel Phelps*, 2018).

Hensel Phelps is significant beyond the three states of the Fifth Circuit, because with that decision the last federal circuit resisting the doctrine fell into line, and the multi-employer policy now operates across the entire federal OSHA program. For the construction bar, the message was blunt: a general contractor with supervisory authority over a site cannot escape citation by pointing out that the exposed workers were a subcontractor's. What matters is control.

What the circuit decisions settled, however, was the doctrine's validity, not its reach in any given case. How much a controlling employer must actually do to discharge its duty, and how much evidence the Secretary must marshal to show it fell short, has been policed since 2019 by the Occupational Safety and Health Review Commission in a line of decisions beginning with *Suncor Energy*. Section IV takes up that line, because it is where the citation defense now lives.

III. Contract Versus Practice: How Control Is Actually Proven

The most litigated question in this area is not whether the doctrine is valid but whether a given employer is a controlling employer at all, and here the gap between the paper and the jobsite is where cases are won and lost. OSHA's own policy states that control can be established either by contract or, in the absence of explicit contractual provisions, by the actual exercise of control in practice (OSHA, 1999). The agency's illustrative examples drive the point home: in one, an employer's contract expressly disclaimed authority over safety, yet the combination of rights the contract did grant gave the employer broad responsibility over the site and the practical ability to direct actions affecting safety — making it a controlling employer notwithstanding the disclaimer (OSHA, 1999).

This is the single most important practical lesson for counsel on either side, and it is fundamentally a factual and technical inquiry rather than a contractual one. The subcontract's indemnity clause and its recitation that "each subcontractor is solely responsible for the safety of its own employees" do not decide the control question. What decides it is the evidence of how the site actually ran: Who set the schedule and the production pace? Who ran the coordination meetings? Who had authority to stop work? Who walked the site, and did their walk-throughs actually look for hazards? Who could direct a subcontractor to fix a condition, and had they done so before? A general contractor whose superintendent conducted daily walk-throughs, issued safety directives, and had a graduated enforcement mechanism has exercised control, whatever the contract says; a general whose "supervision" was purely administrative may not have.

IV. Control Is Not Liability: The *Suncor* Line

Establishing that an employer is a controlling employer answers only the classification question. The reasonable-care standard applied to a controlling employer is not the same as the duty an employer owes its own workers. OSHA's policy expressly provides that the measures a controlling employer must take to satisfy its duty to prevent and detect violations are less than what is required of an employer protecting its own employees (OSHA, 1999). The relevant questions are how frequently and closely the controlling employer inspected, whether it had a system for enforcing the correction of hazards it found, and whether it exercised reasonable diligence in the circumstances, not whether it guaranteed a hazard-free site. In the directive's illustrative example, a general contractor that failed to maintain a graduated system of enforcement to require its subcontractors to correct hazards fell short of that reasonable-care duty even though it conducted adequate inspections (OSHA, 1999).

For two decades that language sat in an enforcement directive, where it bound OSHA only as a matter of internal policy. Since 2019, the Review Commission, the independent adjudicative body that decides contested citations, has converted it into precedent with teeth, repeatedly reversing controlling-employer citations where the Secretary could not show the employer's diligence actually fell short.

A. Suncor Energy (OSHRC 2019): One Lapse Is Not Enough

In *Suncor Energy (U.S.A.) Inc.*, the Commission vacated a fall-protection citation issued to a refinery owner as the controlling employer during a thirty-day maintenance turnaround involving multiple contractors (*Suncor Energy (U.S.A.) Inc.*, 2019). Guardrails had been removed from the third level of a scaffold inside a fired heater, and two employees of a subcontractor retained to inspect welds climbed to that level after obtaining a safe-work permit from *Suncor*: a permit that, in error, did not specify personal fall protection. The administrative law judge found that *Suncor* had missed opportunities to discover the condition and therefore had constructive knowledge of it. The Commission reversed. It emphasized that a controlling employer occupies a secondary safety role and that its duty of reasonable care is less than the duty it owes its own employees; it then held that a single error on one permit, measured against a comprehensive safety program that exceeded OSHA's minimum requirements and an enforcement regime that removed workers from the site for a single fall-protection infraction, could not establish a failure of reasonable care — particularly where *Suncor* was justified in relying on the subcontractor's demonstrated competence to recognize when fall protection was required, and where the violative condition had existed for only a few hours (*Suncor Energy (U.S.A.) Inc.*, 2019).

The decision was widely read as a rebuke of what practitioners had called a "scattershot" approach, citing every employer on the site once a hazard was found, and as a holding that the multi-employer doctrine is not a form of strict liability against controlling employers who have performed their high-level safety duties. After *Suncor*, the Secretary must carry the burden of showing, on objective evidence, that the controlling employer failed to take reasonable measures to prevent or detect the violation; the mere existence of the hazard, and even the employer's conceded controlling status, are not enough.

B. Summit Contracting (OSHRC 2022): Hazards in Plain View

The Commission extended the framework to the construction general contractor — the paradigm controlling employer — in *Summit Contracting Group, Inc.* (*Summit Contracting Grp., Inc.*, 2022), after an intermediate step in StormForce of Jacksonville, LLC (StormForce of Jacksonville, LLC, 2021) applying the same reasonable-care test to a roofing contractor's site. Summit was the general contractor on a multi-family project in Ponte Vedra, Florida, where three employees of framing subcontractors — subcontractors retained by Summit's trade contractor, with no direct contractual relationship to Summit — were observed working without fall protection, in plain view. The Commission nonetheless vacated the citation. Summit showed that it conducted daily walk-throughs of the site, retained an outside safety consultant for monthly inspections, and could point to the framing contractors' own fall-protection plans, training programs, and inspections; on that record, the Commission held that Summit exercised the reasonable care its secondary safety role required and lacked knowledge of the violation (*Summit Contracting Grp., Inc.*, 2022).

Summit Contracting also closed off an analytical shortcut the Secretary had urged: the Commission refused to treat asserted inadequacies in the general contractor's own safety program as an independent basis for imputing constructive knowledge of a subcontractor's violation. The adequacy of the program is evidence bearing on reasonable care; it is not a substitute for proof that reasonable care would have detected the specific violative condition (*Summit Contracting Grp., Inc.*, 2022). (A note on names, since the coincidence invites confusion: *Summit Contracting* Group, the 2022 Commission decision, is unrelated to *Summit Contractors*, the 2009 Eighth Circuit decision discussed in Section II.)

C. The Caveat: OSHA Still Cites the General Contractor First

None of this has stopped the citations. Field enforcement continues to name the general contractor when a subcontractor's crew is found exposed, and the *Suncor* framework must frequently be litigated rather than merely invoked. *Fama Construction, LLC* is the cautionary example: there the administrative law judge recited the controlling-employer framework but then held the general contractor to the more stringent standard applicable to an employer whose own employees are exposed, and the Commission had to remand with instructions to apply the correct, lesser reasonable-care test (Fama Constr., LLC, 2023). The practical lesson is twofold. The reasonable-care defense is real and, on the right record, dispositive, but it is an evidentiary defense, and the general contractor that has not built the *Suncor*-style record of inspections, enforcement, and justified reliance before the compliance officer arrives will have nothing to litigate with.

V. Where This Migrates Into the Civil Case

For plaintiff- and defense-side construction litigators, the multi-employer analysis matters most because of what it enables in the civil forum. Workers' compensation exclusivity ordinarily bars the injured worker's negligence suit against the direct employer, but it does nothing to bar suit against other parties on a multi-employer site, and those parties are precisely the controlling and creating employers the OSHA analysis identifies. The general contractor that controlled the site, the subcontractor whose crew created the hazard, and the equipment supplier whose defect contributed to it are all third parties as to the injured worker, and all are outside the exclusivity bar. That proposition is statutory before it is decisional: the compensation acts bar the employee's action against the *employer* and, in most states, against co-employees, and they expressly preserve the action against others. California's formulation is representative: an employee's compensation claim against the employer does not affect "any right of action against a third person," and the employer's carrier may join or intervene to protect its lien (Cal. Lab. Code § 3852; *Privette v. Superior Court*, 5 Cal. 4th 689, 697 (1993)). New York's is parallel

(N.Y. Workers' Comp. Law § 29(1)), as is Illinois's (820 Ill. Comp. Stat. 305/5(b)) and Texas's (Tex. Lab. Code § 417.001). Counsel should read the operative section rather than a summary of it, because the states diverge on three points that matter to the case: whether the bar extends to co-employees and to the employer's other subsidiaries, whether a statutory-employer or borrowed-servant doctrine can pull a nominal third party back inside the bar, and how the carrier's lien is calculated against a recovery.

Into that third-party case comes the OSHA citation. In most jurisdictions, an OSHA standard violation is admissible as evidence of the standard of care and, where the violated standard was designed to prevent the type of injury suffered, can support a negligence-per-se or presumption-of-negligence theory. The interaction with state law is not uniform — California, for example, worked through a specific statutory history before its Supreme Court confirmed in *Elsner v. Uveges* that Cal/OSHA provisions may be admitted in third-party negligence actions to establish the standard of care (*Elsner*, 2004), so counsel must check the local rule. But the general pattern holds: the same control analysis that made a general contractor a controlling employer for citation purposes makes it a defendant owing a duty of reasonable care for civil purposes, and the citation is evidence that the duty was breached.

The *Suncor* line runs into the civil case as well, and it runs in the other direction. The same documentary record that defeats the citation — the inspection logs, the graduated enforcement history, the evidence of justified reliance on a competent subcontractor — is the record a general contractor will offer a civil jury to show that whatever duty its control created, that duty was met. A vacated citation is not a shield against the negligence suit, but the facts that vacated it are the defense's proof on breach, just as a sustained citation is the plaintiff's.

The technical expert's role in this posture is to reconstruct not the physics of the accident alone but the architecture of control on the site: to map the actual lines of supervisory authority against the contract documents, the meeting minutes, the daily reports, the toolbox-talk sign-in sheets, and the testimony of who told whom to do what. That reconstruction is often more decisive than the accident reconstruction itself, because it answers the question the jury will actually be asked, not what went wrong, but who had the authority and the duty to prevent it.

VI. Practice Implications

Several implications follow for counsel handling construction citations and the injury suits that trail them.

For defense counsel representing a general contractor, the instinct to rely on the subcontract's safety-responsibility and indemnity language is a trap. Those clauses allocate risk between the parties; they do not determine OSHA liability, and 29 U.S.C. § 654(a)(2) imposes a non-delegable duty to comply with standards that contract language cannot erase. The defense that works is evidentiary, and after *Suncor* and *Summit Contracting* it has adjudicative precedent behind it: a documented record of reasonable inspection, a functioning graduated enforcement system, and demonstrated, justified reliance on the subcontractor's own safety competence showing the controlling employer exercised the diligence its secondary safety role requires. Where OSHA has cited the general contractor reflexively, as it still frequently does, that record is also the basis for contesting the citation on the ground that the Secretary cannot carry his burden under the Commission's reasonable-care framework.

For plaintiff's counsel, the opportunity is to build the control record early, while the site documentation still exists. The daily reports, the coordination-meeting minutes, and the enforcement history establish control far more reliably than the organizational chart, and they are the records most likely to disappear as the project closes out and the trailers leave the site. And after *Suncor*, the plaintiff's work does not end with proving control: the record must also show that the general contractor's diligence fell short of even the reduced, secondary-role standard — the missed walk-throughs, the enforcement system that existed on paper only, the reliance on a subcontractor whose safety failures the general had already witnessed.

For both sides, the lesson of *Summit Contractors* and *Hensel Phelps* is that the doctrine is settled; the lesson of *Suncor* and *Summit Contracting* Group is that the doctrine is not strict liability. The fight has moved to classification and to reasonable care, and both are proved the same way. The party that can show — from the site's actual operation rather than its paperwork — who set the pace, who ran the walk-throughs, who could stop the work, and what the party with that authority actually did with it, is the party that controls the answer to the only question that matters: who controlled the hazard, and whether the diligence its control demanded was exercised.

References

Acosta v. Hensel Phelps Constr. Co., 909 F.3d 723 (5th Cir. 2018).

Cal. Lab. Code §§ 3600, 3602, 3852; N.Y. Workers' Comp. Law § 29(1); 820 Ill. Comp. Stat. 305/5(b); Tex. Lab. Code § 417.001.

Elsner v. Uveges, 34 Cal. 4th 915 (2004).

Fama Construction, LLC, No. 19-1467 (O.S.H.R.C. Mar. 29, 2023) (remand order).

Melerine v. Avondale Shipyards, Inc., 659 F.2d 706 (5th Cir. 1981).

Occupational Safety and Health Act of 1970, 29 U.S.C. § 654(a)(2).

Occupational Safety and Health Administration. (1999). Multi-employer citation policy (Directive No. CPL 02-00-124). U.S. Department of Labor. https://www.osha.gov/enforcement/directives/cpl-02-00-124

Privette v. Superior Court, 5 Cal. 4th 689 (1993).

Safety and health regulations for construction; excavations, requirements for protective systems, 29 C.F.R. § 1926.652(a)(1).

Solis v. Summit Contractors, Inc., 558 F.3d 815 (8th Cir. 2009).

StormForce of Jacksonville, LLC, No. 19-0593, 2021 WL 2582530 (O.S.H.R.C. Mar. 8, 2021).

Summit Contracting Group, Inc., No. 18-1451, 2022 WL 1572848 (O.S.H.R.C. May 10, 2022).

Suncor Energy (U.S.A.) Inc., No. 13-0900, 2019 WL 654129 (O.S.H.R.C. Feb. 1, 2019).

CHAPTER 12

State Plans vs. Federal OSHA

When the Stricter Standard Governs the Case

Whether your worksite sat under federal OSHA's General Duty Clause or a state plan's prescriptive standard can change everything downstream: the negligence-per-se theory, the reporting deadline that was missed, and the settlement value. The applicable-standard question belongs at intake, not at trial.

I. Two Regulatory Systems Occupying the Same Country

A worker injured on a jobsite in Nevada, another injured performing identical work across the line in Arizona, and a third in Texas are governed by three meaningfully different bodies of occupational safety law: even though all three states sit under the same federal statute. The reason is the state-plan structure Congress built into the Occupational Safety and Health Act. Under Section 18 of the Act, a state may operate its own occupational safety and health program if OSHA approves the program as "at least as effective" as federal enforcement. The governing text is Section 18 itself, and it repays reading. Section 18(b) permits a state to assume responsibility for standards development and enforcement by submitting a plan for approval; Section 18(c)(2) conditions approval on the state's standards and enforcement being "at least as effective in providing safe and healthful employment" as the federal program; and Section 18(f) obliges OSHA to monitor continuing conformity and to withdraw approval where the state falls short (29 U.S.C. § 667(b), (c)(2), (f); implementing regulations at 29 C.F.R. pt. 1902). Twenty-two states and territories currently operate approved plans covering private-sector as well as state and local government employment, and six states and one territory operate plans limited to state and local government employees (OSHA, State Plans, 29 C.F.R. pt. 1952).

The "at least as effective" floor is the key concept, and it runs in one direction. A state plan may not fall below the protection federal OSHA provides, but it is free to exceed it: adopting standards that are more stringent, or regulating hazards on which federal OSHA has no standard at all (29 U.S.C. § 667(c)(2); 29 C.F.R. §

1902.3(c)). Several states have done exactly that, and the divergence is widest precisely in the areas most relevant to construction and injury practice: heat illness, ergonomics, workplace violence, wildfire smoke, and injury reporting. The thesis of this chapter is procedural but consequential: the applicable-standard determination is not a background detail; it drives the negligence-per-se analysis, the reporting and citation-appeal timelines, and the settlement calculus, and it must be resolved at intake because the differences that matter most are baked in before the first pleading.

II. Where the Standards Diverge

A. Heat Illness: The Sharpest Current Divide

Because federal OSHA still has no heat-specific standard and must litigate heat under the General Duty Clause, the states that have adopted their own heat rules present the starkest example of the stricter standard governing the case. Seven states enforce their own heat illness prevention standards, and the operative citations, not the commentary about them, are what counsel needs at intake: California's outdoor rule at Cal. Code Regs. tit. 8, § 3395 and its indoor rule at § 3396, alongside the Injury and Illness Prevention Program requirement at § 3203; Oregon's at Or. Admin. R. 437-002-0156, reaching both indoor and outdoor work; Washington's outdoor rule at Wash. Admin. Code 296-62-095 through -09560; Minnesota's indoor standard at Minn. R. 5205.0110; Colorado's agricultural provisions under the Agricultural Workers' Rights statute, Colo. Rev. Stat. § 8-3.5-201 et seq., and 7 Colo. Code Regs. 1103-15; Maryland's rule at Md. Code Regs. 09.12.32; and Nevada's at Nev. Admin. Code § 618.900 et seq. They differ substantially in scope and stringency. Maryland's and Oregon's reach most indoor and outdoor work; Washington's centers on outdoor work above defined temperature triggers; Minnesota's is an indoor-only ventilation-and-temperature rule of long standing; Colorado's operates in the agricultural context; and Nevada's is a written-program requirement that leaves more of the substantive protection to employer discretion than the others do. The differences are not academic. A plaintiff's negligence-per-se theory rises or falls on whether the governing rule prescribed shade at a stated index, a written acclimatization protocol, or merely a plan.

The litigation consequence is direct. Where a prescriptive state standard applies, the plaintiff can point to a specific, numeric or procedural requirement the employer violated — a missing shade structure, an absent acclimatization protocol, a water-provision failure — rather than assembling a General Duty Clause hazard from exposure science. A defined standard supports a negligence-per-se or presumption-of-negligence theory in a way an amorphous general duty cannot, and it changes what the technical expert must prove: not that the hazard existed as a matter of physiology, but that a written, enforceable requirement was not met.

B. Ergonomics, Workplace Violence, and Wildfire Smoke

Beyond heat, several state plans regulate hazards federal OSHA does not address by standard. California maintains the only enforceable general-industry ergonomics standard (Cal. Code Regs. tit. 8, § 5110, triggered by two or more repetitive-motion injuries in a twelve-month period) and, since July 1, 2024, a workplace-violence-prevention plan requirement for most employers (Cal. Lab. Code § 6401.9). California and Oregon both regulate wildfire-smoke exposure, Cal. Code Regs. tit. 8, § 5141.1 and Or. Admin. R. 437-002-1081 — requiring air-quality monitoring, communication, and respiratory protection above defined AQI thresholds; Washington's rule appears at Wash. Admin. Code 296-820. For a worker whose injury arises from a repetitive-motion hazard, a workplace assault, or wildfire-smoke exposure, the existence of a governing state standard can be the difference between a cognizable standard-of-care violation and a hazard with no regulatory yardstick at all.

C. Reporting and Appeal Timelines

The procedural divergences are less visible but can be dispositive. State plans administer their own systems for issuing, appealing, and litigating citations, and the deadlines differ from federal practice and from one another. California requires an employer to report a serious injury, illness, or death immediately and in no event more than eight hours after knowledge of it (Cal. Lab. Code § 6409.1(b); Cal. Code Regs. tit. 8, § 342), against the federal twenty-four-hour window for in-patient hospitalization, amputation, and loss of an eye (29 C.F.R. § 1904.39(a)(2)). Oregon allows thirty days to appeal a citation (Or. Rev. Stat. § 654.078; Or. Admin. R. 438-085-0015), where federal practice and most state plans follow the fifteen-working-day period of 29 U.S.C. § 659(a). A defense team that applies the federal appeal clock to an Oregon citation, or misses California's eight-hour reporting trigger, has made an error that no amount of merits work will cure. Even on recordkeeping, where state plans must be substantially identical to Part 1904's recording criteria (29 C.F.R. § 1904.37(b)(1)), states may impose *more* stringent reporting requirements — § 1904.37(b)(2) says so expressly.

Table 12-1 collects the variables counsel actually needs at intake for the state plans that generate the most litigation. It is a starting point rather than a substitute for reading the operative section: state plans amend their rules on their own schedules, and the citation, not the summary, is what governs.

Table 12-1. Selected State Plan Variables at Intake

Plan	Private sector	Stricter standards	Report deadline	Appeal window
California	Yes (Cal/OSHA)	Outdoor heat (§ 3395); indoor heat (§ 3396); IIPP (§ 3203); ergonomics (§ 5110); workplace violence (Lab. Code § 6401.9); wildfire smoke (§ 5141.1)	8 hours, serious injury or death (Lab. Code § 6409.1(b))	15 working days
Oregon	Yes (Oregon OSHA)	Heat, indoor and outdoor (437-002-0156); wildfire smoke (437-002-1081)	8 hours (fatality/catastrophe)	30 days
Washington	Yes (DOSH)	Outdoor heat (296-62-095); wildfire smoke (296-820)	8 hours (fatality/hospitalization)	15 working days
Michigan	Yes (MIOSHA)	Extensive independent general-industry and construction rule set	8 hours (fatality)	15 working days

Plan	Private sector	Stricter standards	Report deadline	Appeal window
Minnesota	Yes (MNOSHA)	Indoor heat (5205.0110); A Workplace Accident and Injury Reduction Act program requirement	8 hours (fatality)	15 working days
Maryland	Yes (MOSH)	Heat (09.12.32)	8 hours (fatality)	15 working days
Nevada	Yes (Nevada OSHA)	Heat, written program (NAC 618.900 et seq.)	8 hours (fatality)	15 working days
Federal OSHA	(default)	None beyond the CFR; heat and ergonomics enforced under § 5(a)(1)	8 hours (fatality); 24 hours (hospitalization, amputation, eye loss)	15 working days

Note. Reporting triggers and appeal windows are stated as of July 2026 and must be verified against the current state rule; several plans define "serious injury" differently from the federal in-patient-hospitalization trigger, and at least one — California's — counts the eight hours from the employer's knowledge rather than from the event.

III. Preemption: Does Federal Wipe Out the State Standard?

A recurring source of confusion, and one that will intensify if OSHA ever finalizes its stalled federal heat rule, is whether a federal standard displaces a state's stricter rule. The governing decision is *Gade v. National Solid Wastes Management Ass'n*, 505

U.S. 88 (1992), and it is worth stating with more care than it usually receives, because the shorthand version is misleading in both directions.

Gade concerned Illinois statutes licensing hazardous-waste workers. Illinois did not have an approved state plan. The plurality held that Section 18 of the Act impliedly preempts *any* state law that regulates an occupational safety or health issue for which a federal standard is in effect, unless the state has obtained approval of a plan under Section 18(b) — reasoning that Section 18(b)'s submission requirement would be meaningless if a state could regulate the same issue without submitting anything (505 U.S. at 99–102). Justice Kennedy concurred in the judgment on express-preemption grounds. Two limits on the holding matter in practice and are routinely lost in the summaries. First, preemption reaches only state law that *regulates occupational safety and health*; a law with a different, legitimate purpose that incidentally touches worker safety is not preempted merely because it has that effect (505 U.S. at 107). Second, and expressly, the Act does not preempt state tort law: Section 4(b)(4) provides that nothing in the Act enlarges, diminishes, or affects the common-law or statutory rights, duties, or liabilities of employers with respect to employee injuries (29 U.S.C. § 653(b)(4)). That second point is why the negligence-per-se and standard-of-care arguments of Chapters 11 and 23 survive at all, and why *Gade* is not the obstacle to civil use of OSHA standards that defendants sometimes suggest.

The Section 18(b) exception is the whole ballgame for state-plan states. Because an approved plan is the mechanism through which a state exercises authority the Act otherwise reserves, a plan state may maintain a heat, ergonomics, or workplace-violence standard more stringent than any federal counterpart. It must remain at least as effective (29 U.S.C. § 667(c)(2)), and once a federal standard issues on the same issue, the state must adopt a standard identical to or at least as effective as the federal rule within six months of the federal promulgation date (29 C.F.R. § 1953.5(a)). What it need not do is repeal the protections it already has. The practical upshot: if OSHA finalizes a federal heat standard, the seven states with existing heat rules do not lose them; they retain provisions at least as protective and conform only where they fall short. For counsel, this means the state standard is not a temporary artifact to be swept away by federal rulemaking: it is a durable feature of the landscape that must be identified worksite by worksite.

There is a third preemption question the chapter should flag because it arises in nearly every construction case in a plan state, and *Gade* does not answer it: whether a state-plan standard can be used in a civil action as evidence of the standard of care or as the predicate for negligence per se. Section 4(b)(4) preserves the question for state law, and the states answer it differently: some treating an OSHA or state-plan violation as negligence per se, some as evidence of negligence, and some excluding it where the plaintiff is not the defendant's own employee. Chapter 11 §V develops the point; it belongs in the intake analysis because it determines what the state standard is actually worth in the case.

IV. What This Changes at Intake

The applicable-standard determination should be among the first analytical steps in any workplace injury or OSHA matter, because it conditions nearly everything that follows.

It fixes the standard of care. Whether the case will be built on a prescriptive state standard or a General Duty Clause hazard determines what the plaintiff must prove and what the defense must rebut. A jurisdiction with a specific heat, ergonomics, or guarding standard offers the plaintiff a negligence-per-se pathway; a General Duty Clause jurisdiction requires the fuller technical showing of hazard, recognition, and feasible abatement. The two cases are built differently from the first interview.

It sets the clock. The reporting and appeal deadlines that govern the matter are the state plan's, not the federal defaults, wherever a full state plan operates. Those deadlines must be calendared correctly at the outset, because several of them, California's eight-hour reporting trigger most notably, run from the incident, not from counsel's involvement.

It shapes the settlement calculus. A defined standard violation, admissible to establish the standard of care, is a more concrete liability exposure than a contested General Duty Clause theory, and it prices differently. Enforcement posture matters too. Inspection activity is not distributed evenly across the plan states, and the differences are documentable rather than anecdotal: OSHA's annual State Plan Performance Reports and the enforcement data published under 29 C.F.R. § 1902.35 report inspection counts, citation rates, and penalty totals plan by plan, and counsel valuing a case in an unfamiliar jurisdiction should consult them rather than rely on reputation. The larger plans — California, Oregon, Washington, and Michigan among them — conduct enough programmed inspection activity that a citation is both more probable and, once issued, more likely to be defended by an agency with the resources to litigate it.

V. Practice Implications

Three implications follow for counsel handling matters that may cross jurisdictional lines.

First, determine the governing regime for each worksite before building the theory of the case. A contractor running crews across state lines may be under a state plan in one location and federal OSHA in the next, with different standards, deadlines, and negligence-per-se exposure at each. The determination is site-specific and cannot be assumed from the company's headquarters or the litigator's home forum.

Second, do not treat state standards as provisional. The *Gade* framework and the "at least as effective" principle mean state-plan standards survive federal rulemaking rather than yielding to it. A stricter state heat or ergonomics standard is a durable basis for liability that will not evaporate when, or if, a federal counterpart is finalized.

Third, recognize that this is where legal analysis and technical proof meet. The applicable-standard question is answered by law, but its consequences are executed through the technical record: which measurements, which protocols, and which documented practices the governing standard actually required. The attorney who resolves the standard question at intake, and then builds the technical case to the specific requirements that standard imposes, has aligned the two halves of the case from the start, and avoided the far more expensive discovery that the standard was stricter, or the deadline shorter, than the federal default assumed.

References

Cal. Lab. Code §§ 6401.9, 6409.1(b); Cal. Code Regs. tit. 8, §§ 342, 3203, 3395, 3396, 5110, 5141.1.

Gade v. Nat'l Solid Wastes Mgmt. Ass'n, 505 U.S. 88 (1992).

Minn. R. 5205.0110; Md. Code Regs. 09.12.32; Nev. Admin. Code § 618.900 et seq.; Colo. Rev. Stat. § 8-3.5-201 et seq.; 7 Colo. Code Regs. 1103-15.

Natural Resources Defense Council. (2026, May 19). States can act now to create workplace heat standards. https://www.nrdc.org/media/states-canact-now-create-workplace-heat-standards Occupational Safety and Health Act of 1970, 29 U.S.C. §§ 653(b)(4), 659(a), 667.

Occupational Safety and Health Administration. (n.d.). State plans. U.S. Department of Labor. https://www.osha.gov/stateplans

Occupational Safety and Health Administration. State Plan Performance Reports and enforcement activity data published under 29 C.F.R. § 1902.35. U.S. Department of Labor.

Or. Rev. Stat. § 654.078; Or. Admin. R. 437-002-0156, 437-002-1081, 438-085-0015.

State plan approval and monitoring, 29 C.F.R. pts. 1902, 1952, 1953 (particularly § 1902.3(c) and § 1953.5(a)); recording and reporting, 29 C.F.R. §§ 1904.37, 1904.39.

State recordkeeping regulations, 29 C.F.R. § 1904.37.

Wash. Admin. Code 296-62-095 to -09560, 296-820.

CHAPTER 13

Defending the Retaliation Claim

The OSHA Whistleblower Case When the Adverse-Action Record Does Not Support the Complaint

Retaliation is the one claim in this book that cannot be photographed, measured, or sampled. It alleges a reason rather than a condition, and a reason leaves its traces only in a documentary sequence — the complaint, the knowledge, the adverse action, and the interval between them — that a qualified safety professional can reconstruct and test against what the employer did in every comparable case it never litigated. This chapter maps that sequence, the points at which it comes apart, and the contemporaneous records that decide it in either direction.

I. The Retaliation Claim Rests Entirely on the Causal Chain

A citation alleges a condition; a retaliation complaint alleges a motive. Nothing on the floor of the plant can be measured to establish why a supervisor signed a termination notice on a Tuesday in March, and the government does not attempt it directly. What it does instead is build the motive by inference, from timing, from comparators, and from the employer's own explanation, and an inference assembled from records is an inference that can be tested against those same records. That is where the defense of these cases lives. When an employee files a complaint under Section 11(c) of the Occupational Safety and Health Act, 29 U.S.C. § 660(c), alleging discharge or discrimination for exercising safety rights, the complainant and ultimately the Secretary must prove a specific sequence: that the employee engaged in protected activity, that the employer knew of it, that the employee suffered an adverse employment action, and that a causal connection links the protected activity to the adverse action. That causal connection is not an observation. It is an inference constructed from a documentary record — complaint logs, discipline files, timing, comparator treatment, supervisor communications, and the employer's stated reasons — and the courts and the Department of Labor's adjudicators have made clear, in decisions spanning four decades, that a break anywhere in that chain can be fatal to the claim.

Counsel defending these claims naturally reach first for the legal architecture: the statutory filing deadline, the definition of adverse action, the shifting burdens of proof. Beneath that architecture sits a layer that is factual and often technical — whether the safety complaint was genuine and reasonable, whether the employer's disciplinary practice was consistent, whether the hazard the employee refused to face actually presented the danger claimed — and it cannot be evaluated from the complaint narrative or the investigator's summary alone. It requires someone who has personally run safety programs, investigated hazard complaints, audited discipline systems, and applied the same standards OSHA's whistleblower investigators are instructed to apply. This chapter describes where the retaliation case breaks, how a technical expert finds the breaks in the record both sides built in real time, and what the reported decisions teach about which defenses succeed and which fail.

II. Where the Retaliation Case Breaks: Five Vulnerabilities

OSHA has published, in considerable detail, how its own personnel are expected to evaluate these claims. The Whistleblower Investigations Manual (CPL 02-03-011) governs the intake, investigation, and merit determination of complaints under every statute the Directorate of Whistleblower Protection Programs administers; the regulations at 29 C.F.R. Part 1977 interpret Section 11(c), including the narrow right to refuse hazardous work; and 29 C.F.R. § 1904.35 and its interpretive guidance extend anti-retaliation principles to injury reporting, post-incident drug testing, and safety incentive programs. These documents are the yardstick. Every gap between what the complainant must prove and what the contemporaneous record actually shows is a fact the defense is entitled to develop, and the gaps cluster in five places.

A. Protected Activity and Its Documentation

The claim begins with what the employee actually did. Section 11(c) protects filing a complaint with OSHA, instituting or testifying in a proceeding, and exercising any right afforded by the Act, which the regulations and case law have read to include internal safety complaints to the employer, participation in inspections, and, in defined circumstances, refusing to perform a task. But the activity must exist as an identifiable event, and its character matters. A generalized grievance about workload or a personality dispute with a supervisor, later recharacterized as a safety complaint, is not protected activity; a complaint about a hazard is. The defense audit therefore starts with the record of the complaint itself: when it was made, to whom, in what words, and whether it appears in the safety suggestion log, the near-miss reporting system, the toolbox-talk minutes, or the maintenance work-order queue, or only in the complainant's recollection after the discharge. Where the alleged complaint left no contemporaneous trace in systems that reliably capture other complaints, its existence and its safety character are both legitimately contestable, and the employer's own reporting infrastructure becomes defense evidence.

B. Employer Knowledge and the Decision-Maker

Retaliation requires a retaliator who knew. The causal element cannot be satisfied unless the person who made the adverse decision knew — or, under the case law, at least suspected — that the employee had engaged in protected activity. The knowledge inquiry is granular: not whether anyone in the company knew of the OSHA complaint, but whether the decision-maker did, and when. OSHA does not disclose a complainant's identity, and where the complaint was filed anonymously, the question becomes what the decision-maker suspected and on what basis: an issue the Eighth Circuit resolved against the employer in *Reich v. Hoy Shoe Co., Inc.*, discussed in Section IV, holding that discharge based on a mistaken or unconfirmed suspicion of protected activity violates the Act just as surely as discharge based on certain knowledge. The lesson cuts both ways. For the defense, a clean record showing the decision-maker had no knowledge and no articulable basis for suspicion severs the chain at its second link; a record showing supervisors speculating about who called OSHA welds the chain shut.

C. Adverse Action, Timing, and Temporal Proximity

The third element is the action itself and, critically, its position on the calendar. Discharge, demotion, suspension, and pay reduction are unambiguous adverse actions; the harder disputes involve schedule changes, transfers, discipline short of discharge, and — under § 1904.35 — policies that penalize injury reporting itself, such as automatic post-incident drug testing untethered to any suspicion of impairment or incentive programs that pay workers for silence. Timing is the complainant's favorite proof: an adverse action following close on the heels of a complaint invites the inference that the one caused the other. But temporal proximity is a two-edged instrument, and the Supreme Court's admonition in *Clark County School District v. Breeden* — that proximity alone must be very close to carry the inference, and that action taken after the employer contemplated it anyway proves nothing — is the doctrinal anchor of the timing defense. The defense audit reconstructs the disciplinary timeline from the records: when the performance problem was first documented, when the termination decision was actually made, and whether the sequence shows a decision already in motion before the protected activity occurred. A discipline file that predates the complaint is the single most valuable document in the case.

D. The Work Refusal and Its Objective Reasonableness

A distinct and recurring species of protected activity is the refusal to work, and it is where the technical expert earns the retention. The Supreme Court in *Whirlpool Corp. v. Marshall* upheld the regulation, 29 C.F.R. § 1977.12(b)(2), that protects an employee who refuses a task, but only within its stated limits: the employee must reasonably and in good faith believe the task presents a real danger of death or serious injury, the danger must be too urgent to abate through normal statutory channels, and the employee must have sought correction from the employer where possible. Each limit is a defense inquiry with technical content. Was the condition the employee refused actually hazardous, measured against the applicable standards, the equipment's guarding and lockout status, the exposure data, and the industry's own practice? Would a reasonable employee, with the training this employee had received, have perceived imminent serious danger? Did the employee raise the condition before refusing, and was a correction available? These are questions of hazard assessment, not credibility, and an expert who reconstructs the condition from photographs, maintenance records, and the physics of the operation can show a tribunal that the refusal fell outside the regulation's boundaries, or, on honest review, squarely within them.

E. Causation, Pretext, and the Consistency Audit

The final vulnerability is the one many claims never survive: the employer's legitimate, non-retaliatory reason and the consistency of its application. Under the framework applied to 11(c) claims, once the complainant establishes the prima facie elements, the employer articulates its reason — the attendance record, the safety violation, the reduction in force — and the fight becomes whether that reason is pretext. Under the newer statutes OSHA administers, the structure is tighter still, and defense counsel who treat all whistleblower claims as McDonnell Douglas cases are preparing for the wrong burden. In the contributing-factor statutes the complainant need show only that protected activity was a contributing factor in the adverse action — any factor that, alone or in combination, tended to affect the outcome in any way — after which the employer must prove by clear and convincing evidence that it would have taken the same action in the absence of the protected activity. The difference is enormous: the complainant's burden falls to something close to a threshold showing, the employer's rises above a preponderance, and the pretext framework's comfortable middle ground disappears.

The statutes that use this framework, which the reader should identify at intake rather than at the hearing, include the Surface Transportation Assistance Act (49 U.S.C. § 31105), the Sarbanes-Oxley Act (18 U.S.C. § 1514A), the Federal Railroad Safety Act (49 U.S.C. § 20109), the National Transit Systems Security Act (6 U.S.C. § 1142), the Consumer Product Safety Improvement Act (15 U.S.C. § 2087), the Food Safety Modernization Act (21 U.S.C. § 399d), the Affordable Care Act's employee-protection provision (29 U.S.C. § 218c), the Consumer Financial Protection Act (12 U.S.C. § 5567), the Seaman's Protection Act (46 U.S.C. § 2114), the Pipeline Safety Improvement Act (49 U.S.C. § 60129), the Moving Ahead for Progress in the 21st Century Act's motor-vehicle-safety provision (49 U.S.C. § 30171), and the Taxpayer First Act (26 U.S.C. § 7623(d)). Section 11(c) itself is *not* among them: it retains the traditional burden-shifting framework, which is why the same conduct can produce very different outcomes depending on which statute the complaint invokes.

Two further features of the modern statutes change defense strategy and are routinely missed. Most of them carry kick-out provisions: where the Secretary has not issued a final decision within a stated period — 210 days under Sarbanes-Oxley, 210 days under FRSA and several others, and there is no delay due to the complainant's bad faith — the complainant may withdraw from the administrative process and file a de novo action in federal district court, frequently with a jury-trial right. An employer that has built its strategy around a favorable OSHA investigative finding can lose that advantage entirely when the complainant elects the kick-out, and the calendar should be diaried from the filing date rather than discovered when the complaint arrives. And several of the statutes carry filing periods materially longer than 11(c)'s thirty days — 180 days under most of them, 90 under STAA — so a claim time-barred under 11(c) may be perfectly timely under the statute the complainant actually pleaded. Under either regime, the decisive evidence is comparative. The defense that survives is the one that can produce the discipline matrix: every employee who committed the same infraction in the relevant period, and what happened to each. Where the complainant was treated as the policy and the comparators were treated the same way, causation collapses. Where the complainant alone was discharged for conduct routinely tolerated, no articulation of policy will save the decision. That consistency audit is a records exercise — discipline logs, HRIS extracts, prior warnings — and it should be run by the defense before OSHA runs it, because the Whistleblower Investigations Manual instructs the investigator to run it in every case.

III. Reading the Complaint Against the Prior Record

A complaint discloses none of this. They live in the parallel records both parties created before anyone contemplated litigation: the safety reporting logs, training rosters, hazard assessments, maintenance and work-order history, discipline files, shift schedules, e-mail and messaging threads, and the OSHA investigative file counsel obtains during the investigation or in discovery. The expert's first task is a document-level reconstruction: place every relevant event on a single timeline — each safety complaint, each documented performance issue, each supervisory communication, the inspection if there was one, and the adverse action — and test the complainant's narrative against it. The reconstruction is conducted against OSHA's own published investigative criteria, which is what gives it force: the defense is not asking the tribunal to adopt a novel theory of the case, but to apply the elements the agency's own manual says every merit finding requires.

The second task is context. A retaliation complaint presents the adverse action as a response to one event, and the employer's broader record frequently shows it was nothing of the kind: a progressive discipline sequence that began months earlier, a documented reduction in force with objective selection criteria, a safety infraction by the complainant that the employer disciplines uniformly. The safety professional adds a dimension counsel cannot: an assessment of the employer's safety culture as it actually operated — whether employees raised concerns freely and without consequence, whether the reporting systems show complaints answered rather than punished, whether the complainant's own prior complaints were addressed on the merits. An employer that can demonstrate a functioning, non-punitive reporting system is presenting character evidence of the most persuasive institutional kind, and the expert who audits that system credibly is the one who can present it.

The third task is restraint, and it separates the expert worth retaining from the one worth impeaching. Not every record supports a defense, and an overreached one damages the credible arguments — Section IV's final case shows a tribunal dismantling an employer's explanation that the documents contradicted. The expert's honest early read — this discipline file was built after the complaint; this refusal was objectively reasonable; this comparator analysis cuts against us — is worth as much to case strategy as any dispositive motion, because it redirects the defense toward early resolution on the items the record will not carry, and concentrates the fight on the elements the complainant actually cannot prove.

IV. Two Cases and a Warning from the Courts

Whirlpool, 445 U.S. 1 (1980), the work refusal, bounded and blessed. Two maintenance employees refused to walk on a mesh guard screen suspended beneath an overhead conveyor, twelve days after a co-worker had fallen through the same screen to his death. The employer sent them home without pay and placed written reprimands in their files, and the Secretary sued under Section 11(c). The Supreme Court unanimously upheld the regulation protecting the refusal, holding that an employee confronted with a choice between performing an assigned task and facing what he reasonably believes to be a real danger of death or serious injury, with no time to pursue normal enforcement channels, is protected when he declines, and that even the modest sanctions of lost pay and a file reprimand were actionable discrimination. The case remains the charter of the refusal doctrine, but its boundaries are the defense's map: the belief must be objectively reasonable as well as sincere, the danger urgent, and the employer must have been given its chance to correct. A refusal case is won or lost on the hazard facts, which is precisely where the technical expert operates.

Hoy Shoe, 32 F.3d 361 (8th Cir. 1994), suspicion is knowledge enough. An employee raised safety concerns about skiving-machine guarding, an anonymous complaint reached OSHA, an inspection followed, and shortly afterward the employee was discharged. The employer's defense was elegant in theory: the complaint was anonymous, so the decision-maker could not have known who filed it, and without knowledge there is no causation. The Eighth Circuit rejected the syllogism, holding that an employer who discharges an employee because it suspects, even mistakenly, that the employee filed the OSHA complaint has discriminated because of protected activity within the meaning of Section 11(c). Any other rule, the court reasoned, would leave anonymous complainants, the employees most in need of protection, wholly unprotected. The holding reshapes the defense inquiry: the question is not what the decision-maker knew but what the record shows the decision-maker suspected, and supervisory speculation about the identity of the complainant — in e-mails, in meetings, in remembered remarks — is the evidence that converts an anonymous complaint into a causal link.

Breeden, 532 U.S. 268 (2001), the warning. The complainant alleged she was transferred in retaliation for filing suit, pointing to the closeness in time between the employer's learning of the suit and the transfer. The Supreme Court, in a per curiam opinion since imported into retaliation analysis across statutes including the whistleblower provisions OSHA administers, rejected the inference twice over: cases accepting temporal proximity alone as sufficient require the proximity to be very close, and — decisively — an employer that was already contemplating the adverse action before learning of the protected activity proves nothing by proceeding with it, because employers need not suspend previously planned decisions upon learning that a complaint has been filed. The warning runs to both sides. For complainants, timing without more is a theory, not a case. For employers, the shield works only if the contemplation was documented before the protected activity — a decision memorandum, a succession plan, a discipline file with dates — because a previously planned action that exists only in post-hoc testimony will be heard as pretext. The defense that wins the timing fight is the one built in the ordinary course, before there was anything to defend.

V. The First Thirty Days After the Complaint

The retaliation defense is perishable, and the work belongs at the front of the case. Section 11(c) complaints must be filed within thirty days of the adverse action, and OSHA's notification to the employer opens a short window in which the position statement will be drafted, the investigator interviewed the employer's witnesses, and the file's trajectory largely set. Within that window, counsel should assemble and preserve the complete contemporaneous record: the complainant's full personnel and discipline file, the safety reporting logs and any entries attributable to the complainant, the comparator discipline data for the same infraction, the communications of every participant in the adverse decision, and the documentation showing when the decision was first contemplated. Litigation holds should issue before routine retention schedules and messaging auto-deletion erase the timeline. The retained safety professional should reconstruct the chronology, audit the discipline system for consistency, assess any work refusal against the *Whirlpool* criteria and the actual hazard record, and evaluate the reporting culture the tribunal will eventually be asked to judge.

The output of that audit is the informed position statement, and the informed settlement posture: which elements the complainant cannot establish, which the record genuinely supports, and where early resolution is cheaper than the answer discovery will produce. Note what makes that audit credible rather than merely useful. The disciplines described here are not devices for defeating meritorious complaints; they are the same elements OSHA's own manual requires its investigators to prove, applied evenhandedly. An expert who audits the employer's record against the agency's criteria, and who says so when the record fails, is the witness a tribunal believes when the record holds. In the retaliation docket the claim is only as good as the causal chain, and the chain is built out of dates. Whoever establishes the sequence first — when the complaint was made, when the decision-maker learned of it, when the discipline was first documented — owns the inference the rest of the case runs on. Counsel who leaves that chronology to be assembled by the investigator has conceded the narrative before answering the complaint.

References

Clark Cnty. Sch. Dist. v. Breeden, 532 U.S. 268 (2001) (per curiam).

Discrimination against employees exercising rights under the Williams-Steiger Occupational Safety and Health Act of 1970, 29 C.F.R. Part 1977, including § 1977.12 (exercise of rights; refusal to work).

Injury and illness recordkeeping — employee involvement and anti-retaliation, 29 C.F.R. § 1904.35; Improve Tracking of Workplace Injuries and Illnesses, 81 Fed. Reg. 29,624 (May 12, 2016).

Occupational Safety and Health Act of 1970 § 11(c), 29 U.S.C. § 660(c).

Occupational Safety and Health Administration. (2016). Interpretation of 29 C.F.R. § 1904.35(b)(1)(i) and (iv): Post-incident drug testing and safety incentive programs (memorandum, as clarified Oct. 11, 2018). U.S. Department of Labor.

Occupational Safety and Health Administration. (2022). Whistleblower Investigations Manual (CPL 02-03-011). U.S. Department of Labor.

Occupational Safety and Health Administration. (n.d.). The Whistleblower Protection Program. U.S. Department of Labor. https://www.whistleblowers.gov

Reich v. Hoy Shoe Co., 32 F.3d 361 (8th Cir. 1994).

Whirlpool Corp. v. Marshall, 445 U.S. 1 (1980).

PART III

THE SCIENCE OF EXPOSURE

CHAPTER 14

From Exposure to Injury

General and Specific Causation, and the Line Between the Industrial Hygienist and the Physician

An occupational disease case is a bridge between two numbers: the dose the worker absorbed and the diagnosis the worker received. The industrial hygienist builds one end of the bridge and the physician the other, and the cases that fail almost always fail at the join: where an exposure estimate is asked to carry a medical conclusion it cannot support, or a diagnosis is offered without the exposure to ground it. This chapter maps the causation framework the courts actually apply, and the division of expert labor that keeps an opinion on the admissible side of it.

I. Two Questions the Law Keeps Separate

Toxic-tort and occupational-disease law divides causation into two questions that the technical witness must never conflate, because the courts do not. General causation asks whether the substance is capable of causing the disease at issue — can respirable crystalline silica cause this fibrosis, can benzene cause this leukemia, can the solvent cause this encephalopathy — at all, in anyone. Specific causation asks whether it did cause the disease in this plaintiff, given this exposure. A plaintiff who cannot establish general causation never reaches the specific question; a plaintiff who establishes only general causation has proven that the agent can hurt someone, not that it hurt this someone. The two questions carry different proof, different witnesses, and different failure modes, and the reconstruction of exposure that Chapter 16 describes feeds the specific question while the epidemiology feeds the general one.

The industrial hygienist owns neither question outright. What the hygienist owns is the exposure — the dose, its route, its duration, its uncertainty — which is a necessary input to both questions and a sufficient answer to neither. The failure the reported decisions punish most reliably is the hygienist who crosses from characterizing the exposure to diagnosing its effect, and the physician who opines on a dose no hygienist reconstructed. The opinion that survives is the one that respects the boundary the next sections draw.

II. General Causation and the Epidemiology

General causation is established primarily through epidemiology, and a legal reader does not need to become an epidemiologist, but does need to read a study the way a court does. The measures that matter are few: the relative risk or odds ratio, which expresses how much more often the disease appears in the exposed than the unexposed; the confidence interval, which expresses the precision of that estimate and, when it crosses 1.0, signals a result that may be chance; and the dose-response relationship, which — when risk rises with exposure — is among the strongest available evidence of a true effect. Courts have absorbed a rough heuristic that a relative risk above 2.0 implies it is more likely than not that a given exposed plaintiff's disease was caused by the exposure, and while the inference is cruder than the epidemiology deserves, counsel should know it because opponents will invoke it.

The framework courts use to move from association to causation is the set of considerations articulated by Sir Austin Bradford Hill — strength of the association, consistency across studies, specificity, temporality, biological gradient (dose-response), plausibility, coherence, experimental evidence, and analogy — not as a checklist to be scored but as a structured argument for or against a causal reading. The weight-of-the-evidence methodology that applies the Hill considerations was accepted as reliable in *Milward v. Acuity Specialty Products Group, Inc.*, 639 F.3d 11 (1st Cir. 2011), and it is the method a well-prepared general-causation opinion will name and follow. The opponent's attack, correspondingly, is that the witness cherry-picked the studies, ignored the null results, or dressed a conclusion as a method: the unsupported assertion the court rejected in *General Electric Co. v. Joiner*, 522 U.S. 136 (1997).

III. Specific Causation and Differential Etiology

Specific causation is usually established through differential etiology: the physician's method of listing the possible causes of this patient's disease and systematically ruling out the ones the evidence excludes, until the exposure remains as a substantial contributing cause. The method is accepted, but it is only as good as two things: the completeness of the initial list, and the reliability of the rule-outs.

The decisions that exclude differential-etiology opinions cluster on those two failures. In *Tamraz v. Lincoln Electric Co.*, 620 F.3d 665 (6th Cir. 2010), a specific-causation opinion linking welding fume to Parkinsonism was excluded as speculation because the ruling-in step rested on hypothesis rather than established science; in *Pluck v. BP Oil Pipeline Co.*, 640 F.3d 671 (6th Cir. 2011), the differential diagnosis failed because the expert could not reliably rule out alternative causes and had no reliable exposure quantification to rule benzene in. The lesson runs directly to the industrial hygienist: a differential etiology is only as sound as the exposure estimate that lets the physician rule the occupational agent in, which is why the specific-causation opinion and the exposure reconstruction must be built to the same standard and must actually connect.

IV. *Daubert, Frye,* and the Split Over Differential Etiology

The method described in the preceding section is not received the same way in every courtroom, and the difference is large enough to change which opinions a case can plead. Federal courts and most states apply Rule 702 and the *Daubert* line. A shrinking but still significant minority retain *Frye v. United States*, 293 F. 1013 (D.C. Cir. 1923), or a descendant of it, New York, Pennsylvania, Illinois, and Washington among them, with California applying its own variant alongside the gatekeeping duty of *Sargon Enterprises, Inc. v. University of Southern California*, 55 Cal. 4th 747 (2012). The map moves: Florida completed its switch to *Daubert* in 2019, and Maryland abandoned *Frye*-Reed in *Rochkind v. Stevenson*, 471 Md. 1 (2020).

The formal difference between the two tests is narrow and the practical difference is not. *Frye* asks whether the methodology is generally accepted in the relevant scientific community; it does not, in terms, ask whether that methodology was reliably applied to these facts. Differential etiology is generally accepted, so a mechanical *Frye* analysis clears it at the threshold, where amended Rule 702(d) is aimed precisely at the two failure points the preceding section identified. The gap is the difference between a court that examines the list and the rule-outs and one that satisfies itself a recognized method was used.

New York shows how a *Frye* jurisdiction reaches the federal destination by another road. Its courts police specific causation through the foundation inquiry of *Parker v. Mobil Oil Corp.*, 7 N.Y.3d 434 (2006), which requires evidence of exposure to a level of the agent known to cause the harm claimed. *Nemeth v. Brenntag North America*, 38 N.Y.3d 336 (2022), overturned a verdict because the expert worked backward from the diagnosis to an assumed sufficient dose, and *Sean R. v. BMW of North America, LLC*, 26 N.Y.3d 801 (2016), rejected an inference the scientific community did not accept. Pennsylvania has pulled the other way: *Walsh v. BASF*

Corp., 234 A.3d 446 (Pa. 2020), held that a *Frye* court may not parse an expert's conclusions to reject a generally accepted methodology.

The same report can therefore be admissible in one forum and excluded across the river, and the divergence is widening, because the 2023 amendment targets application while *Frye* addresses method. Build to the stricter of the two: a rule-out analysis that satisfies Rule 702(d) will satisfy *Frye* anywhere, and the reverse is not true.

V. Latency and Temporality

Nothing separates the disease case from the injury case more sharply than the interval between exposure and diagnosis, and nothing is more often missing from the reconstruction. Latency is a property of the agent and the disease rather than of the plaintiff: mesothelioma commonly appears twenty to fifty years after first exposure, asbestosis and chronic silicosis after ten to thirty, benzene-related acute myeloid leukemia after an interval usually measured in single-digit years, and acute silicosis after months of extraordinary exposure.

Temporality is the legal edge of that fact. It is the one Bradford Hill consideration that operates as a necessary condition rather than as a weight, because the exposure must precede the disease by an interval the biology allows. Two failure modes follow. The first is the interval too short: a diagnosis four years after first exposure to an agent with a twenty-year minimum makes no case against that employer whatever the dose, and this is the defense argument most likely to resolve a disease claim before trial. The second treats the interval as having no upper bound, sweeping every exposure in a forty-year work history into the causal account without asking whether agent, dose, and interval fit together.

The engineered-stone silicosis docket of Chapter 28 is the reason the second error now cuts both ways. Accelerated silicosis in dry countertop fabrication has appeared after a handful of years of very high-intensity exposure, and a defense expert reasoning from chronic silicosis timelines will call the interval too short when the literature says otherwise. Latency is agent-specific, disease-specific, and sensitive to intensity, and it must be sourced rather than assumed.

For the industrial hygienist the operational lesson is that the reconstruction must be dated, not merely quantified. A cumulative dose with no chronology attached cannot answer the temporality question, cannot support apportionment among employers, and cannot address the limitations defense, which in most jurisdictions runs from the date the plaintiff knew or should have known of both the injury and its cause. *Urie v. Thompson*, 337 U.S. 163 (1949), is the origin of that discovery rule in occupational disease; it exists because latency would otherwise extinguish these claims before anyone could bring them. The chronology of jobs, tasks, products, and controls is an element of the opinion, not background to it.

VI. Dose, Thresholds, and the "Every Exposure" Problem

Between general and specific causation sits the question that has reorganized toxic-tort law over the past two decades: how much exposure is enough, and what must the plaintiff prove about the dose? The proposition that every exposure to a toxicant, however small, is a substantial contributing cause of the resulting disease — the "each and every exposure" theory, born in the asbestos docket — has been rejected as unreliable by a broad and still-growing set of courts, and the reasoning travels to every agent this Part covers. The theory fails as science because it reads the absence of a demonstrated safe threshold as proof that no threshold exists, and fails as law because it makes the causation element vanish: if every exposure is substantial, nothing is. *Bostic v. Georgia-Pacific Corp.* states the modern rule with unusual clarity — proof of some exposure is not enough; the plaintiff must establish a dose, at least approximately, and connect that dose to the disease through the epidemiology — and the same demand now confronts the silica, solvent, and ethylene-oxide plaintiff that first confronted the asbestos plaintiff.

The doctrinal counterweight is the substantial-factor framework California built in *Rutherford v. Owens-Illinois*, under which the plaintiff need not trace the fibers or prove that a particular defendant's product was the but-for cause, but must show that the exposure attributable to the defendant was, in reasonable medical probability, a substantial factor in contributing to the aggregate dose and hence the risk. *Rutherford* is not the every-exposure theory in different clothing — it still requires an exposure assessment with enough resolution to say something meaningful about this defendant's contribution — and the difference between the two is precisely the industrial hygienist's work product. A task-based reconstruction that quantifies frequency, intensity, and duration by product and by period gives the physician a dose to reason from and gives the court a basis to find substantiality; the affidavit that says only that the worker was exposed and the disease is dose-responsive gives neither, and it is the affidavit the modern decisions strike.

One further distinction organizes the proof and belongs in counsel's intake analysis: the signature disease against the multifactorial one. Silicosis, asbestosis, and mesothelioma are, for practical litigation purposes, specific to their agents: the diagnosis itself carries most of the general-causation freight, and the fight moves to exposure and apportionment. Lung cancer, leukemia, kidney disease, and the neurological syndromes are multifactorial: the diagnosis proves nothing about the cause, the background rate supplies the defense its alternative explanation, and the relative-risk and dose-response evidence must do work the signature-disease case never asks of it. The two case types demand different reconstructions — the signature case needs the dose apportioned among sources; the multifactorial case needs the dose established at a level the epidemiology associates with a risk large enough to support specific causation at all — and the expert team that recognizes at intake which case it is holding builds the record the theory will eventually require.

VII. Biomarkers and Biomonitoring as an Exposure Input

Air sampling and reconstruction estimate what the worker breathed. Biomonitoring measures what reached the body, and it arrives with increasing frequency as an exposure input in the cases this Part covers. It is powerful in one direction and misleading in the other.

Three categories must be kept apart, and courts routinely conflate them. A biomarker of exposure is the agent or its metabolite measured in blood, urine, breath, or tissue. A biomarker of effect is an early biological change the exposure produces, such as the beta-2-microglobulin elevation the cadmium standard tracks. A biomarker of susceptibility describes a host characteristic that changes the risk. Only the first is a measure of dose; the second belongs to the physician's side of the line drawn below, and the third proves nothing about exposure.

The strength of a biomarker of exposure is that it integrates every route and every source, capturing inhalation, dermal absorption, and incidental ingestion in a way no air sample does. That is also its weakness, because it does not know which source it came from: it is the best available answer to whether this plaintiff was exposed and among the worst to whether this defendant caused the exposure. Interpretation requires a comparator: the ACGIH Biological Exposure Indices, which express the concentration expected in a worker exposed at the corresponding airborne limit, and the population distributions the CDC publishes from its national biomonitoring program. A value called elevated without reference to either is an adjective, not a measurement.

The error that decides most challenges is the half-life. A marker with a short biological residence, most solvents and most urinary metabolites among them, is cleared in hours to days and describes recent exposure only; it says nothing about the exposure decades earlier that caused the disease. Markers that persist will support retrospective work: lead in bone, cadmium in the kidney, and the serum PFAS concentrations whose multi-year half-lives make historical inference possible at all. An expert who uses a recent urinary result to establish a historical dose has made an error a competent cross will find in three questions.

Where a standard requires biological monitoring, lead, cadmium, and benzene being the principal examples, the employer already holds the results, discoverable and retained for thirty years under the exposure-and-medical-records rule. Counsel who does not ask for them leaves the most direct exposure evidence in the case in a filing cabinet. The point carries forward: because essentially the entire population now carries a measurable serum PFAS concentration, the PFAS litigation of Chapter 28 turns not on whether the plaintiff was exposed but on whether a defendant's contribution can be distinguished from background.

VIII. Apportionment and Multiple Employers

Occupational disease rarely arrives from a single clean exposure. The worker was exposed to the agent across employers, alongside other agents, and against a background of non-occupational risk: the smoker with silica exposure, the painter exposed to three solvents across two decades and four jobsites. Apportionment is the attempt to divide causal responsibility among these contributors, and it is where the technical and legal frameworks grind against each other. Some agents act independently and some synergistically — the multiplicative interaction of asbestos and smoking on lung-cancer risk is the classic example — and the hygienist's reconstruction must characterize each contributing exposure well enough to support whatever apportionment the governing law requires, which varies sharply by jurisdiction and by whether the theory sounds in workers' compensation, product liability, or negligence. Counsel should resolve the apportionment framework at intake, because it determines what the exposure reconstruction must be built to prove: a last-injurious-exposure rule, a several-liability regime apportioning by contribution, and a joint-and-several regime each demand a different reconstruction.

IX. The Division of Labor Between Experts

The admissible causation case is a relay, and the baton passes cleanly. The industrial hygienist reconstructs and quantifies the exposure — dose, route, duration, uncertainty — and may opine on how that exposure compares to the levels the literature associates with the disease, but stops short of the medical diagnosis and the clinical judgment that this patient's disease was caused by it. The occupational physician or toxicologist takes the exposure as an input, addresses general causation through the epidemiology, and performs the differential etiology that yields specific causation. When the hygienist reaches for the diagnosis or the physician invents the dose, the opinion is vulnerable at exactly the seam the opponent will target. The strongest occupational-disease presentations pair the two experts so that each testifies to the matters within their qualification and each relies on the other's work in the way Rule 703 permits: the hygienist's reconstruction becoming the physician's foundation, disclosed as such, defensible because the reliance is exactly the reliance the disciplines practice.

The relay does not stop at the physician, and the second handoff is the one this book's two audiences most often leave unmanaged. Where the causation case succeeds, the findings pass again — to the life-care planner, the vocational rehabilitation expert, and the economist, and what they receive from the safety and industrial hygiene side is frequently less than they need and occasionally more than the opinion supports. The magnitudes involved make this worth a page. In the anchor case of Chapter 6, more than half of a nine-figure verdict was future care and future loss of enjoyment, which is to say that the majority of the number was produced downstream of the experts this book is about, from inputs those experts supplied.

Three transfers actually matter, and each has a failure mode. The first is the physical-capacity and restriction record. The life-care planner and the vocational expert build from the functional limitations the treating and examining physicians document, and the safety professional's contribution is the job itself: the actual physical demands of the position, the actual exposures it entails, and whether a restriction is accommodatable in that workplace rather than in the abstract. A job description written for a personnel file is not a physical-demands analysis, and the vocational opinion built on one is attackable at exactly the seam Chapter 4 teaches. The second is the exposure history, which the life-care planner needs in a form the medical monitoring recommendation can rest on: not the conclusion that exposure exceeded a limit, but the agent, the dose, the period, and the surveillance the literature associates with that dose. The third is causation-limited apportionment — where the reconstruction supports attributing part of the impairment to a non-occupational or prior-employer source, the economist cannot apply that division unless someone states it in a form that can be applied.

The discipline governing all three is the one this chapter has already stated in the other direction. The exposure expert supplies inputs and does not supply conclusions belonging to the downstream discipline. An industrial hygienist who opines that the worker will require lifetime pulmonary surveillance costing a stated amount has crossed into the life-care planner's lane as surely as one who diagnoses the disease has crossed into the physician's, and the crossing is punished the same way. Conversely, a life-care plan whose surveillance schedule rests on an exposure characterization no hygienist has stated, or a vocational opinion whose physical demands were assumed rather than analyzed, has a foundation problem that a competent defense will find before trial. The instruction for counsel is to run the handoff explicitly: identify, at the point the causation experts are retained, which findings the damages experts will need, and have them stated in the reports of the experts qualified to state them. Chapter 8 §VII discusses what these numbers do to reserves, mediation posture, and coverage; the point here is that they begin as technical findings, and a finding that never got stated in a form the downstream expert could use is a finding that will not appear in the verdict.

X. Practice Implications

For counsel, causation is where occupational-disease cases are won and lost, and the work belongs at intake: identify the general-causation literature and whether it will bear the weight, resolve the apportionment framework the jurisdiction imposes, and pair the exposure and medical experts so the bridge has both ends and a sound join. For the safety and industrial hygiene professional, the discipline is restraint: build the exposure reconstruction to the standard Chapter 16 demands, state honestly what it can and cannot support, and hand the physician a foundation rather than a conclusion. The exposure number and the diagnosis are different opinions held by different experts under different qualifications; keeping them distinct, and connecting them with a reconstruction built to carry the weight, is what survives the motion the opponent has already drafted.

References

In re Amendments to the Florida Evidence Code, 278 So. 3d 551 (Fla. 2019).

American Conference of Governmental Industrial Hygienists. Biological Exposure Indices (current ed.). ACGIH.

Bostic v. Georgia-Pacific Corp., 439 S.W.3d 332 (Tex. 2014).

Centers for Disease Control and Prevention. National report on human exposure to environmental chemicals (National Health and Nutrition Examination Survey biomonitoring data). U.S. Department of Health and Human Services.

Federal Judicial Center. (2011). Reference manual on scientific evidence (3d ed.). National Academies Press.

Fed. R. Evid. 702; Fed. R. Evid. 703.

Frye v. United States, 293 F. 1013 (D.C. Cir. 1923).

Gen. Elec. Co. v. Joiner, 522 U.S. 136 (1997).

Hill, A. B. (1965). The environment and disease: Association or causation? Proceedings of the Royal Society of Medicine, 58(5), 295–300.

Lead, 29 C.F.R. § 1910.1025; cadmium, 29 C.F.R. § 1910.1027; benzene, 29 C.F.R. § 1910.1028; access to employee exposure and medical records, 29 C.F.R. § 1910.1020.

Milward v. Acuity Specialty Prods. Grp., Inc., 639 F.3d 11 (1st Cir. 2011).

Milward v. Acuity Specialty Prods. Grp., Inc., 969 F. Supp. 2d 101 (D. Mass. 2013), aff'd sub nom. Milward v. Rust-Oleum Corp., 820 F.3d 469 (1st Cir. 2016) (specific-causation exclusion following the general-causation reversal).

National Research Council. (2006). Human biomonitoring for environmental chemicals. National Academies Press.

Nemeth v. Brenntag N. Am., 38 N.Y.3d 336 (2022).

Parker v. Mobil Oil Corp., 7 N.Y.3d 434 (2006).

Pluck v. BP Oil Pipeline Co., 640 F.3d 671 (6th Cir. 2011).

Rochkind v. Stevenson, 471 Md. 1 (2020).

Rothman, K. J., Greenland, S., & Lash, T. L. (2008). Modern epidemiology (3d ed.). Lippincott Williams & Wilkins.

Rutherford v. Owens-Illinois, Inc., 16 Cal. 4th 953 (1997).

Sargon Enters., Inc. v. Univ. of S. Cal., 55 Cal. 4th 747 (2012).

Sean R. v. BMW of N. Am., LLC, 26 N.Y.3d 801 (2016).

Tamraz v. Lincoln Elec. Co., 620 F.3d 665 (6th Cir. 2010).

Urie v. Thompson, 337 U.S. 163 (1949).

Walsh v. BASF Corp., 234 A.3d 446 (Pa. 2020).

CHAPTER 15

You Took Three Samples. Now What?

Turning Limited Air Monitoring Data Into a Defensible Decision

A sampling report with a few numbers under the exposure limit is not the same thing as a decision about whether the workforce is protected. This chapter explains the difference, and it is the one chapter in the book that speaks to a third audience.

A note on this chapter's address. *This chapter addresses the employer who purchases industrial hygiene services, and it stays in the second person throughout. Counsel should read it as a checklist of the decisions an opposing expert's client made before any of this became a case.*

I. The Question Behind the Sampling Report

An employer pays someone to measure the air in its plant. Several weeks later a report arrives. It contains a handful of numbers, all of them below the applicable exposure limit, and a sentence saying exposures were within acceptable levels. The natural response is relief, followed by filing the report and moving on.

Sometimes that response is correct. Often it is premature, and occasionally it is expensive. The measurements in that report answer a narrow question: what was in the air on the particular days somebody happened to be wearing a sampling pump. The question that actually needs answering is broader. Is a worker who does this job every day, for years, likely to breathe more than the limit on a meaningful number of those days?

Those are different questions, and the second one cannot be answered by reading the numbers off the page. It requires a method. What follows explains what that method is, why it exists, and what a competent industrial hygienist should be delivering. It contains no formulas and assumes no background in statistics. The goal is that you finish it able to tell a rigorous exposure assessment from a superficial one, and able to ask the three or four questions that reveal which kind you have received.

II. Why "All Results Were Below the Limit" Is Not an Answer

Consider a simple analogy. Suppose you want to know whether a stretch of road floods. You drive it on three days, chosen partly by convenience, and it is dry each time. Have you established that the road does not flood? Obviously not. You have established that it was dry on three days. Whether that tells you anything useful depends on how often the road floods, whether you drove during the season it floods, and how much water it takes to matter.

Workplace air behaves the same way, for three reasons that operate together.

Exposures vary enormously from day to day. The same worker, doing the same task, in the same building, can be exposed to concentrations that differ several fold from one shift to the next. Production rate changes. A bay door is open or closed. A ventilation hood is working well or is partly blocked. Someone runs a second machine nearby. A batch of material has a different composition. None of this is unusual or a sign of a badly run operation. Variability is simply a property of workplace air.

A few samples usually miss the high days. If the worst days are relatively uncommon, and you collect three samples, the odds are good that you sampled ordinary days rather than bad ones. The results will look reassuring precisely because the sampling was too sparse to find the exposures that matter most. This is the single most common way a well-intentioned monitoring program produces a false sense of security.

An exposure limit describes a lifetime, not an afternoon. For substances that cause chronic disease (silica, solvents, welding fumes, isocyanates, and most of the agents that generate long term liability), the limit represents a concentration a worker can breathe repeatedly across a working career without unacceptable risk. The relevant question is therefore about the pattern of exposure over years, not about the value recorded on one Tuesday in March.

Put those three together and the conclusion follows. A report that says all results were below the limit has reported a fact. It has not yet made a decision.

III. How Workplace Exposures Actually Behave

Air sampling results for a given job almost never spread out evenly around an average. They bunch up at the low end, with a long tail of occasional high values stretching out to the right. Most days are unremarkable. A few days are much worse than the rest. This shape is so consistent across industries and substances that it has been the accepted working model in the field for decades.

The practical consequence is that the average of a set of results is a poor guide to risk. Two jobs can have identical averages while one produces occasional exposures at five times the limit and the other never exceeds half of it. If you judge both by their averages, they look the same. They are not the same, and the workers doing them do not face the same risk.

This is why hygienists focus on the upper end of the range rather than the middle of it. The convention is to estimate what the high days look like, specifically the level that only about one shift in twenty exceeds, and compare that to the exposure limit. If the high days stay comfortably below the limit, the job is controlled. If the high days routinely exceed it, the job is not controlled, no matter how good the average looks. When a hygienist tells you they are evaluating the upper end of the exposure distribution rather than the average, they are not overcomplicating the matter. They are looking at the part of the picture where the health risk lives.

There is a second consequence, less obvious and more important for your purposes. Because the high days are the uncommon ones, a small number of samples is unlikely to capture them. So a set of low results carries less reassurance than it appears to. The fewer the samples, the truer this is.

IV. What the Hygienist Is Actually Deciding

A rigorous exposure assessment does not produce a verdict of pass or fail. It produces two things, and you should expect both.

A. Conclusion About Where the Job Stands

The industrial hygiene profession uses a graded scale rather than a single line. In broad terms, a job is rated somewhere between highly controlled, where even the bad days are a small fraction of the limit, and poorly controlled, where the high days exceed it. The middle categories cover jobs where exposures are real but well managed, and jobs that are technically compliant but running close enough to the limit that any slip in controls will push them over.

That gradation matters to you commercially. A job in the middle of the scale is compliant today and vulnerable tomorrow. Knowing that lets you plan a ventilation upgrade on your schedule instead of discovering the problem during an inspection, a workers compensation claim, or a lawsuit filed fifteen years from now by someone who no longer works for you.

B. Statement of How Confident That Conclusion Is

This is the part most often missing, and it is the part that separates a defensible assessment from a superficial one. Any conclusion drawn from limited data carries uncertainty, and the professional obligation is to quantify that uncertainty and state it plainly rather than bury it.

A competent report will tell you not only where the job sits but how firmly it sits there. High confidence means the data are consistent and the conclusion would not change much with more sampling. Low confidence means the data are compatible with the job being fine and also compatible with the job being a problem, and the assessment cannot presently distinguish between those possibilities. Both are legitimate findings. Only one of them justifies closing the file.

If there is a single practical point to carry out of this chapter, it is this one. A report that states a conclusion without stating its confidence has delivered half of what was paid for, and it is the half that will not hold up when someone challenges it, which is also, exactly, the half Chapter 17 §VI teaches an opposing expert to go looking for.

V. Small Data Sets and Professional Judgment

It is worth understanding that small datasets are the ordinary condition of this work, not a sign that your consultant cut corners. Sampling is constrained by cost, by production schedules, by the number of workers who actually perform the task, by shortened shifts, and by the routine attrition of samples that fail laboratory quality checks. A plan for ten samples that yields three or four valid results is an unremarkable outcome.

The important question is therefore not how the hygienist avoids small datasets, but how the hygienist reasons from them. Here the research is uncomfortable and worth knowing about, because it explains why the answer cannot simply be to trust an experienced professional's instincts.

Investigators have repeatedly tested how well industrial hygienists can judge exposure levels when relying on experience and observation without analytical tools. The results are consistent across studies, and the number that matters has to be stated against its baseline. In the desktop studies, hygienists sorted exposures into four AIHA exposure-rating categories — the four-band collapse those studies used, rather than the five-band scheme (Categories 0 through 4) that the AIHA strategy itself sets out and that Appendix G applies — so random assignment scores twenty-five percent. Unaided judgments were correct roughly a third of the time: an improvement over chance too small to rely on, and in several of the follow-up studies not statistically distinguishable from it (Logan et al., 2009; Vadali et al., 2012; Arnold et al., 2016). AIHA's own framing of this body of work is the careful one — that unaided professional judgment is sometimes not statistically different from random chance, and the careful framing is the one to use, because an opposing expert who has read the underlying papers will read an overstatement aloud. The errors also ran in one direction: hygienists tended to underestimate exposure (Logan et al., 2009). Follow-up work found that seniority and credentials offered surprisingly little protection against this pattern (Logan et al., 2011).

The encouraging half of the same research is that the problem is fixable, and cheaply. Structured methods correct it. Training in data interpretation raised accuracy from roughly half to two thirds of judgments and eliminated the tendency to guess low (Vadali et al., 2012); the higher baseline in that study reflects its different design, in which participants worked from sampling data rather than from observation alone, and the improvement rather than the absolute figure is the comparable result. A structured checklist, used during the walkthrough, nearly doubled the accuracy of qualitative judgments and shifted the remaining errors toward the cautious side (Arnold et al., 2016).

The lesson for you as a purchaser is direct. Experience is necessary and it is not sufficient. What you are buying, and what you should insist on, is a documented method applied to your data, not a seasoned opinion delivered with confidence. When a hygienist walks your floor, looks at a process, and pronounces it fine, the research says there is roughly a one in three chance they are right against a scheme in which guessing scores one in four — and that if they are wrong, they are probably wrong in the direction that leaves your workers overexposed and you exposed to liability. The margin over guessing is real. It is not large enough to buy.

VI. How Good Assessments Get More From Less

A skilled hygienist does not treat three samples as though they were the only information in existence. Your three results arrive in a world that already contains a great deal of relevant knowledge: measurements from comparable operations, published industry survey data, government databases of monitoring results, the documented performance of the specific ventilation system installed in your building, and validated models that predict airborne concentrations from how much of a material is used and how well the space is ventilated.

Modern practice combines that existing knowledge with your new measurements through a formal, transparent procedure. The technical name is Bayesian analysis, and you do not need to understand its mathematics. You need to understand what it does and what protects you from its misuse.

What it does is straightforward. It starts with what was reasonably believed about the job before sampling, updates that belief using the measurements actually collected, and reports the result as a probability. Instead of a bare number, the deliverable is a statement such as: there is a seventy-eight percent probability this job is well controlled, a fifteen percent probability it is only marginally controlled, a five percent probability that exposures routinely approach the limit, and a two percent probability that they exceed it. That is a form of answer a manager can act on. It tells you both what is most likely true and how much risk of being wrong you are carrying.

The protection you should insist on concerns where the starting assumptions came from. The method is powerful precisely because it lets the hygienist bring outside knowledge to bear, which means it can be abused by an analyst who selects convenient assumptions to produce a comfortable answer. The safeguard is disclosure. A defensible report states exactly what prior information was used, where it came from, and why it applies to your operation. If the report cannot show you that, the conclusion is not verifiable. When genuinely comparable data do not exist, the honest approach is to start from neutral assumptions, let your three samples carry the full weight, and accept the wider uncertainty that results.

VII. A Worked Example (Three Samples, One Over the Limit)

A case adapted from the professional literature illustrates why method matters more than sample count (Hewett et al., 2006). A hygienist collects three full shift samples on a task. One of the three exceeds the exposure limit.

A superficial reading declares the job noncompliant and stops there. A different superficial reading averages the three results, finds the average below the limit, and declares the job fine. Both readings are indefensible, and they reach opposite conclusions from identical data, which should tell you something about the value of reading numbers without a method.

What resolves the case is context, handled openly. Suppose this operation is one of many similar operations, and dozens of full shift measurements from comparable processes under comparable conditions have consistently fallen well below the limit. That body of evidence is legitimate, documentable prior knowledge. Combined formally with the three new results, it can support a conclusion that the job is most probably well controlled, with the single high reading treated as an unusual event to be investigated for a specific cause: a failed hood, an unusual batch, a temporary change in the work method.

Now change one fact. Suppose no such history exists, and the operation is new or unique. The same three results, one of them over the limit, should worry you considerably. There is no basis for treating the high value as an anomaly, and the responsible conclusion is that control has not been demonstrated and further work is required.

Identical data, opposite conclusions, and the difference lies entirely in evidence outside the three numbers. That is the work you are hiring someone to do. A report that shows you this reasoning is defensible. A report that announces a conclusion without it is an opinion in a binder.

VIII. What a Defensible Report Contains

The elements below appear in an annotated model report at Appendix G, together with an eight-element test that can be run against any deliverable in about five minutes.

You do not need technical training to audit an exposure assessment report. You need to check that certain elements are present. A report that contains all of them was probably produced by a rigorous process. A report missing several was probably not.

A clear description of who was assessed and why they were grouped together. Sampling results are used to draw conclusions about a group of workers whose exposures should be similar. The report should identify that group and justify it. If a full time production operator and a maintenance technician who visits the area twice a week were lumped into one group, every conclusion downstream is suspect.

What was sampled, how, and against which limit. The substance, the sampling and analytical method, the duration of each sample, and the specific exposure limit applied, including whether it is an OSHA permissible limit, a more protective consensus limit, or an internal standard. These differ, sometimes by a great deal, and which one was used should never be ambiguous.

It is worth being explicit about why that choice matters beyond bookkeeping, because it is a recurring battleground in litigation. Many OSHA permissible exposure limits are the 1971 adoptions of consensus values from the 1960s and have never been updated; the ACGIH Threshold Limit Value and the NIOSH Recommended Exposure Limit for the same substance are frequently far lower, and in some cases lower by an order of magnitude. Compliance with the PEL is therefore not the same proposition as control of the hazard, and the two propositions are argued by different parties for different purposes: the plaintiff will offer the TLV or REL as the standard of care a competent professional actually applies, and the defense will answer that the PEL is the only limit with the force of law and that consensus values are advisory, unenforceable, and set with safety margins that are not injury thresholds. Both characterizations are partly right, which is precisely why the assessment should state which limit it applied, why, and what the conclusion would be under the others. An assessment that quietly benchmarks to the most permissive available number has made an advocacy choice before any lawyer was involved.

Noise is the clearest illustration available, and it will recur through the next three chapters as a second running example alongside silica, because it exercises the same method with entirely different physics. OSHA's permissible exposure limit for noise is 90 A-weighted decibels as an eight-hour time-weighted average, computed with a five-decibel exchange rate — meaning the permitted exposure time halves for every five-decibel increase. NIOSH's recommended exposure limit is 85 dBA with a three-decibel exchange rate, which is the physically correct doubling of sound energy. Those are not two ways of saying nearly the same thing. On an eight-hour shift at 90 dBA, a worker is at 100 percent of the OSHA allowable dose and at roughly 300 percent of the NIOSH one; at 95 dBA the same shift is 200 percent of the OSHA dose and roughly 1,000 percent of the NIOSH one. An assessment can therefore report, accurately, that a job is in compliance and, equally accurately, that the same job carries a substantial risk of noise-induced hearing loss, and the only thing distinguishing those two reports is the criterion each benchmarked to.

Two features make this the most useful benchmark example in the book. The first is that OSHA itself operates both criteria: the 90 dBA PEL sets the compliance line, while the 85 dBA action level of 29 C.F.R. § 1910.95(c) triggers the monitoring, audiometric testing, training, and hearing-protector obligations of the hearing conservation program. An employer can be simultaneously compliant with the PEL and in violation of the action-level program requirements, and the assessment that reports only the first has answered half the question the employer asked. The second is that the exchange rate is not a matter of policy preference at the margin: an assessment that uses a five-decibel exchange rate is understating the dose of any worker whose exposure varies across the shift, and the more variable the exposure the larger the understatement. A report that does not name its exchange rate has left out an input that can change the answer by a factor of three.

The practical instruction for the purchaser is short. Require the report to state the criterion, the exchange rate, the threshold below which sound was not integrated, and whether the result is expressed as a TWA in dBA or as a percentage dose — and require it to report the result against both the OSHA and the NIOSH criteria when the two diverge. That second report costs the consultant one additional line of output from the same dosimeter data. It is the line that will matter ten years later, when someone asks what the employer knew.

A conclusion expressed as a category, not a verdict. Where does this job sit on the scale from well controlled to poorly controlled? A bare statement that results were below the limit is not this.

An explicit statement of confidence. How certain is that placement, and what would change it? As discussed above, this is the element most often absent and the one that matters most.

Disclosure of any outside information used. If comparable data, published studies, or exposure models informed the conclusion, the report should say so and identify them.

A recommended action tied to the finding. Accept and monitor periodically, improve controls, or collect additional data. The recommendation should follow logically from the confidence level. A low confidence finding paired with a recommendation to take no further action is internally inconsistent.

A plain statement of limitations. What this assessment does not cover: other tasks, other shifts, upset conditions, maintenance activities, changes in production rate. Honest limits are a sign of quality, not weakness.

IX. Questions to Ask, Before and After

Four questions will tell you most of what you need to know about the quality of the work you are buying.

Ask before you hire: how will you decide, not just how will you sample? A consultant who can only describe the field work has told you about the easy half of the job. Ask what decision framework they use and how they handle limited data. The answer should reference an established professional strategy rather than personal experience alone.

Ask when the report arrives: how confident are you, and what would change your mind? A professional who cannot answer this in plain language either has not done the uncertainty analysis or cannot explain it, and neither is acceptable.

Ask: what did you rely on besides my samples? The answer should be specific and documented. Vagueness here is the warning sign that matters most.

Ask: would this conclusion survive being read aloud by opposing counsel in ten years? Long latency diseases mean today's assessment may be examined long after the plant has changed, the records have thinned, and the people involved have moved on. A report written to withstand that scrutiny costs no more to produce than one that will not.

X. What Is Actually at Stake

It is tempting to treat exposure assessment as a compliance formality, a document generated to satisfy a regulation and then filed. Three consequences argue otherwise.

The first is the obvious one. Workers who are quietly overexposed for years develop diseases that appear decades later, and no settlement repairs that. The engineered stone countertop industry provides a recent and grim demonstration of how quickly a poorly assessed exposure can produce severe disease in young workers.

The second is financial and delayed. Occupational disease claims and toxic tort suits are litigated largely on the question of how much the person was exposed to and for how long. When that question is asked, your monitoring records and the assessments built on them become the evidence. An assessment that documented its method, stated its uncertainty, and drove corrective action is a strong record. A thin file of numbers with a conclusory sentence attached is not, and its weakness will be apparent to everyone in the room.

The third runs the other way and is often overlooked. Poor assessment also causes overspending. Organizations install unnecessary ventilation, impose respirator programs that carry their own costs and compliance burdens, and restrict operations, all because nobody established what the actual exposure was. Rigor cuts both ways. It identifies the hazards that need money spent on them and it releases you from the ones that do not.

Three samples will never carry the confidence of thirty, and no method invents information that was never collected. But the number of samples is rarely what determines whether an exposure decision holds up. What determines it is whether the group was defined sensibly, the right question was asked of the data, the uncertainty was stated honestly, and the reasoning was written down where someone else can check it. Those are the things to require from the professional you hire, and they are the things worth paying for.

References

29 C.F.R. § 1910.95 (occupational noise exposure; hearing conservation program, monitoring, audiometric testing, and recordkeeping requirements, including the age-correction tables of Appendix F).

American Industrial Hygiene Association Guideline Foundation. (2019). Principles of good practice for exposure assessment strategies (Section 2). American Industrial Hygiene Association.

Arnold, S. F., Stenzel, M., Drolet, D., & Ramachandran, G. (2016). Using checklists and algorithms to improve qualitative exposure judgment accuracy. Journal of Occupational and Environmental Hygiene, 13(3), 159–168.

European Committee for Standardization. (2018). Workplace exposure: Measurement of exposure by inhalation to chemical agents, strategy for testing compliance with occupational exposure limit values (EN 689:2018). European Committee for Standardization.

Hewett, P., Logan, P., Mulhausen, J., Ramachandran, G., & Banerjee, S. (2006). Rating exposure control using Bayesian decision analysis. Journal of Occupational and Environmental Hygiene, 3(10), 568–581.

Jahn, S. D., Bullock, W. H., & Ignacio, J. S. (Eds.). (2015). A strategy for assessing and managing occupational exposures (4th ed.). American Industrial Hygiene Association.

Lavoué, J., Joseph, L., Knott, P., Davies, H., Labrèche, F., Clerc, F., Mater, G., & Kirkham, T. (2019). Expostats: A Bayesian toolkit to aid the interpretation of occupational exposure measurements. Annals of Work Exposures and Health, 63(3), 267–279.

Logan, P., Ramachandran, G., Mulhausen, J., & Hewett, P. (2009). Occupational exposure decisions: Can limited data interpretation training help improve accuracy? Annals of Occupational Hygiene, 53(4), 311–324.

Logan, P. W., Ramachandran, G., Mulhausen, J. R., Banerjee, S., & Hewett, P. (2011). Desktop study of occupational exposure judgments: Do education and experience influence accuracy? Journal of Occupational and Environmental Hygiene, 8(12), 746–758.

National Institute for Occupational Safety and Health. (1998). Criteria for a recommended standard: Occupational noise exposure (revised criteria 1998) (DHHS (NIOSH) Publication No. 98-126). U.S. Department of Health and Human Services.

Rappaport, S. M., & Selvin, S. (1987). A method for evaluating the mean exposure from a lognormal distribution. American Industrial Hygiene Association Journal, 48(4), 374–379.

Vadali, M., Ramachandran, G., Mulhausen, J. R., & Banerjee, S. (2012). Effect of training on exposure judgment accuracy of industrial hygienists. Journal of Occupational and Environmental Hygiene, 9(4), 242–256.

CHAPTER 16

Reconstructing the Past

How an Industrial Hygienist Rebuilds a Worker's Exposure Years — or Decades — After the Fact

> *Silica, solvent, and welding-fume cases are rarely won on the medicine alone. They are won, or lost, on whether the exposure estimate behind the diagnosis can survive a Rule 702 challenge.*

I. The Problem: A Diagnosis Today, an Exposure Long Gone

A fabricator walks into a pulmonologist's office at 38 with progressive massive fibrosis. A machinist develops a blood cancer at 55 after thirty years around degreasing tanks. A boilermaker who welded stainless steel through the 1990s is diagnosed with a neurological disorder. In each case, the medicine is only half the battle. Before any physician can tie the disease to the workplace, someone has to answer a harder question: how much of what was this person actually breathing, and for how long? The answer almost never exists as a tidy set of air-sampling records. It has to be rebuilt.

That rebuilding is called retrospective exposure assessment (REA), and it is the discipline on which most toxic tort and occupational disease cases quietly turn. REA is the estimation of the exposure associated with a person's work history, and its central difficulty is converting fragmentary, largely qualitative information — job titles, tasks, materials, ventilation, the era's work practices — into a defensible quantitative estimate of dose (Sottas et al., 2020). Get it right and the causation opinion has a foundation. Get it wrong, or get it right but fail to document the method, and the whole case can collapse at a *Daubert* hearing before a jury ever hears the word "silicosis."

This chapter walks through how a qualified industrial hygienist actually performs that reconstruction, and, just as important for counsel, where opposing experts' reconstructions tend to fail under the reliability standards that now govern expert testimony in federal court and most states. The thesis is simple: in long-latency disease litigation, everything turns on the exposure estimate, and its methodology is what the court will scrutinize first.

II. Why These Cases Are Different: Latency and Data

Occupational disease litigation has a structural feature that acute-injury cases do not: a latency period that can span decades between exposure and diagnosis. Chronic silicosis typically develops after 10 or more years of exposure to respirable crystalline silica, and many solvent- and metal-related diseases have comparably long or longer intervals (NIOSH, 2002). By the time the disease appears, the employer may be defunct, the plant demolished, the ventilation long since changed, and contemporaneous air monitoring — if it ever existed — lost or never collected at all.

The reliability of a past-exposure estimate is precisely what matters most in these long-latency diseases, and it is precisely what is hardest to establish (Sottas et al., 2020). This is why the same technical exercise decides both the science and the litigation. The plaintiff must prove that the exposure was sufficient to cause the disease (general and specific causation); the defense must show that it was not, or that the estimate offered is too speculative to support any opinion at all. Both sides are litigating the same reconstruction: they are simply arguing about its inputs, its method, and its uncertainty.

Chapter 2 sets out the 2023 amendment to Rule 702 in full; the reconstruction problems below assume it.

III. The Methodology: Building the Reconstruction

A defensible reconstruction is not an educated guess dressed up in credentials. It follows the structured framework that the American Industrial Hygiene Association has codified across four editions of its consensus text, A Strategy for Assessing and Managing Occupational Exposures (Jahn et al., 2015), adapted to look backward rather than forward. The steps below are the ones I walk through in every case, and the ones counsel should expect any competent expert, on either side, to be able to defend line by line.

A. Basic Characterization: Reconstructing the Workplace

The first task is to rebuild the workplace as it existed during the exposure period. That means assembling the worker's complete occupational history — every employer, job title, department, and task, with dates — and then characterizing the agents, processes, materials, and controls associated with each. The specific determinants that drive a quantitative estimate are well established: the work period, the type and frequency of each task, the physical setting, how conditions changed over time, and the presence or absence of engineering controls and personal protective equipment (Sottas et al., 2020). In practice the raw material comes from depositions, co-worker testimony, personnel and payroll records, product safety data sheets, purchasing records, photographs, union records, and OSHA inspection files. The quality of this stage sets a ceiling on everything that follows; a reconstruction built on a vague job title and nothing else is a reconstruction built on sand.

B. Similar Exposure Groups: Grouping Like With Like

Because no one measured this particular worker on the days that mattered, the hygienist reasons from groups of workers whose exposures were governed by the same processes, tasks, and conditions: what the AIHA framework calls similar exposure groups, or SEGs (Jahn et al., 2015). If a body of measurements exists for edge-grinders in stone fabrication, or for vapor-degreaser operators in a given decade, those data can be brought to bear on an individual who did the same work under comparable conditions. The defensibility of the whole exercise depends on getting these groupings right; workers placed in a group whose exposures were not truly similar are, in the language of the discipline, misclassified, and misclassification is one of the most productive lines of cross-examination available against a sloppy opponent.

C. Filling the Data Gap: Analogues, Models, Judgment

With the workplace characterized and the groups defined, the hygienist estimates the actual airborne concentrations. Three sources of information are used, in descending order of preference. First, direct historical measurements: the worker's own employer records, or published industry surveys and government datasets (for example, OSHA's Chemical Exposure Health Database or NIOSH health hazard evaluations) for the same task and era. Second, validated physical-chemical models: mathematical exposure models that estimate airborne concentration from emission rates, room volume, and ventilation, which are especially useful for solvent vapors where the governing physics is well understood. Third, and only where the first two are thin, structured professional judgment. Quantitative, data-driven approaches are preferred wherever the records permit, precisely because they are more transparent and reproducible than judgment alone (Sottas et al., 2020). A well-built reconstruction is explicit about which of the three it is relying on at each step, and honest about the uncertainty each introduces.

D. From Concentration to Dose — and to a Regulatory Yardstick

An airborne concentration becomes a dose only when combined with the duration and frequency of exposure, expressed as an 8-hour time-weighted average and, where the disease depends on cumulative burden, integrated over the years of work. That figure is then compared against the relevant occupational exposure limits, both current and historical. For respirable crystalline silica, for instance, OSHA's 2016 final rule set a permissible exposure limit of 50 micrograms per cubic meter as an 8-hour TWA, with an action level of 25, effective June 23, 2016 (Occupational Exposure to Respirable Crystalline Silica, 2016). Crucially, the analysis cannot simply apply today's limit to yesterday's work; the reconstruction has to identify the limit in force during each exposure period and account for how both the standards and the control technology evolved over the worker's career.

IV. Two Case Studies in Why This Matters

A. Engineered-Stone Silicosis

No current example illustrates the stakes better than the epidemic of silicosis among engineered-stone countertop fabricators. Engineered stone is a manufactured quartz composite that can contain more than 90 percent crystalline silica — far more than granite (roughly 30 percent) or marble (under 10 percent) — so cutting, grinding, and polishing it can release extreme concentrations of respirable silica (Fazio et al., 2025). The result has been severe, accelerated disease in unusually young workers. In a California case series of 52 fabrication workers, the median age at diagnosis was just 45; more than a third already had progressive massive fibrosis, and roughly one in five had died (Fazio et al., 2023). By November 2024, California alone had identified 219 cases, including at least 14 deaths and 26 lung transplants (Fazio et al., 2025).

For litigation, these cases compress the reconstruction problem into a few brutal years rather than decades, but every methodological demand still applies. The material's silica content must be established from the specific product used; the tasks (dry versus wet cutting, the presence or absence of local exhaust ventilation and respirators) must be reconstructed from testimony and shop records; and the resulting exposures must be quantified against the PEL in force at the time. Where a California worker's exposures during a documented inspection were found to reach many times the applicable limit — in one reported instance, dozens of times the limit then in effect (Rose et al., 2019) — the reconstruction turns an abstract diagnosis into a concrete, defensible exposure narrative. That is the difference between a case that settles well and a case that never survives summary judgment.

B. Noise-Induced Hearing Loss

The silica case compresses the reconstruction into a few years. The second example stretches it across a working life, and it is worth developing because it exercises every step of the method above on an agent whose measurement physics is nothing like a respirable dust. Noise-induced hearing loss is, by claim volume, among the largest occupational disease categories in the country, and its reconstruction problem is unusually tractable: the exposure is continuous rather than episodic, the standard governing it is prescriptive, and the standard generates records.

Basic characterization proceeds as it does for any agent, with one advantage. Because 29 C.F.R. § 1910.95 has required monitoring since 1983 wherever exposures may equal or exceed the 85 dBA action level, an employer in a covered industry either has area and personal dosimetry results for the relevant departments or has a documented failure to have obtained them, and the second is as useful to a reconstruction as the first. The records the standard compels — the monitoring results, the audiometric test records the employer must retain for the duration of employment, the training documentation, the hearing-protector attenuation selections, and the notifications required when a standard threshold shift is recorded — are precisely the discoverable paper trail this book keeps telling counsel to demand. A hearing-loss case in which no one has asked for the § 1910.95 file has not been worked.

Similar exposure groups form on the same logic as in the foundry example, and they fail in the same way. A grinder and a crane operator in the same bay are not one group; a maintenance millwright who moves through six departments is not a member of any of them without a task-time analysis. Where measurements are missing for the early years, and for a claimant who started in 1978 they always are, the reconstruction fills the gap from the same three sources in the same order of preference: the employer's own historical surveys where they exist, published task-based and industry-based noise data where they do not, and characterized professional judgment last. Machine specifications and sound-power data from the equipment manufacturers are frequently obtainable and are commonly overlooked; a documented equipment inventory with dates of installation and replacement will often support a defensible period-by-period profile where nothing else will.

The dose step is where noise diverges from a dust, and where the reconstruction is most often attacked. Cumulative noise exposure is not a simple product of concentration and years, because the metric is logarithmic: the conventional expression is a noise immission level, which integrates the eight-hour equivalent level across the exposure years, and it must be computed with a stated exchange rate. Chapter 15 §VIII sets out why the choice between the five-decibel OSHA rate and the three-decibel NIOSH rate is not a rounding decision; here it is worth adding that the two also imply different hearing-protector adequacy conclusions for the same job, since the derated attenuation of a protector is applied against a dose that differs by the exchange rate before the protector is considered at all. The estimate is then compared against the criterion in force during each period — which for a career beginning before 1983 means a period in which the hearing conservation amendment did not yet exist, and the anachronism failure of Section V below is committed the moment an expert applies the modern program requirements to work performed before they applied.

For counsel the practical value of the noise example is that it is the cleanest available demonstration that this method generalizes. The agent, the instrumentation, the units, the biological mechanism, and the governing standard all differ from the silica case. The reconstruction does not. It is still basic characterization, still similar exposure groups, still a documented hierarchy of data sources, still a dose integrated over time, still a period-matched yardstick, and still an uncertainty statement. An expert whose method changes when the agent changes has not been applying a method.

V. Where Opposing Experts Fail Under *Daubert*

The modern reliability framework descends from the Supreme Court's "*Daubert* trilogy", *Daubert v. Merrell Dow Pharmaceuticals, Inc.* (1993), *General Electric Co. v. Joiner* (1997), and *Kumho Tire Co. v. Carmichael* (1999) — which together charge the trial judge with a gatekeeping duty and extend that duty to technical and experience-based testimony, not just conventional science (*Kumho Tire*, 1999). *Daubert* itself identified non-exclusive reliability factors: testability, peer review and publication, known or potential error rate, and general acceptance in the relevant field (*Daubert*, 1993). The 2023 amendment to Rule 702 then tightened the proponent's burden to satisfy those reliability requirements by a preponderance of the evidence. Against that backdrop, retrospective exposure opinions tend to fail in five recurring ways.

First, the ipse dixit estimate. An expert asserts a number — "his exposures exceeded the PEL" — without showing the data, model, or reasoning that produced it. *Joiner* squarely authorizes exclusion where too great an analytical gap separates the data from the opinion (*Joiner*, 1997); a conclusion connected to the facts only by the expert's say-so is the paradigm case for exclusion under amended Rule 702.

Second, misclassification through bad grouping. An expert borrows measurements from a similar exposure group whose tasks, materials, or controls were not actually similar to the plaintiff's: lumping an intermittent maintenance worker with full-time production operators, say. Because the SEG concept is the recognized backbone of the methodology (Jahn et al., 2015), a demonstrably ill-fitting group undermines the reliable-application prong directly.

Because the chapter has called misclassification one of the most productive lines of cross-examination available, it should show one. The example below is composite and simplified, but the shape is the shape these attacks actually take.

The opinion. A plaintiff's hygienist reconstructs a maintenance mechanic's respirable crystalline silica exposure at a foundry between 1996 and 2008. No measurements exist for the mechanic. The hygienist works from thirty-one personal samples the employer collected on shakeout and finishing operators during the period, and reports them in this order: geometric mean 0.06 mg/m^3, geometric standard deviation 3.7, arithmetic mean 0.14 mg/m^3. He adopts the arithmetic mean, assigns it to the mechanic at full-time occupancy — 2,000 hours a year, an occupancy fraction of 1.0 — across thirteen years, and reports a cumulative exposure of 0.14 × 1.0 × 13, or roughly 1.8 mg/m^3-years: comfortably within the range the epidemiology associates with silicosis.

The attack, in four moves. The first is **task-time**: the mechanic's own work orders, which the hygienist did not obtain, show a median of 6.2 hours per week in the shakeout and finishing areas, with the balance of his time in the maintenance shop, the yard, and the electrical room. The SEG the hygienist borrowed was defined by *area and task*, and the mechanic was in it for roughly fifteen percent of his hours. The second is **task-identity**: within those hours the mechanic was not performing the operators' tasks. He was troubleshooting and replacing components on stopped equipment, sometimes during production and sometimes not, and the generation mechanism for a silica exposure during a shakeout cycle is not the generation mechanism during a bearing replacement on an idle line. The third is **the wrong statistic**: exposure data are approximately lognormal, and for a lognormal distribution the arithmetic mean exceeds the geometric mean by a factor of $\exp(\sigma^2/2)$, where σ is the natural logarithm of the geometric standard deviation. The report's own figures are internally consistent on this point: an arithmetic mean of 0.14 mg/m^3 against a geometric mean, and median, of 0.06 is a ratio of 2.3 — exactly the factor a geometric standard deviation of 3.7 implies. That is a wide distribution, and it means the arithmetic mean the hygienist chose sits near the seventy-fifth percentile rather than the middle of the operators' own exposures.

The choice between the mean and the median is a methodological decision with a factor-of-two consequence, and the report never disclosed that it was making one. The fourth is **the missing control history**: the employer installed local exhaust ventilation on the shakeout in 2003, and the borrowed dataset pools pre- and post-installation samples, so the single mean assigned uniformly across thirteen years describes neither period.

What the corrected analysis produces. The rebuttal does not stop at criticism, which is the discipline Chapter 17 §VII insists on. It re-derives the estimate: the mechanic's actual hours in each area from the work orders, a task-specific exposure profile assembled from the subset of samples collected during maintenance activity plus published task-based data for maintenance in ferrous foundries, and separate pre- and post-2003 control periods, 2003 being the only control breakpoint that falls inside the 1996–2008 exposure window. The corrected arithmetic is stated so that it can be checked in a single reading. The occupancy fraction is 6.2 hours a week against a 40-hour week, or 0.15. The pre-LEV period is 1996 through 2002 — seven years at a maintenance-task arithmetic mean of 0.12 mg/m^3, giving $0.12 \times 0.15 \times 7 = 0.13$ mg/m^3-years. The post-LEV period is 2003 through 2008 — six years at 0.05 mg/m^3, giving $0.05 \times 0.15 \times 6 = 0.045$ mg/m^3-years. The corrected cumulative estimate is therefore roughly 0.17 mg/m^3-years against the original 1.8: an order of magnitude below it, and in a range where the specific-causation opinion built on the original does not stand.

Note. Cumulative exposure = $\Sigma(C_i \times f_i \times t_i)$, where C_i is the mean airborne concentration during period i, f_i is the fraction of full-time hours the worker spent in the exposed task during that period, and t_i is the length of the period in years. Reporting the three inputs separately for each period, rather than only their product, is what makes a reconstruction checkable by the other side — and by the court.

The lesson, in both chairs. Note what the attack did not do. It did not argue that borrowing data is impermissible; borrowing is the method, and Section III of this chapter endorses it. It argued that *this* group was not this worker's group, and it proved the point from the defendant's own work orders rather than from the expert's skepticism. That is the misclassification attack in its effective form, and it is available to the plaintiff's side just as often: against a defense hygienist who assigns a worker to a well-controlled SEG on the strength of a job title while the maintenance log shows him inside the enclosure three times a week.

Third, anachronistic standards. The expert applies a current exposure limit, or current control assumptions, to a work period governed by different rules and different technology, or vice versa. A reconstruction that ignores how the PEL and the state of ventilation practice changed over a multi-decade career is not a reliable application of the method to the facts.

Fourth, cherry-picked inputs. The expert selects the measurements or model assumptions that favor the retaining party and omits contrary data without explanation. Federal courts applying amended Rule 702 have treated selective use of the evidentiary record and unexplained changes in analysis as grounds for exclusion, not merely impeachment (Advisory Committee on Evidence Rules, 2023).

Fifth, false precision, or none at all. A reconstruction that reports a single confident number while concealing the uncertainty around it is as vulnerable as one that offers no quantification. Because different defensible approaches applied to the same scenario can yield materially different results, the discipline itself insists that uncertainty be characterized rather than hidden (Sottas et al., 2020). An expert who cannot state the confidence bounds on the estimate has not finished the job.

VI. Practice Implications: What Counsel Should Demand

For plaintiff's counsel, the reconstruction is the load-bearing wall of the case, and it should be engaged early, ideally before the causation experts are retained, so that the medical opinion rests on a quantified exposure rather than the reverse. Insist that your own expert document every input, disclose the data-versus-model-versus-judgment basis for each step, and state the uncertainty. An opinion built that way is far harder to exclude, because it can demonstrate reliable application by a preponderance of the evidence exactly as amended Rule 702 requires.

For defense counsel, the same framework is a checklist for challenge. Probe how the similar exposure groups were formed and whether they truly match the plaintiff's work; test whether historical standards and control technology were correctly time-matched; ask what data were considered and rejected, and why; and press for the error bounds the opinion conveniently omits. Under the current Rule 702, many of these points are no longer mere cross-examination fodder: they go to admissibility, and a well-framed motion can keep an unreliable estimate away from the jury entirely.

For both sides, the lesson is the same one that runs through every long-latency disease case: the fight that looks medical is usually, at its foundation, a fight about exposure reconstruction. The party whose expert can build, and defend, the more rigorous reconstruction is usually the party that controls the outcome. The estimate is the case.

References

29 C.F.R. § 1910.95 (occupational noise exposure; hearing conservation program requirements are annotated in full in the Chapter 15 reference list).

Advisory Committee on Evidence Rules. (2023). Report of the Advisory Committee on Evidence Rules (Committee note to the 2023 amendment to Fed. R. Evid. 702). Committee on Rules of Practice and Procedure of the Judicial Conference of the United States.

Daubert v. Merrell Dow Pharms., Inc., 509 U.S. 579 (1993).

Fazio, J. C., Gandhi, S. A., Flattery, J., Heinzerling, A., Cummings, K. J., & Harrison, R. J. (2025). Deadly countertops: An urgent need to eliminate silicosis among engineered stone workers. American Journal of Public Health, 115(4), 490–493.

Fazio, J. C., Gandhi, S. A., Flattery, J., Heinzerling, A., Kamangar, N., Afif, N., Cummings, K. J., & Harrison, R. J. (2023). Silicosis among immigrant engineered stone (quartz) countertop fabrication workers in California. JAMA Internal Medicine, 183(9), 991–998.

Fed. R. Evid. 702.

Gen. Elec. Co. v. Joiner, 522 U.S. 136 (1997).

Jahn, S. D., Bullock, W. H., & Ignacio, J. S. (Eds.). (2015). A strategy for assessing and managing occupational exposures (4th ed.). American Industrial Hygiene Association.

Kumho Tire Co. v. Carmichael, 526 U.S. 137 (1999).

International Organization for Standardization. (2013). ISO 1999: Acoustics — Estimation of noise-induced hearing loss. ISO. (Adopted in the United States as ANSI/ASA S3.44.)

National Institute for Occupational Safety and Health. (2002). Health effects of occupational exposure to respirable crystalline silica (DHHS [NIOSH] Publication No. 2002-129). U.S. Department of Health and Human Services.

National Institute for Occupational Safety and Health. (1998). Criteria for a recommended standard: Occupational noise exposure (revised criteria 1998) (DHHS (NIOSH) Publication No. 98-126). U.S. Department of Health and Human Services.

Occupational Exposure to Respirable Crystalline Silica, 81 Fed. Reg. 16285 (Mar. 25, 2016) (codified at 29 C.F.R. pts. 1910, 1915, 1926).

Rose, C., Heinzerling, A., Patel, K., Sack, C., Wolff, J., Zell-Baran, L., Weissman, D., Hall, E., Sooriash, R., McCarthy, R. B., Bojes, H., Korotzer, B., Flattery, J., Weinberg, J. L., Potocko, J., Jones, K. D., Cummings, K. J., Materna, B. L., & Harrison, R. J. (2019). Severe silicosis in engineered stone fabrication workers — California, Colorado, Texas, and Washington, 2017–2019. Morbidity and Mortality Weekly Report, 68(38), 813–818.

Sottas, P.-E., Lavoué, J., & Vernez, D. (2020). Retrospective exposure assessment methods used in occupational human health risk assessment: A systematic review. International Journal of Environmental Research and Public Health, 17(17), 6190.

CHAPTER 17

Attacking the Exposure Opinion

Auditing an Opposing Expert's Sampling and Reconstruction Under Rule 702

Chapter 10 takes apart the government's measurement; this chapter takes apart the retained expert's judgment. In the civil case the exposure number is rarely handed down by a compliance officer: it is built by an industrial hygienist for one side, from a sampling design that chose what to measure and a reconstruction that supplied the rest. Both are arguments dressed as data, and both can be audited. This chapter is the audit: how to read an opposing exposure opinion for the design decisions, the borrowed data, the misapplied method, and the buried uncertainty that decide whether the number reaches the jury at all.

I. The Three Attacks, and Why This One Is Different

Everything in this book about contesting an exposure number converges here, so it is worth fixing the boundaries at the outset. Chapter 4 teaches the deposition and cross-examination of an opposing expert: the litigation craft of the campaign, applicable to any witness in any field. Chapter 10 teaches the attack on the Secretary's sampling in an OSHA overexposure case: a measurement performed by a compliance officer under the Field Operations Manual, audited against the Commission's case law and the agency's own methods. This chapter teaches the third and, in civil litigation, the most common attack: the audit of a retained expert's exposure opinion — the plaintiff's or the defendant's industrial hygienist who designed a sampling campaign, or reconstructed a dose that no one measured, and then asked a court to treat the result as reliable science.

The distinction is not academic, because it changes the governing standard and the target. The compliance officer's sampling is attacked against a prescriptive protocol and a body of Review Commission decisions; the retained expert's opinion is attacked against Rule 702 and the *Daubert* trilogy, in a forum where the trial judge is the gatekeeper and the proponent must establish reliability by a preponderance of the evidence. The compliance officer measured what was in front of him on the day of the inspection; the retained expert, especially in a long-latency disease case, is usually reconstructing an exposure that ended years or decades before anyone thought to sample it. The government's hygienist is defending a number; the retained hygienist is defending a set of judgments about how to build a number. Those judgments — which samples, which group, which model, which assumptions, which uncertainty — are the surface this chapter maps.

Chapter 16 has already named the five recurring ways a reconstruction fails under *Daubert*: the ipse dixit estimate, misclassification through bad grouping, anachronistic standards, cherry-picked inputs, and false precision. This chapter does not repeat that catalogue; it supplies the audit that produces the findings behind it, and then shows where each finding goes to work — into the Rule 702 motion built in Chapter 2, the deposition scripted in Chapter 4, and the rebuttal report written under the discipline of Chapter 3. The catalogue is the diagnosis; what follows is the examination.

II. Reading the Opinion Before You Attack It

An exposure opinion cannot be audited until it has been disassembled into its load-bearing parts, and the first task on receiving an opposing report is to separate what the expert measured from what the expert assumed, modeled, or borrowed. Every quantitative exposure opinion is some combination of four kinds of input: primary data, meaning samples actually collected on this plaintiff or this operation; analogous data, meaning measurements imported from some other worker, task, or facility asserted to be similar; modeled values, meaning concentrations calculated from a physical or statistical model rather than measured; and professional judgment, meaning the expert's unquantified estimate filling a gap the other three could not. A defensible opinion states, for every number it advances, which of the four produced it. An opinion that blurs the four together, presenting a modeled or borrowed value with the same confidence as a measured one, has concealed the very seam where it is weakest.

The Rule 26 report, dissected as Chapter 3 describes, is where this separation is forced. The report must disclose the bases and reasons for each opinion and the facts or data the expert considered; the audit reads it for what it conspicuously does not disclose. Where did each measurement come from, and was it collected under a method the expert will name? For every borrowed value, what is the asserted basis of similarity, and is it documented or merely asserted? For every modeled value, what model, what inputs, and what validation? Where the report supplies a single number without a stated basis, the audit has found either an undisclosed opinion, inadmissible on its own terms, or an unsourced estimate, and the deposition of Chapter 4 exists to make the witness say which.

The most useful early product of the audit is a simple inventory: a line for every exposure value in the opinion, coded by input type, with the source cited or marked absent. That inventory is the spine of the deposition outline and, later, of the motion. It converts a fluent narrative report into a ledger of testable claims, and it almost always reveals that the opinion's headline number rests on far less primary data than its confidence implied.

III. The Sampling-Strategy Critique: The AIHA Framework

Where the opposing expert collected data, the attack is not on the numbers the laboratory returned but on the design that decided which numbers would ever exist. Sampling is a choice before it is a measurement — of whom to monitor, on what day, for how long, under what conditions — and a campaign designed to find a particular answer will usually find it. The recognized standard against which a design is judged is the American Industrial Hygiene Association's exposure assessment strategy, which frames the professional obligation to define similar exposure groups, characterize the full exposure profile rather than a convenient slice of it, and rate the result against the limit with its uncertainty stated (Jahn et al., 2015). That framework is the yardstick the audit applies, and departures from it are the findings.

Four design questions carry most of the weight. First, representativeness: were the sampled days, tasks, and workers typical of the exposure being characterized, or were they selected — a low-production day, a well-ventilated bay, a cautious operator — to bias the profile? The plaintiff's expert who sampled only peak tasks and the defense expert who sampled only the quiet shift have made the mirror-image error, and each is vulnerable to the same critique. Second, sample number and duration: does the campaign contain enough full-shift samples to support a statistical statement about the exposure profile, or is a single grab sample being asked to stand for a career? Chapter 15 sets out what a handful of samples can and cannot support; the audit measures the opposing design against that ceiling. Third, group construction: is the similar exposure group genuinely homogeneous, or has the expert pooled workers whose tasks, materials, or controls differ enough that the group's mean describes no one in it? Fourth, the measured-agent question: did the sampling capture the exposure that matters — the right analyte, the right fraction, the right averaging period — or did it measure something adjacent and reason across the gap?

Each of these is a reliable-application question under amended Rule 702, not a matter of weight, because each goes to whether the method the expert claims to have followed was actually followed on these facts. An expert who invokes the AIHA strategy as the basis for the opinion and then violated its representativeness or grouping requirements has supplied the cross-examiner with the most powerful kind of impeachment: a departure from the witness's own stated standard.

IV. Misapplied Methods: Right Reference, Wrong Way

A distinct and often decisive line of attack is the misapplication of a named sampling or analytical method. Experts cite authority for their procedures — a NIOSH Manual of Analytical Methods number, an OSHA sampling and analytical method, a substance-specific standard's mandatory appendix — and the citation lends the opinion the appearance of rigor. The audit checks whether the method was actually satisfied, because a method invoked is not a method followed. Analytical methods prescribe minimum and maximum sampling volumes, flow-rate tolerances, medium types, storage and shipping conditions, and limits of detection and quantification; a result generated outside those bounds is not supported by the authority the expert cited for it, however confidently the number is reported.

Three misapplications recur. The wrong method for the matrix: an analyte quantified by a procedure validated for a different sampling medium or a different co-contaminant environment, so that the reported concentration carries an error the method was never characterized to bound. The out-of-range result: a value derived from a sample below the method's limit of quantification, or from a sampling time outside the validated window, reported as though it were a quantified measurement rather than an estimate the method cannot support. And the imported limit: an exposure compared against a permissible limit or an averaging convention that did not govern the analyte, the industry, or the period at issue — the anachronism Chapter 16 names, seen here at the level of the method rather than the standard. In each, the expert has borrowed the credibility of a published procedure without meeting its conditions, and the OSHA Technical Manual and the method documents themselves supply the neutral text that proves the gap.

A fourth misapplication deserves separate treatment, because it is the most common defect in the exposure literature this book audits and the one least often examined at deposition: what the analysis did with the results the laboratory could not quantify. Almost every real sampling campaign returns some proportion of results reported as less than a stated value: non-detects, in the vernacular; left-censored data, in the statistics. The laboratory has not measured zero. It has measured something it cannot distinguish from zero at the method's limit, and the difference between those two statements is where a large share of exposure opinions quietly acquire their answer.

The conventional handling is substitution: every non-detect is replaced by a single number — zero, half the limit of detection, the limit divided by the square root of two, or the full limit — and the dataset is then treated as though it had been measured throughout. Substitution is defensible only when the censored fraction is small, and it is defensible then largely because nothing much turns on it. As the censored fraction rises, three consequences follow, and an expert who cannot describe them has not understood the data. The first is that the choice of substitute moves the estimate, and it is a discretionary choice the report usually does not disclose. The second is less intuitive and more damaging: substituting a constant for many observations compresses the apparent variability of the distribution, so the rule that raises the arithmetic mean lowers the estimated upper percentile, which means the substitution that looks conservative on the average is the anti-conservative one on exactly the part of the distribution where the health risk and the compliance question live. The third is that substitution produces a point estimate and nothing else. It cannot generate a confidence interval, because it has silently converted an unknown into a datum.

The methods that handle censoring honestly are established, published, and available in ordinary industrial hygiene software: maximum-likelihood estimation, which fits a lognormal distribution to the detected values while treating each non-detect as the information it actually is — a value somewhere below the reporting limit; regression on order statistics, which is more stable at small sample sizes; and the Bayesian treatments that carry censoring through to a posterior distribution and therefore report the uncertainty rather than absorbing it. None of these is exotic, and their availability is what converts an undisclosed substitution from a defensible simplification into a reliable-application problem. The audit therefore asks for five things, and their absence is the finding: the censored fraction; the limit of detection and the limit of quantification for the analyte, stated per sample rather than in general, because both depend on the air volume actually collected; the substitution rule, named; the reason it was chosen; and a sensitivity run showing what the conclusion does under the alternatives.

A worked example shows the shape of the attack, and it is the companion to the misclassification example of Chapter 16 §V. The composite facts are these. A hygienist reconstructs a solvent exposure for a similar exposure group from twenty-four full-shift samples, of which fourteen were quantified and ten were reported as below a limit of detection of 0.5 parts per million, a censored fraction of forty-two percent. The report states that non-detects were "treated as one-half the detection limit in accordance with standard practice," and reports a single number.

Run the same data four ways. Substituting half the limit returns a geometric mean of 0.86 ppm, a geometric standard deviation of 3.54, and an estimated ninety-fifth percentile of 6.87 ppm. Substituting the limit divided by the square root of two returns 0.99, 3.07, and 6.27. Substituting the full limit returns 1.15, 2.68, and 5.82: note that the substitution producing the highest geometric mean has produced the lowest upper percentile, which is the compression effect described above, visible in four numbers. A maximum-likelihood fit to the identical data, treating the ten results as censored rather than as measurements, returns a geometric mean of 0.74, a geometric standard deviation of 4.45, and a ninety-fifth percentile of 8.60 ppm: roughly half again the figure the most permissive substitution produced, and the only one of the four accompanied by an interval. Nothing about the workplace changed across those four lines. What changed was a methodological decision the report did not disclose and the expert was never asked to defend.

The attack proceeds in three moves and belongs in the deposition before it belongs in the motion. Establish the censored fraction, which the report frequently does not state and which the laboratory package always does. Establish that the substitution was a choice among recognized alternatives rather than a requirement, and that the alternatives were available to the witness: the AIHA strategy and the published literature on censored occupational data are the neutral authority, and the witness will concede them because they are her own field's. Then establish the sensitivity: ask what the estimate becomes under each rule and under a maximum-likelihood fit, and whether the witness ran any of them. An expert who ran the alternatives and reported the range has done the work and should be left alone; an expert who ran one rule, did not name it, and cannot say what the others return has supplied an undisclosed discretionary choice at the precise point where the data were weakest.

Two disciplines keep this line honest, and the second is the one that keeps the expert who uses it credible. The first is that the mirror error exists and is equally available: the defense hygienist who substitutes zero for every non-detect has understated the dose by exactly the mechanism the plaintiff's hygienist used to overstate it, and a report that treats every censored result as an absence of exposure is making an affirmative claim the laboratory never made. The second is a point about sample volume that decides more cases than it should. Because the detection limit for a given analyte is a mass divided by the volume of air drawn, a short sample can carry a detection limit above the exposure limit itself. A non-detect on such a sample establishes nothing whatever about compliance, the method could not have seen an overexposure had one occurred, and a report that lists it among results "below the permissible exposure limit" has not merely simplified. It has reported as evidence of control a measurement that was incapable of detecting its absence, and that sentence, put to the witness with the calculation behind it, is the most efficient cross-examination in this chapter.

A fifth misapplication belongs here because it is where the noise example of Chapters 15 and 16 becomes an audit rather than a reconstruction. The hearing conservation standard does not merely require measurement; it requires a series of interpretive determinations, each of which is a contested methodological choice made by the employer's own contractor and each of which is discoverable. The audit runs to four of them.

The first is the standard threshold shift determination. A recordable shift under 29 C.F.R. § 1910.95(g)(10) is an average change of ten decibels or more at 2000, 3000, and 4000 hertz in either ear relative to the baseline audiogram. Three inputs decide whether a shift is found: which audiogram is being used as the baseline, whether the comparison averaged the correct frequencies, and whether a retest was performed within the thirty days the standard permits and, if so, which of the two results was carried forward. A pattern of shifts that appear on the annual test and vanish on the retest is a finding about the retest protocol, not about the ear.

The second is age correction. The standard permits, but does not require, the employer to subtract the portion of a shift attributable to aging using the tables in its Appendix F. The permission is frequently exercised silently. Age correction applied without disclosure converts a recordable shift into a non-recordable one and removes an entry from the OSHA 300 log, and the audit question is not whether the correction is legitimate, it is, but whether it was applied, disclosed, and applied consistently across the workforce rather than selectively to the cases near the threshold. An expert offering an opinion on the employer's program who has not asked whether the audiometric database was age-corrected has audited the output of a process without knowing the process.

The third is baseline revision. The standard allows the audiologist or physician to revise a worker's baseline when a shift is persistent, and revising the baseline resets the comparator so that the same ear must lose another ten decibels before anything is recorded again. Serial revision is the single most effective way to make a deteriorating audiometric record look stable, and it is usually visible on the face of the database as a stairstep. Ask for the complete audiometric history for the claimant and for the department, ask which baselines were revised and on whose written determination, and ask what the record would show against the original baseline. That last calculation takes an afternoon and it is frequently the case.

The fourth is the hearing-protector adequacy calculation. A protector's Noise Reduction Rating is derived under laboratory conditions that real workplaces do not reproduce, and the standard's own appendix and NIOSH's guidance both derate it substantially before it is applied. An assessment that subtracts the full labeled NRR from the measured exposure has made the same error as the reconstruction that credits a respirator at its assigned protection factor without a fit-test record, a program, or evidence of use — and the parallel is exact enough that the discovery list in Section V of this chapter transfers with only the nouns changed. The questions are whether the protector was selected against the actual measured exposure, whether the derating method is named, whether workers were trained and fitted, and whether anyone verified that the protector was worn in the noise, which for a device the worker can remove between tasks is a question the attenuation label cannot answer.

V. The Analogous-Data Leap

The heart of most retained exposure opinions in a disease case is not measurement at all but reconstruction, and the heart of the reconstruction is the analogous-data leap: the inference that a measurement taken somewhere, sometime, on someone describes this plaintiff's exposure closely enough to carry a causation opinion. Reconstruction is a legitimate and often unavoidable discipline — the worker's exposure ended before anyone sampled it, and the estimate must be built from the surviving record — but its legitimacy depends entirely on the defensibility of the analogies and assumptions that bridge the gap. The audit's task is to expose each bridge and test whether it bears the load the opinion places on it.

The reconstruction rests on a chain of determinants of exposure — the material's composition, the task and its duration, the ventilation and enclosure, the use and efficiency of respiratory protection, the work rate, the room volume and air exchange — and the expert has supplied a value, measured or assumed, for each. The attack proceeds determinant by determinant, asking of each assumed value: what is its source, how sensitive is the final estimate to it, and what happens to the opinion if it is wrong? An expert who assumed the local exhaust ventilation always functioned, or that a respirator was worn and fitted whenever the task was performed, or that the historical product contained the silica fraction of a modern one, has made a choice the record may not support, and if the estimate collapses when the assumption is relaxed, the opinion was resting on the assumption rather than on the data.

Respiratory protection deserves separate scrutiny, because it is the determinant most often credited at full value on the thinnest record. A reconstruction that divides the airborne concentration by an assigned protection factor has made three assumptions at once: that a respirator of that class was issued, that it was actually worn for the full duration of the exposure, and that it performed at its assigned factor on this wearer. The assigned protection factor is a program value, not a device property: it presupposes the written program, the medical evaluation, the fit testing, the cartridge change schedule, and the facial-seal conditions that the respiratory protection standard requires, and it degrades sharply when any of them is absent. The audit therefore asks for the program records rather than the respirator: the fit-test history for this worker, the training documentation, the issuance and maintenance records, and the evidence of use during the specific task. Where those records are missing, an opinion that nonetheless applies the full assigned protection factor has reduced the estimated dose by a factor the record does not support, and the reduction is exactly the kind of unexamined discretionary choice the reliable-application inquiry now reaches. The mirror error belongs to the other chair: the expert who assumes no protection at all, where the program records show a functioning one, has inflated the dose by the same mechanism.

Where the reconstruction uses a physical model — a well-mixed-room or two-zone mass-balance model, or a statistical or Bayesian treatment of sparse data — the model's inputs are assumptions in mathematical dress, and the same sensitivity discipline applies. A modeled concentration is only as reliable as its generation rate, its ventilation assumption, and its mixing assumption, and a model run at a single favorable set of inputs, without a sensitivity or uncertainty analysis, has produced a number with an unstated and possibly enormous error. The reconstruction that survives is the one that states its determinants, discloses which were measured and which assumed, and shows how the estimate moves across the plausible range of each; the one that fails presents a single confident figure and hopes no one asks what it was built from.

VI. Uncertainty and Bias: The Analysis the Opinion Omits

The final and most technical stratum of the audit is the one opposing opinions most often skip entirely: the honest characterization of uncertainty and the search for systematic bias. Exposure science is not a discipline of single true numbers; it is a discipline of distributions, and any defensible estimate carries a range that reflects measurement error, model error, and the variability of the exposure itself. Because different defensible approaches applied to the same scenario can yield materially different results (Sottas et al., 2020), an opinion that reports a single point without its bounds has not merely been imprecise: it has withheld the information a court needs to judge the opinion's reliability at all.

Uncertainty and bias are distinct, and both belong in the audit. Uncertainty is the width of the estimate, the confidence bounds around the number, and an expert who cannot state it has not finished the analysis. Bias is the direction of the error: whether the design, the borrowed data, the assumptions, and the model selection all lean the same way, toward the answer the retaining party needed. A campaign that sampled the peak task, borrowed data from a dustier analogue, assumed ventilation failure, and modeled with a conservative generation rate has not made four independent judgment calls; it has made one directional decision four times, and the accumulated bias is a proper subject of both the motion and the cross. The tools that discipline this — sensitivity analysis that reveals which inputs drive the result, and a frank statement of the estimate's confidence bounds — are exactly what a reliable opinion contains and an advocacy opinion omits.

This is where the audit most often crosses from impeachment to inadmissibility. Under the pre-2023 practice, an unstated uncertainty was cross-examination fodder, a point for the jury to weigh. Under amended Rule 702, the proponent must show reliable application by a preponderance, and an expert who cannot characterize the uncertainty of an estimate has not carried that burden. The federal courts applying the amended rule have treated the selective use of the evidentiary record and unexplained analytical choices as grounds for exclusion rather than mere weight (Advisory Committee on Evidence Rules, 2023), and a directional bias documented across every discretionary step of a reconstruction is precisely such a choice, multiplied.

VII. Where the Findings Go: Motion and Deposition

An audit that stays in the expert's file changes nothing; its value is realized only when its findings are routed into the three instruments that actually move an exposure opinion out of the case or blunt its force. The first is the Rule 702 motion built in Chapter 2. The strongest findings — a number with no disclosed basis, a similar exposure group that groups dissimilar work, a method invoked but violated, a reconstruction whose estimate collapses when a single unsupported assumption is relaxed, an uncertainty never characterized — are reliable-application defects, and under the amended rule they are matters of admissibility. Framed in writing on a motion in limine, they ask the gatekeeper to keep the number from the jury entirely, and the audit inventory is the exhibit that shows the court how little supports it.

The second instrument is the deposition of Chapter 4, and the audit is its script. The examination does not argue with the witness; it walks the inventory, establishing for each value which of the four inputs produced it, pinning the witness to a stated basis or extracting the concession that none exists, and locking the sensitivity of the estimate to each contested assumption before the witness appreciates where the questions lead. A deposition run from the audit produces the admissions the motion needs and forecloses the reconstruction the expert would otherwise attempt at trial. It is Chapter 10's cross-examination of the measurement chain, turned on a retained witness and conducted under Rule 702 rather than before the Commission.

The third instrument is the rebuttal report, written under the discipline Chapter 3 imposes on every expert report. The rebuttal does not merely list the opposing opinion's flaws; it models the corrected analysis — the properly constituted group, the method applied within its bounds, the reconstruction with its determinants disclosed and its uncertainty stated — and in doing so it demonstrates the reliable application the opposing opinion lacked. A rebuttal built that way survives the identical audit run against it, which is the point: the expert who takes apart an opposing exposure opinion must build one that cannot be taken apart the same way, because the record is indifferent to who commissioned it, and the more rigorous reconstruction is the one that wins. That symmetry is the discipline of this chapter and the reason its method belongs to whichever party builds it honestly.

References

29 C.F.R. § 1910.95 (occupational noise exposure; hearing conservation program requirements are annotated in full in the Chapter 15 reference list).

Advisory Committee on Evidence Rules. (2023). Report of the Advisory Committee on Evidence Rules (Committee note to the 2023 amendment to Fed. R. Evid. 702). Committee on Rules of Practice and Procedure of the Judicial Conference of the United States.

Daubert v. Merrell Dow Pharms., Inc., 509 U.S. 579 (1993).

Fed. R. Civ. P. 26(a)(2), (b)(4).

Fed. R. Evid. 702.

Gen. Elec. Co. v. Joiner, 522 U.S. 136 (1997).

Jahn, S. D., Bullock, W. H., & Ignacio, J. S. (Eds.). (2015). A strategy for assessing and managing occupational exposures (4th ed.). American Industrial Hygiene Association.

Kumho Tire Co. v. Carmichael, 526 U.S. 137 (1999).

National Institute for Occupational Safety and Health. (1998). Criteria for a recommended standard: Occupational noise exposure (revised criteria 1998) (DHHS (NIOSH) Publication No. 98-126). U.S. Department of Health and Human Services.

National Institute for Occupational Safety and Health. (2020). NIOSH manual of analytical methods (NMAM) (5th ed.). U.S. Department of Health and Human Services.

Nicas, M., & Jayjock, M. (2002). Uncertainty in exposure estimates made by modeling versus monitoring. American Industrial Hygiene Association Journal, 63(3), 275–283.

Occupational Safety and Health Administration. (1999, as updated). OSHA technical manual (TED 01-00-015). U.S. Department of Labor.

Ramachandran, G. (2005). Occupational exposure assessment for air contaminants. CRC Press.

Sottas, P.-E., Lavoué, J., & Vernez, D. (2020). Retrospective exposure assessment methods used in occupational human health risk assessment: A systematic review. International Journal of Environmental Research and Public Health, 17(17), 6190.

CHAPTER 18

The Dose That Didn't Come From Work

The Workplace Boundary, the Non-Occupational Exposure, and the Rule-Out That Has to Survive Rule 702

Every chapter of this Part has taught the reconstruction of a workplace exposure. A large share of exposure cases are decided on the other side of that line — on whether the harm was occupational at all, and on whether the alternative cause was ever characterized to the standard the workplace estimate was held to. The hobby woodshop, the prior employer, and the ambient background are not footnotes to the causation analysis; they are frequently the whole of it.

I. The Line That Decides the Case Before Causation Does

Chapters 14 through 17 have taught one direction of travel: the exposure happened at work, and the task is to quantify it, connect it to the disease, and defend or attack the estimate. That direction assumes an answer to a question the forums actually litigate first. Before anyone asks whether this exposure caused this condition, someone asks whether this was a workplace exposure at all, and three different bodies of law ask it in three different vocabularies, at three different levels of rigor, with three different consequences.

The recordkeeping rule asks whether the case is work-related and therefore recordable. The workers'-compensation act asks whether the injury arose out of and in the course of employment. The tort case asks whether the occupational exposure was a substantial factor in producing the disease, on a record that must satisfy Rule 702. These are not three phrasings of one standard. They ascend in rigor, they are decided by different people at different times, and the answer to one does not establish the answer to another. The professional or lawyer who treats them as interchangeable makes the single most common analytical error in this territory, and Section II is written to prevent it.

The second half of the boundary problem arrives even where the exposure is indisputably occupational. The defense will locate the dose somewhere else — the home shop, the second job, the renovation, the neighborhood — and the plaintiff's specific-causation opinion will stand or fall on whether those sources were characterized or merely mentioned. Chapter 14 named the governing authority: a differential etiology is only as good as the completeness of the initial list and the reliability of the rule-outs, and in *Pluck v. BP Oil Pipeline Co.*, 640 F.3d 671 (6th Cir. 2011), the opinion failed at both ends. It is worth noticing what kind of case *Pluck* was. The plaintiffs had bought a home above ground contaminated by pipeline releases and used the well to drink, wash, and irrigate; the disease was non-Hodgkin's lymphoma; and the alternative sources the expert did not adequately account for were the ordinary ones — the court observed that benzene is ubiquitous in ambient air and a constituent of vehicle exhaust and cigarette smoke. The case this book has cited for the discipline the workplace estimate requires is, on its own facts, a case about the discipline the other sources require. Both halves of that lesson are this chapter.

II. Was It a Workplace Injury at All? § 1904.5

The recordkeeping rule begins with a presumption, and counsel should read it before arguing about any entry. Under 29 C.F.R. § 1904.5(a), an injury or illness is work-related if an event or exposure in the work environment either caused or contributed to the condition, or significantly aggravated a pre-existing condition, and work-relatedness is presumed for injuries resulting from events or exposures occurring in the work environment. The work environment is defined broadly, reaching the establishment and any other location where one or more employees are working or are present as a condition of their employment. The presumption is defeated only through the exceptions the rule itself supplies, and the list is exclusive rather than illustrative.

Those exceptions are where the boundary is actually fought, and they are more particular than their reputation. A case is not recordable when the employee was present as a member of the general public rather than as an employee; when the symptoms surface at work but result solely from a nonwork event or exposure — the provision that carries most of the argument, because it distinguishes a condition merely manifesting at work from one the work contributed to; when the injury arises from voluntary participation in a wellness program, or a recreational or social activity; when it results from eating, drinking, or preparing food for personal consumption; when it results from personal tasks at the establishment outside assigned working hours; when it arises from personal grooming, self-medication for a nonwork condition, or intentional self-inflicted injury; when it results from a motor vehicle accident in a company parking lot or on a company access road during the commute; and for the common cold or flu, and for mental illness absent a supporting opinion from a licensed health care professional. The rule then supplies its own boundary geography for travel status and for work performed at home, where an injury is work-related if it occurs while the employee is performing work for pay or compensation in the home and is directly related to the performance of that work rather than to the home environment: a provision written for a rarity that is no longer rare, and one that will do steadily more work as the telework record grows.

What matters for litigation is the character of the decision the rule produces. The recordability call is made by a safety professional, contemporaneously, often years before any claim, on a causal standard — caused or contributed to — substantially lower than anything a tort plaintiff must prove. That makes the entry, and equally the conspicuous absence of one, evidence of how the employer characterized the event when it had no litigation reason to characterize it either way. It is not an admission of legal causation, and counsel who argues it as one will be corrected. Chapter 23 develops the log as an evidentiary instrument: what it establishes about knowledge, and how to read it against its own gaps; this section supplies the boundary rule that generates the entry in the first place.

The compensation forum asks a parallel question in its own language: did the injury arise out of and in the course of the employment. Arising out of addresses the causal connection to the work; in the course of addresses time, place, and circumstance. The doctrines that populate that boundary are the ones every claims professional knows and every technical witness should: the going-and-coming rule and its exceptions for special errands, dual-purpose trips, and employer-furnished transportation; the premises and parking-lot rules; the personal-comfort doctrine; deviation and horseplay; the idiopathic fall; and, increasingly, the telework claim that asks where the premises are when the premises are a spare bedroom. The standard is no-fault and generally liberal, but it is not the recordkeeping standard, and an entry on the log neither establishes compensability nor forecloses a contest of it.

The phrase does decisive work outside the compensation forum as well, and in a direction that surprises counsel who have only ever used it defensively. In *Schneider Electric USA, Inc. v. Williams*, decided by the Supreme Court of Kentucky on March 19, 2026, the employer argued that a household member's asbestos disease fell within the compensation bargain because she had briefly worked at the plant herself. The court held that the compensation scheme cannot convert non-occupational exposure into employer immunity: where the injury did not arise out of or in the course of the employment, the exclusivity provision supplies no shield, and the experts on both sides agreed the disease came from decades of household contact rather than from the summer job. The same six words that channel a worker's claim into the compensation system left the employer fully exposed in tort to a plaintiff whose exposure happened at home. Chapter 20 takes up that case and the docket it belongs to.

The three standards are best held in the mind at once. Recordability asks whether work caused or contributed, is decided by the employer at the time, and produces a document. Compensability asks whether the injury arose out of and in the course of employment, is decided by the compensation forum, and produces a benefit or a denial. Tort causation asks whether the occupational exposure was a substantial factor on a dose the record can support, is decided by a court applying Rule 702, and produces a judgment. Each is a higher bar than the one before it, which yields the practical rule counsel should carry into every intake: an entry on the log does not establish compensability, a compensable claim does not establish tort causation, and — the point most often missed — the absence of an entry establishes the negative of none of them.

III. Reconstructing the Non-Occupational Exposure

The alternative cause the defense invokes and the plaintiff must survive is not a suggestion. It is an exposure estimate, and it is subject to every discipline Chapter 16 imposed on the workplace one. This is the reconstruction the literature of this field has largely left unwritten: the reader has been taught to rebuild a dose from a job that ended twenty years ago, and has been told that the unexamined alternative cause is grounds for exclusion, but has never been shown how a hygienist actually characterizes the woodshop, the kiln, the renovation, or the ambient background so that the rule-out survives Rule 702. It is the same method turned ninety degrees.

A. The Symmetry Principle

The governing rule can be stated in one sentence: the non-occupational estimate is held to the standard the occupational estimate is held to, by whichever party offers it. For the plaintiff, the consequence is the one *Pluck* delivers: the expert who cannot reliably exclude the alternative source has an inadmissible opinion however well the workplace exposure was reconstructed. For the defense, the consequence is less often noticed and at least as sharp: an alternative cause is an affirmative technical proposition, and asserted without characterization it is the ipse dixit of General Electric Co. v. *Joiner*, 522 U.S. 136 (1997), wearing the defense's colors. The expert who testifies that the plaintiff kept a home woodshop, offers no task, no material, no hours, no ventilation, and no era, and concludes that the disease is therefore attributable to it, has done precisely what the reconstruction chapters condemn, with the direction of the advocacy reversed.

The asymmetry is itself the vulnerability, and it is visible on the face of a report. An opinion that reconstructs the workplace dose to three significant figures across four job titles and two decades, then disposes of twenty years of avocational exposure in a subordinate clause, has announced that the method was selected for the answer rather than brought to the question. It is the pattern the Introduction described, assumptions that all lean one direction, and it is the pattern the amended Rule 702 instructs courts to look for.

B. The Source Inventory Comes Before the Method

Completeness is the first failure point, so the inventory is built before it is known which entries help. The categories are stable enough to work from a fixed protocol: prior and concurrent employment, including moonlighting, self-employment, seasonal work, and military service; avocational sources — woodworking, ceramics and kiln firing, stained glass and soldering, indoor firing ranges, automotive restoration and brake work, casting and forge work, and art materials; domestic renovation and maintenance, which reaches disturbance of lead paint in pre-1978 housing, asbestos in flooring, joint compound, and pipe insulation, and respirable silica from concrete, masonry, and tile cutting; consumer products, particularly the solvent-bearing strippers, adhesives, and cleaners whose historical formulations bear no resemblance to their current ones; the household environment, including active and passive tobacco use, wood and coal heat, radon, well water, and attached garages; and community background — proximity to a permitted industrial source, a traffic corridor, or a regional airshed with a history of its own. The exposure a worker carries home to someone else belongs on the same list, and Chapter 20 takes it up as its own subject.

The discipline is that the inventory is compiled from the protocol rather than from the theory of the case. A source list containing no entry adverse to the retaining party is not a finding; it is a tell, and it is the first thing a competent opponent will count.

C. The Method Is the Same Method

The architecture is the task-based reconstruction of Chapter 16: frequency, intensity, and duration, assembled task by task and period by period, with the assumptions stated and the uncertainty carried forward. What differs is the evidentiary base, and the difference is severe: there is no employer, no sampling program, no written exposure control plan, and no thirty-year record file under 29 C.F.R. § 1910.1020 to anchor the estimate.

Two consequences follow. The first is that history-taking becomes the primary instrument, and it is self-reported evidence gathered in litigation, subject to recall failure and to incentive in both directions: the plaintiff who genuinely cannot recall how many weekends the shop ran, and the plaintiff who recalls it as fewer. Structured elicitation is the available answer: task, material, method, frequency, duration, enclosure and ventilation, respiratory protection, and era, asked source by source rather than gathered from an open-ended question about hobbies. The second is that documentary anchoring matters more, not less, than in the workplace reconstruction. Purchase and tool receipts, building permits and contractor invoices, photographs and home video, club and range membership records, prior employers' own § 1910.1020 exposure records — which prior employers owe for the same thirty years — prior compensation claims, and military service and veterans' exposure files are all obtainable, and Section VI develops them as discovery targets for each side. The rule of practice is to anchor every self-reported frequency to a document where one exists, and to say plainly, in the report, where none does.

D. The Published Data the Estimate Runs On

The estimate is defensible only if its inputs are published, cited, and independently checkable, and the sources exist. Activity patterns, time-use, and inhalation rates by age and activity are compiled in the Environmental Protection Agency's Exposure Factors Handbook. Community background at a given residence and era can be approached through the agency's ambient monitoring networks and its air toxics screening assessments, and through state agency monitoring where it was conducted. Population body burden — the comparator that answers whether this plaintiff's measured burden is distinguishable from the general population's at all — is available through the Centers for Disease Control and Prevention's National Report on Human Exposure to Environmental Chemicals and the biomonitoring series behind it. Hobby and small-shop exposures have been characterized directly in NIOSH Health Hazard Evaluations, which have sampled indoor firing ranges, art and ceramics studios, and small woodworking operations and supply the task-level data a residential exposure otherwise lacks, alongside a peer-reviewed avocational literature that is thinner than the occupational one but is not empty. Period-correct safety data sheets and formulation records establish what the consumer product actually contained in the year it was used.

Where no measurement exists for the source at issue, the mathematical models the profession already uses for workplace estimation apply without modification to a garage, a basement shop, or a spare room: the near-field and far-field treatments take a generation rate, a room volume, and an air-exchange rate and produce a concentration with a stated uncertainty. The models are published, the inputs are cited, and the arithmetic can be re-run by the opponent, which is the whole of what Rule 702 asks. Chapter 15's lesson governs the output with more force here than anywhere else in the book: the workplace estimate at least had three samples, and the non-occupational estimate usually has none, so the obligation to state what the data supports rather than what it suggests is correspondingly heavier.

E. The Same Five Failures, in the Other Direction

Chapter 16's catalogue of reconstruction failures transfers intact, and naming each in its non-occupational form is the fastest audit available to either side. The ipse dixit estimate is the source asserted without characterization, he was a woodworker, carrying a causal conclusion no stated method produced. Misclassification through grouping is the borrowed time-weighted average from a production cabinet shop applied to a weekend hobbyist, when the intensity, the duration distribution, and the population are all different. Anachronism runs in both directions: the modern low-solvent consumer formulation applied to a 1985 exposure, the current ambient background applied to a 1970s community, or the historical formulation applied to a product bought last year. Cherry-picked inputs are the inventory of subsection B assembled from the theory rather than the protocol. And false precision is the modeled residential concentration carried to two decimals from a deposition memory of a few weekends a year.

The disclosure obligation closes the method. Rule 26(a)(2)(B) reaches the facts and data considered, which includes the sources considered and rejected, and the reasoned rejection stated in the report is stronger in every forum than the silent omission discovered in deposition.

F. Why This Serves Both Chairs

For the defense the use is obvious: an alternative cause characterized to this standard is the rule-in the exclusion decisions require, and it is the difference between an argument and an opinion.

For the plaintiff the use is counterintuitive and routinely missed. The plaintiff's own affirmative reconstruction of the non-occupational sources is the strongest available form of the rule-out. A source quantified and shown to sit an order of magnitude below the occupational dose is foreclosed as a defense in a way that a source never mentioned never is; it supplies the completeness the differential etiology requires; and it removes from the cross-examiner the single most effective question in this docket, which is not whether the woodshop caused the disease but why the expert never looked. Occasionally the plaintiff's interest runs further still, to ruling the alternative source in rather than out: a real but minor non-occupational contribution, honestly characterized, is the predicate for the apportionment Section IV takes up, not a concession against it.

G. A Worked Structure: Hearing Loss

Noise-induced hearing loss is the textbook application of everything above, and it is worth running through because its alternative causes are unusually well characterized — which cuts in both directions and is the reason the symmetry principle of Section III.A is easier to demonstrate here than anywhere else in the book.

The source inventory has four standing entries beyond the occupational exposure, and each of them is quantifiable rather than merely nameable. The first is age. Presbycusis is not a competing hypothesis; it is a measured population phenomenon with published median and percentile hearing-threshold distributions by age and sex, and the same tables that OSHA permits an employer to use for age correction under Appendix F of § 1910.95 are available to either expert for the causation analysis. The second is recreational and non-occupational noise: firearms use, which produces a characteristic asymmetry — greater loss in the ear opposite the trigger hand, the left ear in a right-handed shooter, because the other is shielded by the head — and is the single most common non-occupational entry; power tools and shop work; motorcycles, small aircraft, and racing; amplified music by attendance and by personal listening device. Each of these has published exposure literature, and each can be reconstructed by the same frequency-intensity-duration protocol used for the workplace.

The third is ototoxicity, which is the entry most often missed by both sides. Aminoglycoside antibiotics, platinum-based chemotherapeutics, high-dose salicylates, loop diuretics, and certain solvents — toluene, styrene, carbon disulfide, and carbon monoxide all affect the auditory system, and several of the solvents are themselves occupational, which means the inventory can produce an occupational alternative cause that the plaintiff's own reconstruction did not consider and that helps the plaintiff. The fourth is the set of non-noise medical and traumatic causes: head injury, acoustic neuroma, Meniere's disease, sudden sensorineural loss, chronic otologic disease, and family history of early loss.

What makes hearing loss the cleanest illustration of the chapter's method, though, is the audiogram itself. Unlike the diseases of Chapters 16 and 17, this injury produces a serial quantitative record of the outcome across the working life. That record does two things no other case in this book offers. It permits the exposure reconstruction to be tested against the observed progression rather than merely asserted — published risk models predict expected threshold shift as a function of noise immission level and age, and an opinion whose reconstructed dose predicts a loss the audiograms do not show has a problem the report should address before the other side does. And it dates the injury, which is frequently the whole case: a claimant whose loss was already established at the pre-placement audiogram did not acquire it at this employer, and a claimant whose thresholds were stable for eleven years and then moved sharply has a question to answer about what changed, in the workplace or outside it.

The five failures of Chapter 16 §V transfer here without translation. The ipse dixit estimate is the defense expert who attributes the loss to "age and recreational noise" without an age-adjusted expected threshold or a single quantified recreational source. Misclassification is the plaintiff's expert who assigns a departmental dosimetry average to a worker whose task-time analysis places him in the loud bay four hours a week. The anachronism is the application of the 1983 hearing conservation requirements to a 1974 work period. The cherry-pick is the report that reproduces the two audiograms bracketing the shift and omits the eleven in between. And false precision is the attribution of a specific decibel quantum to occupational exposure when the underlying models support a range. An expert who runs both the occupational and the non-occupational reconstruction to the same standard has produced the only kind of apportionment opinion that survives, and, as Section III.F says of every agent, has produced it for whichever chair retained her.

IV. Apportionment Across Sources

Chapter 14 introduced apportionment as a division among employers and agents, and named the frameworks that decide what the reconstruction must be built to prove: the last-injurious-exposure rule, the several-liability regime apportioning by contribution, and the joint-and-several regime. Adding the non-occupational share changes the analysis, because that share is usually not a share at all in the legal sense. No party stands behind the woodshop. In most tort regimes the non-occupational contribution operates not as an allocated portion of liability but as a subtraction from the plaintiff's proof: it is the reason the occupational dose may fail to be substantial, or the reason the differential etiology may fail to close. In the compensation forum it reappears in a different guise, as the apportionment between a compensable aggravation and the underlying non-occupational condition. Counsel should know which of those two operations the jurisdiction is actually performing before instructing an expert to apportion anything.

Two technical traps recur. The first is the linear split of a non-linear interaction. Where agents combine multiplicatively rather than additively, the interaction of asbestos and smoking on lung-cancer risk being the canonical example, there is no defensible way to assign a percentage to each contributor, because the excess risk is a product rather than a sum, and the expert who reports a clean sixty-forty division has produced a number the epidemiology does not contain. The honest treatment states the interaction, states that the contributions are not separable in the way the question assumes, and gives the court the bounded statement the science supports. The second is the attributable fraction pressed into individual service. The population attributable fraction derived from a relative risk describes what share of disease in a group is attributable to an exposure; converting it into a statement about this plaintiff imports the same crude inference as the relative-risk-above-two heuristic Chapter 14 flagged, and while opponents will invoke it, the expert who uses it should say what it is.

The signature-versus-multifactorial distinction from Chapter 14 governs how much work the non-occupational source can do. In the signature case — silicosis, asbestosis, mesothelioma — the diagnosis carries most of the general-causation freight and the alternative sources fight over share. In the multifactorial case — lung cancer, leukemia, the neurological syndromes — the background rate is itself the defense's alternative explanation, and characterizing the non-occupational sources is not a subsidiary exercise but the central one. And the substantial-factor framework of *Rutherford v. Owens-Illinois, Inc.*, 16 Cal. 4th 953 (1997), operates on the same material from the other side: an aggregate dose assembled from occupational and non-occupational contributions can still support a finding that the defendant's share was substantial, which is why the plaintiff's expert who characterizes the other sources honestly is frequently strengthening the case rather than conceding it.

V. What the Cases Teach

Pluck is the decision to read first, and to read for its facts as well as its holding. The plaintiff's specific-causation expert was excluded because he formulated a dose opinion without exposure data, having been told only that the plaintiff was heavily exposed to benzene in her water; because he could not explain the calculations behind the dose; because he relied on a no-safe-dose theory other courts had already rejected as a basis for specific causation; and because he did not adequately account for the plaintiff's other sources of benzene exposure in a record where the court noted that the compound is ubiquitous in ambient air and present in vehicle exhaust and tobacco smoke. Two failures, one discipline: the agent could not be reliably ruled in without a dose, and the alternatives could not be ruled out without characterizing them. The plaintiff who reconstructs only the exposure she is suing over has done half the work the method requires.

Tamraz v. Lincoln Electric Co., 620 F.3d 665 (6th Cir. 2010), supplies the ruling-in half in isolation: an opinion excluded because the step that brought the occupational agent into the differential rested on hypothesis rather than established science. *Joiner* supplies the rule that governs both chairs equally, and it is worth restating in this context because it is so often deployed in one direction only: an opinion is not admissible merely because a qualified witness says it, and nothing in that principle exempts an alternative cause offered by the defense. The defense expert who names the hobby and stops has offered the court exactly the analytical gap *Joiner* describes.

Schneider Electric USA, Inc. v. Williams teaches the boundary lesson from Section II in its most consequential form. The employer lost the immunity it assumed it had because the injury at issue did not arise out of or in the course of the employment, and the question of where the disease came from, which the parties had prepared as a causation question, decided the availability of the defense before causation was reached. That is the pattern of this entire chapter: the boundary question is answered with exposure evidence, and the side that has done the exposure work owns the answer.

VI. Discovery Targets and Expert Assignments

For counsel developing or attacking the non-occupational side, the record exists and is largely obtainable. Prior employers owe the same thirty-year retention of exposure and medical records under § 1910.1020 that the defendant employer owes, and their monitoring files, written exposure control plans, and injury logs are reachable by subpoena; prior compensation claims and their medical records establish both exposure history and prior characterizations of causation; military service and veterans' exposure registry records cover a period of many plaintiffs' lives that civilian discovery misses entirely. On the avocational and domestic side, the anchoring documents are building permits and contractor invoices, tool and material purchase records, homeowner's insurance and renovation claims, club and range membership records, and photographs and home video, with the residence history itself supplying the addresses and dates the ambient background analysis needs. The treating medical records carry the social, smoking, and occupational histories the plaintiff gave when no one was litigating, which is precisely what makes them valuable to both sides.

The expert assignments follow the same symmetry the method does. The industrial hygienist reconstructs both the occupational and the non-occupational exposures, to one standard, and states the uncertainty of each; the assignment is written that way in the engagement letter, before anyone knows which estimate will help. The physician performing the differential etiology receives both reconstructions and must list and dispose of each source on the record, because the completeness of the list is the element the exclusion decisions test. Where a consumer product or historical formulation is at issue, someone must do the era-correct research rather than assume the current label describes the 1985 can. And counsel on both sides should resolve at intake which of the two apportionment operations of Section IV the jurisdiction actually performs, because that decision determines what the reconstruction must be built to prove.

The instruction that governs all of it is the one this chapter began with. The source inventory is compiled from a protocol before it is known which entries help, the estimate for each source is built and disclosed to the standard the workplace estimate is held to, and the rejections are stated with reasons. An opinion assembled that way survives the motion the opponent has already drafted. An opinion that reconstructs one side of the line with rigor and waves at the other has told the court which answer it was retained to reach.

References

29 C.F.R. § 1904.5 (determination of work-relatedness).

29 C.F.R. § 1910.1020 (access to employee exposure and medical records).

29 C.F.R. § 1910.95 (occupational noise exposure; hearing conservation program requirements are annotated in full in the Chapter 15 reference list).

American Industrial Hygiene Association. Mathematical models for estimating occupational exposure to chemicals (2d ed.).

Centers for Disease Control and Prevention. National report on human exposure to environmental chemicals. U.S. Department of Health and Human Services.

Federal Judicial Center. (2011). Reference manual on scientific evidence (3d ed.). National Academies Press.

Fed. R. Civ. P. 26(a)(2)(B); Fed. R. Evid. 702.

Gen. Elec. Co. v. Joiner, 522 U.S. 136 (1997).

National Institute for Occupational Safety and Health. Health hazard evaluation program reports. U.S. Department of Health and Human Services.

National Institute for Occupational Safety and Health. (2018). Preventing hearing loss caused by chemical (ototoxicity) and noise exposure (DHHS (NIOSH) Publication No. 2018-124). U.S. Department of Health and Human Services.

International Organization for Standardization. (2013). ISO 1999: Acoustics — Estimation of noise-induced hearing loss. ISO. (Adopted in the United States as ANSI/ASA S3.44.)

Pluck v. BP Oil Pipeline Co., 640 F.3d 671 (6th Cir. 2011).

Rutherford v. Owens-Illinois, Inc., 16 Cal. 4th 953 (1997).

Schneider Electric USA, Inc. v. Williams, No. 2023-SC-0436-DG (Ky. Mar. 19, 2026).

Tamraz v. Lincoln Elec. Co., 620 F.3d 665 (6th Cir. 2010).

U.S. Environmental Protection Agency. Exposure factors handbook. Office of Research and Development.

PART IV

THE INJURY CASE FILE

CHAPTER 19

The Jobsite Is Not Immune

Escaping the Exclusive Remedy, and the Technical Record the Third-Party Case Adds

Scope note. The exclusive-remedy analysis in this chapter applies to land-based employment. Maritime employment under the LHWCA and the Jones Act, and railroad employment under FELA, are governed by separate schemes and are outside the scope of this book; see the Introduction. Chapter 6 set out the method: the physics, the standards matrix, the hierarchy of controls, the leading-indicator record, and the discipline of running all of it before knowing which way it points. This chapter does not repeat that method. It takes up what the third-party posture adds to it: the exclusive-remedy escape routes and what each demands technically, the products-liability overlay and the machine-safety standards that decide guarding cases, the empty chair the immune employer leaves on the verdict form, the staffing agency whose worker belongs to two employers at once, and the role the safety professional and the industrial hygienist play inside the compensation system rather than around it. The method is Chapter 6's. The obstacles are this chapter's.

I. What the Third-Party Posture Adds

The exclusive-remedy provision of every state's workers'-compensation act is the quiet architect of workplace-injury litigation. It bars the injured worker's tort suit against the employer in exchange for no-fault benefits, and in doing so it redirects nearly every serious workplace-injury claim toward the parties the bar does not reach: the general contractor that ran the site, the owner that hired the work, the subcontractor whose crew created the condition, the staffing agency, the manufacturer of the press or the lift, the supplier of the solvent, the maintenance firm that serviced the machine last.

The technical work in those cases is the work Chapter 6 describes, and this chapter assumes it. The standards matrix is built the same way, the hierarchy of controls is run the same way, the leading-indicator record is read the same way, the preservation obligations attach on the same day and are treated at length in Chapter 21, and the amended Rule 702 polices the resulting opinion under the same standard. An expert who has internalized Chapter 6 already knows how to build a third-party opinion.

What the third-party posture adds is a set of obstacles that do not exist in the two-party case, and every one of them is answered with technical evidence rather than argument. Four of them organize this chapter. The compensation bar must be escaped, and each escape route imposes a distinct evidentiary burden. Where the defendant is a manufacturer rather than a site party, a products-liability overlay applies that the site standards do not reach: a design file, a body of machine-safety standards, and an alternative design that must be engineered rather than named. The immune employer's share of the fault does not disappear; in most states it is argued to an empty chair, and in some it is not argued at all, which changes the theory before the complaint is drafted. And on a large and growing fraction of sites the injured worker has two employers at once, which can pull a nominal third party back inside the bar the plaintiff was trying to escape.

There is also a fifth thing worth saying at the outset, because the book has otherwise treated the compensation system as an obstacle to be escaped: the safety professional and the industrial hygienist have substantial work to do *inside* that system, and for a large share of the second audience it is where they will actually encounter litigation. Section VII takes it up.

One structural note before the obstacles. The party with the best records, the employer, is usually not a party at all, which means the most important custodian in the litigation is reached by subpoena and deposition rather than by party discovery, and neither side controls what it preserved. Chapter 6 §III's inventory of demands applies, with one addition that belongs at the top of the plaintiff's list and the defense's alike: the non-party employer's file, subpoenaed early, and the deposition of its safety personnel taken while institutional memory is intact. The employer's investigation, its root-cause analysis, its comp file, its monitoring data under 29 C.F.R. § 1910.1020, and its equipment maintenance history are the spine of the case, and they are held by someone with no stake in either side's theory and no automatic duty to preserve anything for either.

II. The Escape Routes, and What Each One Costs to Prove

The compensation bar is not one rule but a family of them, and the routes around it differ enough that the choice of route determines the expert's assignment. Counsel should resolve which routes are available in the governing jurisdiction at intake, because the answer determines what the technical record has to establish.

A. The Third-Party Route

The ordinary route is not an exception at all. The compensation acts bar the action against the employer and preserve it against everyone else, and they say so expressly (Chapter 11 §V collects the statutes). Nothing needs to be proved to use this route except that the defendant is not the employer, which is why Section V matters, since the borrowed-servant and statutory-employer doctrines exist precisely to make that question contestable.

B. Intentional Tort and Substantial Certainty

Where the plaintiff wants the employer itself in the case, the route runs through the intentional-tort exception, and it is narrow by design. The states cluster into three groups, and the group determines whether the case is viable at all. A minority require a deliberate intent to injure this worker, a standard almost never met outside assault. A larger group applies a substantial-certainty test, asking whether the employer knew the injury was substantially certain to follow from the conditions it maintained. A handful have codified intermediate standards by statute.

The substantial-certainty showing is the one that turns on technical evidence, and Chapter 6 §IV describes how it is built: a specifically identified hazard, dated management knowledge, feasible controls identified by the employer's own program and by consensus standards, and a decision renewed shift after shift to continue the exposure unchanged. The verdict that anchors Chapter 6 was returned on exactly that construction. What deserves emphasis here is the negative: adjectives do not meet this standard, and neither does the severity of the injury. An expert asked to support a substantial-certainty theory should say plainly, and early, when the leading-indicator record will not carry it, because the theory pleaded and lost costs the plaintiff the tolling, the settlement posture, and frequently the credibility of the rest of the case.

C. Dual Capacity

The dual-capacity doctrine allows suit where the employer occupied a second legal persona independent of the employment relationship: most commonly as the manufacturer of the product that injured its own employee, and occasionally as a landowner or a medical provider. It survives in a minority of jurisdictions and has been narrowed or abolished by statute in several. Where it lives, the technical question is precise and worth naming: was the product the employer manufactured placed in the stream of commerce for general sale, or was it built for the employer's own use? An in-house machine, fabricated by the maintenance department for the plant's own line, is generally the employer acting as employer. The same machine sold to twenty other plants is the employer acting as manufacturer, and the design file of Section III becomes discoverable against the employer directly.

D. Fraudulent Concealment and the Uninsured Employer

Two smaller routes deserve a line each because they turn on documents the safety professional generated. A number of states recognize an exception where the employer fraudulently concealed a known injury or a known exposure and thereby aggravated it: a theory that reaches directly into the industrial hygiene file, because the concealed thing is usually a monitoring result, a medical surveillance finding, or a hazard-communication failure, and the § 1910.1020 access rule supplies the duty the concealment breached. And in every state the employer who failed to secure compensation coverage loses the bar entirely, which converts an ordinary negligence case into one where the employer's own safety record is fully in play.

III. The Products-Liability Overlay

When the defendant is a manufacturer, the analysis shifts onto ground the site standards do not cover, and this is where the book has the most to add, because the technical literature on guarding is largely invisible to lawyers and the legal literature on design defect is largely invisible to safety professionals. A substantial fraction of third-party injury cases are guarding cases. They are decided on a body of standards, and on a design record, that a site-safety analysis never touches.

A. The Machine-Safety Standards Matrix

The regulatory floor is OSHA's, and it is thin: the general machine-guarding requirement of 29 C.F.R. § 1910.212, the mechanical-power-press standard at § 1910.217, the specific-machinery provisions of Subpart O, and the energy-control standard at § 1910.147. Those provisions bind the *employer*, they are written in performance terms, and they predate the entire modern discipline. They are almost never sufficient to resolve a design case.

The operative standards sit above them, and an expert in a guarding case should be fluent in all of them. ANSI B11.0 supplies the risk-assessment methodology for machinery — hazard identification, risk scoring by severity and probability of occurrence, risk reduction through the hierarchy, and residual-risk documentation — and the B11 series carries machine-specific standards beneath it (B11.1 for power presses, B11.19 for the performance criteria applicable to safeguarding devices, B11.20 for integrated manufacturing systems, and so on). ISO 12100 is the international companion, stating the general principles for design and risk reduction. ISO 13849-1 and IEC 62061 govern the functional safety of the safety-related control system — the question of how reliably the guard, interlock, light curtain, or two-hand control actually performs its safety function, expressed as a required Performance Level (PLa through PLe) or Safety Integrity Level. NFPA 79 governs the electrical installation of industrial machinery. And ANSI/ASSP Z244.1 addresses control of hazardous energy in terms considerably more current than § 1910.147.

Why this matters to a lawyer, in one paragraph: the modern standards do not ask whether a guard was present. They ask whether the manufacturer performed a documented risk assessment, what hazards it identified, what risk-reduction measures it selected, why it selected them, and whether the safety-related control system was designed to a performance level appropriate to the risk it was mitigating. Each of those questions has a paper answer that exists or does not exist in the manufacturer's file. A defense expert who can produce a B11.0 risk assessment, a stated Performance Level with the calculation behind it, and a validation record has a strong case that the design met the state of the art. A plaintiff's expert who finds no risk assessment at all — which is common, particularly with older equipment and with machines built by small integrators — has found the fracture, and it is a fracture in the manufacturer's own documented process rather than in the expert's opinion about what a guard should have looked like.

The reverse is equally true and belongs in any honest treatment. Where the risk assessment exists, identifies the hazard that injured the plaintiff, and documents a rational basis for the safeguarding actually provided — and where the incident occurred because a downstream party defeated the interlock, removed the guard, or bypassed the light curtain — the design case is over and the case is about the employer's conduct, which is where the compensation bar puts it.

B. The Design File and What to Demand

The discovery follows the standards. For equipment, the demands are: the hazard analysis and risk assessment for the machine and, critically, for the specific configuration sold; the guarding, interlock, and safeguarding options considered, offered as options, standard, or declined, together with the pricing and specification sheets that show what the purchaser was offered; the functional-safety calculations and validation records for the safety-related control system; the testing and standards-compliance documentation, including any third-party certification; the drafting history of the warnings, labels, and manuals, which is where a warnings case is actually decided; service bulletins, retrofit kits, and field modification instructions, which carry post-sale knowledge; and the prior-claims, complaint, and warranty file, which carries the notice question in a products case the way near-miss data carries it in a site case.

Two of those deserve particular emphasis because they are routinely under-demanded. The options file answers the most common defense in a guarding case — that a guard was available and the purchaser declined it — and it answers it with a document rather than a recollection. And the retrofit record answers the post-sale duty question, which several jurisdictions recognize and which turns on what the manufacturer learned after the sale and what it did about it.

C. The Alternative Design Must Be Engineered, Not Named

Most jurisdictions require a plaintiff pursuing a design-defect theory to establish a reasonable alternative design, and the appellate record is unforgiving about what that requires. *Sardis v. Overhead Door Corp.*, 10 F.4th 268 (4th Cir. 2021), is the cautionary case and is treated at Section VI: a multimillion-dollar verdict vacated because the design opinion rested on standards that did not apply and testing that was never done. The discipline the decisions demand is concrete. The alternative must have existed and been feasible at the time of manufacture, not at the time of trial, which makes the dating of the standards, the catalogs, and the competitor equipment a research task rather than an assumption. It must be shown to prevent the mechanism that actually injured the plaintiff, which requires the incident reconstruction Chapter 6 §II.D describes. It must not introduce a greater hazard or defeat the machine's function, which is an engineering question the plaintiff's expert should answer before the defense asks it. And where the discipline permits, it should be tested, modeled, or demonstrated on an exemplar rather than asserted — because *Sardis* is what an untested alternative-design opinion is worth on appeal.

D. The Chemical Supplier

Where the product is a substance rather than a machine, the overlay shifts but the logic holds. The demands are the formulation and its change history; the safety data sheet revision history and the hazard-determination file behind it, which is the supplier's own documented reasoning about what its product does and what it was required to say about it under the hazard communication standard; downstream-user communications, including technical bulletins and industrial hygiene guidance the supplier issued, or considered and withheld; and the prior-claims file. The sophisticated-user and learned-intermediary defenses live in those documents, and so do their answers.

IV. The Empty Chair

In most jurisdictions the defense will ask the jury to assign a share of fault to the immune employer, and the verdict form will let them. In some it will not. This is not a detail to be resolved at the charge conference; it changes the plaintiff's theory, the defense's expert assignment, and the settlement value, and it belongs in the intake analysis alongside the applicable-standard determination of Chapter 12.

The states fall into three groups. A number permit allocation of fault to the immune employer as a nonparty, so that the named defendants pay only their own shares and the plaintiff simply loses the employer's percentage. A number bar it outright, on the reasoning that a party immune from liability cannot be assigned fault, with the consequence that the named defendants absorb the employer's share: a rule that makes the site defendants' exposure substantially larger and their incentive to settle correspondingly greater. And a number permit it only where the nonparty has been formally identified within a statutory deadline, which is a trap for defense counsel who assume the argument is always available. Where the employer's negligence is severe and the site defendants' involvement is peripheral, the difference between the first rule and the second can be the whole case.

The technical consequence for the plaintiff's expert is the one Chapter 6 §IV's discipline anticipates but the third-party case sharpens: generic safety failure drains toward the employer's chair. An opinion that says the site was badly run has handed the defense its allocation argument. The opinion that survives the empty chair isolates the failures that belonged specifically to the named defendants — the rungs of the hierarchy only they could reach, the design choices only the manufacturer made, the sequencing only the controlling contractor could have changed, the knowledge only they held — and concedes candidly what belonged to the employer. The concession is not a weakness. It is what makes the allocation the expert does argue for credible.

The technical consequence for the defense is the mirror, and it is the harder discipline. The employer's conduct is usually the largest single causal factor on the board, and the temptation is to point at the chair and sit down. Pointing is not evidence. What persuades a jury to assign a large share to an absent party is a sequence analysis: which decisions, made by whom, on what day, actually produced the exposure, and which of them no site defendant could have seen or scheduled around. Absent that work, the plaintiff's answer is the hierarchy of controls — that the parties above the employer held the more effective rungs and used none of them — and it is a good answer.

V. Two Employers, One Worker: The Borrowed Servant

A large and growing share of the workforce on any industrial site is employed by someone other than the entity directing its work, and the resulting doctrine can dissolve a third-party case that looked secure at intake.

The general rule in most states is that a temporary worker has two employers — the staffing agency that hired and paid her (the general employer) and the host that directs her work (the special or borrowed employer) — and that both enjoy the compensation bar where the elements of the borrowed-servant relationship are satisfied. The elements vary by state but cluster around a familiar set: whether the worker consented to the special employment, whether the host controlled the details of the work, whether the work performed was essentially the host's, and whether the host had the right to discharge her from the assignment. Some states apply a test weighted almost entirely to control; some ask whether there was an implied contract of hire; a few permit the parties to allocate the status by written agreement.

The practical consequence is that the host manufacturer or contractor the plaintiff sued as a third party may not be a third party at all. Counsel on both sides should therefore treat the staffing arrangement as a discovery target in its own right, and the documents are specific: the staffing agreement itself and any clause purporting to allocate employer status, workers'-compensation responsibility, or indemnity; the client service agreement's description of supervision; the worker's own onboarding, orientation, and assignment records; the host's training records for the worker, which frequently prove the control the host would prefer to deny; the timekeeping and supervision records; and the certificate of insurance identifying which carrier accepted the risk.

For the safety professional, this is one of the places where ordinary program documentation has consequences no one anticipated when it was created. OSHA's temporary-worker enforcement position is that the staffing agency and the host are joint employers with allocable responsibilities, that the host generally owns the site-specific hazard training and the day-to-day conditions, and that the agency owns the general training and the duty to inquire into the conditions its workers will face. A host that documents site-specific training, hazard communication, and supervision is discharging a real safety obligation and simultaneously building the control record that may confer the compensation bar on it. A host that documents none of it has neither.

VI. What the Cases Add

Chapter 6 §VI collects the decisions on the shared method: the Rule 702 gate, the spoliation framework, the one-sided opinion. Chapter 21 collects the preservation cases. What follows are the three decisions that speak specifically to the obstacles above.

Sardis v. Overhead Door Corp. (4th Cir. 2021), the design verdict that evaporated at the gate. A worker was killed unloading a commercial door when the wooden handhold of a shipping crate failed; the jury returned a multimillion-dollar wrongful-death verdict against the manufacturer on design and warning theories, and the Fourth Circuit vacated it. The plaintiff's engineering and warnings experts should never have reached the jury: the design opinion rested on standards that did not apply and testing that was never done, the warnings opinion on the expert's ipse dixit, and the district court had abdicated its gatekeeping by leaving reliability to cross-examination. Without the excluded opinions the claims could not stand. The instruction runs to both chairs. For the plaintiff, an alternative-design opinion must be engineered — tested, tied to the standards that actually govern the machine, quantified — because a verdict built on an inadmissible opinion is a verdict held on appeal bond. For the defense, the Rule 702 challenge must be briefed, ruled on, and preserved, because *Sardis* is precisely what the preserved objection is worth.

Schmid v. Milwaukee Electric Tool Corp. (3d Cir. 1994), the exemplar disassembled and the sanction reversed. *Sardis* is about an opinion that should have been kept from the jury; this is its mirror, and the design chapter needs it because the machine in a guarding case usually has to be taken apart before anyone can say what failed. A worker's hand was severely injured when a circular saw kicked back while a coworker cut wooden forms, and the theory was that debris in the housing had kept the lower blade guard from returning to the closed position. The plaintiff's engineer photographed the saw and then disassembled it, altering the condition the defense wanted to examine, and the district court struck his testimony in its entirety and entered judgment for the manufacturer. The Third Circuit reversed. Striking the testimony was an extreme sanction, and the court set out the framework that still governs the choice: the degree of fault of the party who altered the evidence, the degree of prejudice to the opponent, and whether a lesser sanction will avoid substantial unfairness and, where the offending party is seriously at fault, deter repetition — the instruction being to select the least onerous sanction corresponding to the willfulness of the act and the prejudice actually suffered. Two points follow for the expert. Prejudice is structurally lower in a design case than in a manufacturing-defect or maintenance case, because the opponent can test another machine built to the same design; the sanction analysis is therefore theory-dependent, and an expert should know which kind of case she is in before she picks up a wrench. And the photographs are what saved the opinion. Chapter 21 and Chapter 22 state the protocol — document in place, notify the parties whose interests are known, agree or obtain a protocol, invite attendance — and *Schmid* shows what partial compliance with it is worth: enough to survive, at the cost of an appeal.

Hooker v. Department of Transportation (Cal. 2002) and *Sandoval v. Qualcomm Inc.* (Cal. 2021) — retained control, and the discovery that decides it. California's line is among the most developed statements of the question every site case asks. *Hooker* holds a hirer liable for a contractor's employee's injury only where the hirer retained control over the work and exercised it in a manner that affirmatively contributed to the injury; mere authority, unexercised, is not enough. *Sandoval* reversed a substantial plaintiff's verdict against a facility owner in an arc-flash case, holding that the owner had delegated the work and that its conduct did not amount to the affirmative contribution the doctrine requires. Read as an illustration of a category rather than a national rule — the retained-control doctrines vary, and a few states impose a non-delegable duty on the controlling contractor that *Hooker* would not — the two decisions define the discovery target: not the contract clause, but the field record of control in operation. The directives given, the inspections conducted, the stop-work authority exercised or withheld on the day in question. The plaintiff who builds that record survives the doctrine. The defendant whose file documents genuine delegation has the summary-judgment motion the doctrine was written to reward.

Amorgianos v. National Railroad Passenger Corp. (2d Cir. 2002), the exposure case lost inside the hygienist's spreadsheet. A bridge painter alleged neurological injury from xylene exposure over weeks of enclosed work. His industrial-hygiene expert set out to reconstruct the dose, used inputs the record contradicted, and left the calculation unfinished, while the medical experts relied on an exposure level the hygienist never established. The Second Circuit affirmed exclusion of the entire causation case and, with it, summary judgment: a reliable method unreliably applied is inadmissible, and every link in the exposure-to-disease chain must hold. *Amorgianos* belongs in this chapter rather than in Part III because of what it says about the third-party exposure case specifically. The chemical-supplier defendant is frequently reached only through an exposure reconstruction, and the reconstruction is frequently the only thing standing between the plaintiff and the compensation bar. A half-finished dose estimate does not merely weaken that case. It ends it.

VII. The Expert Inside the Compensation System

This book has treated workers' compensation almost entirely as a wall to be climbed. For a large part of its second audience that is exactly backward, because the compensation system is where the safety professional and the industrial hygienist most often encounter litigation, and the work there is substantial.

Occupational disease claims are the clearest case. A compensable occupational disease requires proof that the condition arose out of the employment, and in a state with a schedule of presumptive diseases the fight is over whether the claimant's exposure fits the presumption; in a state without one, it is a general-causation and exposure question conducted with less procedure and considerably less discovery than a tort case. The hygienist's assignment is the same reconstruction Chapter 16 describes, performed on a thinner record and against a shorter clock.

Apportionment among employers under last-injurious-exposure rules is the second. Where a disease develops across a career, most states assign liability to a single employer by rule — typically the last employer where the claimant was exposed to the hazard for a defined minimum period — and the technical question that decides which employer that is, and whether the exposure at that employer was capable of contributing to the disease at all, is an exposure-reconstruction question. It is frequently the entire case, and it is frequently litigated with no exposure expert on either side, which is a professional opportunity and, more to the point, a reason claims are decided wrongly.

Third, exposure assessment adjacent to the independent medical examination. The physician evaluating causation in a comp claim needs a dose to reason from and usually does not have one. Where a hygienist supplies a defensible exposure characterization — with its uncertainty stated, in the register Chapter 17 §VI demands — the medical opinion improves in both directions, and the claimant with a genuine occupational disease is better served than by a system in which the exposure is assumed by whoever asserts it most confidently.

Two cautions for the professional entering this work. The record is thinner, the procedure is looser, and the temptation to overstate what a sparse record supports is correspondingly greater; the discipline Chapter 15 describes for limited data is the discipline that belongs here. And an opinion given in a compensation proceeding is an opinion given under oath in a proceeding of record, which means it is discoverable, quotable, and reachable in the third-party case that may follow the same injury. The professional who has said one thing to the compensation judge and something else to the district court has supplied the cross-examination Chapter 4 describes, and no explanation of the different procedural postures will repair it.

VIII. What the Third-Party Case Asks of the Expert

The obstacles in this chapter are legal in form and technical in substance, which is the book's argument in its most concrete version. Whether the compensation bar is escaped depends on what the leading-indicator record shows about knowledge, or on whether a machine was built for sale or for the plant's own line. Whether the manufacturer defends the design depends on whether a risk assessment exists in its file and what Performance Level the safety-related control system was designed to. Whether the empty chair absorbs the verdict depends on whether the defense produced a sequence analysis or a gesture. Whether the host contractor is a third party at all depends on training records the safety department created for entirely different reasons. Not one of those questions can be answered from the pleadings, and every one of them is answered from documents that exist before anyone files anything.

That is why the retention timing this book recommends more often than any other recommendation applies with particular force here. The machine is repaired and returned to production. The scene is rebuilt by the next phase of the schedule. Site cameras overwrite on cycles measured in days. The subcontractor demobilizes and its daily logs scatter. And the employer — a non-party, with no automatic duty to preserve anything for anyone — completes its comp-driven investigation and moves on. An expert retained in the first weeks writes the preservation letter that names the systems, the exemplar, and the design file. An expert retained at the report deadline writes about what happens to be left.

The chapter closes on the point that distinguishes it from a defense manual or a plaintiff's manual. Every obstacle above cuts both ways. The substantial-certainty record that supports the plaintiff's escape from the bar is the same record that, honestly read, sometimes shows an ordinary industrial accident and no more. The design file that indicts one manufacturer exonerates the next. The empty-chair analysis that reduces a site defendant's exposure is the same analysis that, run first by the plaintiff, tells her which defendants are worth suing. The technical work does not know which of those results it will produce until it is finished, and the professional who finds out before counsel does, and says so, is the one worth the retention.

References

American National Standards Institute / American Society of Safety Professionals. ANSI B11.0, Safety of machinery — General requirements and risk assessment; ANSI B11.19, Performance requirements for risk reduction measures; ANSI B11.20, Integrated manufacturing systems; ANSI/ASSP Z244.1, Control of hazardous energy — Lockout, tagout and alternative methods.

American Society of Safety Professionals. ANSI/ASSP A10.33, Safety and health program requirements for multi-employer projects.

Amorgianos v. Nat'l R.R. Passenger Corp., 303 F.3d 256 (2d Cir. 2002).

Fed. R. Evid. 702 (as amended Dec. 1, 2023); Fed. R. Civ. P. 26, 30(b)(6), 34, 37(e).

Gen. Elec. Co. v. Joiner, 522 U.S. 136 (1997).

Hooker v. Dep't of Transp., 27 Cal. 4th 198 (2002).

International Electrotechnical Commission. IEC 62061, Safety of machinery — Functional safety of safety-related control systems.

International Organization for Standardization. ISO 12100, Safety of machinery — General principles for design; ISO 13849-1, Safety-related parts of control systems.

Machine guarding, 29 C.F.R. § 1910.212; mechanical power presses, § 1910.217; machinery and machine guarding generally, 29 C.F.R. pt. 1910, subpt. O; control of hazardous energy, § 1910.147; access to employee exposure and medical records, § 1910.1020; hazard communication, § 1910.1200.

National Fire Protection Association. NFPA 79, Electrical standard for industrial machinery.

Occupational Safety and Health Administration & National Institute for Occupational Safety and Health. Recommended practices: Protecting temporary workers (OSHA 3735) and the Temporary Worker Initiative guidance bulletins.

OSHA Multi-Employer Citation Policy, CPL 02-00-124 (Dec. 10, 1999).

Privette v. Superior Court, 5 Cal. 4th 689 (1993).

Restatement (Second) of Torts § 414 (1965); Restatement (Third) of Torts: Products Liability § 2 (1998).

Sandoval v. Qualcomm Inc., 12 Cal. 5th 256 (2021).

Sardis v. Overhead Door Corp., 10 F.4th 268 (4th Cir. 2021).

Schmid v. Milwaukee Elec. Tool Corp., 13 F.3d 76 (3d Cir. 1994).

Solis v. Summit Contractors, Inc., 558 F.3d 815 (8th Cir. 2009).

CHAPTER 20

The Exposure That Goes Home

Take-Home and Community Claims — When the Employer's Own Record Is Read by a Plaintiff Who Never Worked There

Chapter 19 was organized around an escape: the worker's claim is channeled into the compensation system, so the third-party case must find a defendant outside the employment relationship. This chapter inverts that structure. The plaintiff here never worked at the facility and was never counted in any exposure assessment the employer performed — she is the spouse who laundered the clothing, the child who hugged her father at the door, the neighbor a quarter-mile downwind, and her claim is proven or defeated with documents the employer generated for compliance purposes decades earlier.

I. The Inversion of the Exclusive Remedy

Chapter 19 taught the escape. The exclusive remedy bars the injured worker's tort suit against the employer, so the third-party case runs against the general contractor above, the premises owner, or the manufacturer of the machine. The take-home case inverts the structure rather than escaping it. The household plaintiff has no employment relationship to bar and no exclusivity provision to escape, because exclusivity reaches injuries arising out of and in the course of employment and this plaintiff's injury arose at home.

The Supreme Court of Kentucky put the point plainly in *Schneider Electric USA, Inc. v. Williams*, decided March 19, 2026, holding that the coverage available under the compensation act cannot convert non-occupational exposure into employer immunity: on a showing that the injury did not arise out of or in the course of the employment, the employer is not entitled to immunity, and exclusivity does not apply. The facts show why the analysis deserves to be performed rather than assumed. The decedent had herself worked briefly at the plant as a teenager, and the employer argued that this brought her entire indivisible injury inside the compensation bargain. The argument failed because the experts on both sides agreed that her mesothelioma came from decades of household contact with her father's work clothing rather than from her own summer employment. The exclusivity question, in other words, was decided by the exposure evidence: the same evidence the parties had assembled for causation, doing work at the threshold that neither side had planned for it.

The consequence is worth stating without softening: the employer that is immune from the worker it injured may face ordinary tort liability, uncapped and undiverted, from that worker's daughter, on the same facts, from the same dust, proven from the same records. Two qualifications keep the proposition accurate. First, the household member's claim is not freestanding in every sense — New Jersey's foundational decision treats the duty owed to the spouse as derivative of the duty owed to the worker, so that where no duty ran to the worker, none runs to the household. Second, the analysis turns on whose injury and whose employment: a worker's own exposure to his employer's materials at home has been held to remain within exclusivity where a substantial contributing cause still arose from the employment. The threshold question is factual and it is technical, which is the theme of everything that follows.

II. The Duty Split, and What It Actually Turns On

Whether the employer or premises owner owes any duty at all to a household member is the most jurisdictionally fractured question in this book, and counsel must resolve it at intake, because it decides whether there is a case rather than how large one is. The split is genuine and roughly balanced among the states that have reached it, and a substantial number have never addressed it.

The recognizing line reasons from foreseeability: if it was foreseeable that dust carried out on clothing would be inhaled by whoever laundered it, the ordinary duty of reasonable care extends to that person. California so held in *Kesner v. Superior Court*, 1 Cal. 5th 1132 (2016), decided with *Haver v. BNSF Railway Co.* and confining the duty to members of the worker's household. New Jersey reached the spouse who laundered the clothing in *Olivo v. Owens-Illinois, Inc.*, 186 N.J. 394, 895 A.2d 1143 (2006). Tennessee extended it to a worker's daughter in *Satterfield v. Breeding Insulation Co.*, 266 S.W.3d 347 (Tenn. 2008), and Kentucky joined that reasoning in Schneider Electric, bounding the duty to household members who came into regular, repeated, and close contact with contaminated work clothing over an extended period rather than to the general public. Alabama, Indiana, Louisiana, Utah, Virginia, and Washington have recognized comparable duties. Delaware, having once refused, changed course in *Ramsey v. Georgia Southern University Advanced Development Center*, 189 A.3d 1255 (Del. 2018): a reminder that the survey counsel relied on three years ago may no longer describe the jurisdiction.

The rejecting line reasons from relationship and from the limits of liability, holding that foreseeability does not by itself create a duty to persons outside the employment relationship, and declining to expose the employer to a class of claimants it cannot identify or protect. New York took that course in *In re New York City Asbestos Litigation (Holdampf v. A.C.&S., Inc.)*, 5 N.Y.3d 486, 840 N.E.2d 115 (2005). Arizona, Georgia, Iowa, Maryland, Michigan, and Texas have reached comparable results, several of them — Maryland and Texas conspicuously — resting not on any categorical rule but on what was knowable about household exposure in the specific decade at issue, which converts the duty question into the state-of-the-art question this chapter keeps returning to. Two states have removed the question from the courts: Kansas and Ohio each bar premises-owner liability by statute for exposures that did not occur at or near the owner's property, and the Kansas provision reaches silica as well as asbestos.

Three variables move cases within each line, and each is a technical assignment before it is a legal one. The first is the class of plaintiff, where the recognizing jurisdictions themselves divide: *Kesner* confined the duty to household members, while New Jersey in *Schwartz v. Accuratus Corp.*, 225 N.J. 517, 139 A.3d 84 (2016) — a beryllium case arising from a ceramics plant rather than an asbestos case, and involving a girlfriend who later became a spouse — declined to impose any categorical limit and directed a fact-specific inquiry instead. The second is the defendant's posture: an employer sued in negligence, a premises owner sued for conditions on its land, and a product manufacturer sued for failure to warn are governed by three different duty analyses, and the manufacturer's case frequently turns on what it told the purchasing employer about contaminated clothing rather than on any relationship to the plaintiff at all. The third, and the one that most often decides the case, is temporal: what the defendant knew or had reason to know about take-home exposure at the time of the exposure. Schwartz made that structural, listing the employer's knowledge of the dangerousness of the exposure at the time among the factors that fix the scope of duty. It is a state-of-the-art question, it is litigated with the published literature and the regulatory history, and it is where Section VI's expert assignments concentrate.

III. Reconstructing the Household Dose

The plaintiff must still prove an exposure, and the reconstruction is the method of Chapter 18 applied to a residence, with the workplace estimate as its upstream input rather than its subject. The chain runs from the worker's own occupational exposure, established where possible from the employer's monitoring, to the contaminant load deposited on clothing, skin, hair, and tools; to the release of that load into the home during transport, handling, laundering, and ordinary domestic activity; to the household member's own inhalation over the years the practice continued.

Each link has a literature. The asbestos take-home body of work includes simulation studies that measured airborne fiber concentrations generated by handling and shaking contaminated clothing, and those studies are the closest thing this field has to direct measurement of the exposure at issue. The lead literature adds surface-loading and settled-dust data supporting the ingestion pathway the asbestos analogy misses, which matters because the household plaintiff in a lead case is frequently a child. Where the exposure was community-wide as well as domestic, the vermiculite operations at Libby, Montana supply the most thoroughly characterized record in existence: worker exposure, household contamination, and ambient community exposure investigated together, with the federal environmental and health-agency work available to both sides as an independent data source rather than a litigation product.

The honesty standard of Chapter 18 governs with no relaxation. A household reconstruction that assumes the worker's full-shift time-weighted average as the concentration in the family kitchen has not modeled anything; it has restated the workplace estimate in a new location. The defensible version states the transfer assumption explicitly, sources it, propagates the uncertainty of the occupational estimate through it, and reports a range wide enough to be honest about a chain in which every link multiplies the error of the one before it. The defense audit begins at the same seam: the transfer step is where a household opinion is most often assumed rather than derived, and it is the step the opposing hygienist should be asked to defend first.

IV. The Employer's Own Record Decides It

Here the thesis of this book arrives in its most literal form. The claim is brought by a stranger to the workplace, and it is proven or defeated with documents the employer generated for compliance purposes, often decades before this plaintiff was diagnosed and sometimes before she was born.

The substantive standards did the work in advance, because take-home contamination is a hazard the health standards address directly. The asbestos standard at 29 C.F.R. § 1910.1001 and the lead standards at § 1910.1025 and § 1926.62 require change rooms and separate storage for street and work clothing, prohibit the removal of contaminated clothing from the workplace, impose laundering obligations on the employer, and — a provision whose evidentiary value is routinely underestimated — require the employer to inform whoever launders the clothing of the hazard. The silica standards at § 1910.1053 and § 1926.1153 reach the same territory through housekeeping, restricting the dry sweeping and compressed-air practices that redistribute settled dust onto clothing and skin.

Each of those provisions is simultaneously a compliance obligation and, in the household case, evidence that the pathway was recognized, and the courts have used them exactly that way. *Kesner* grounded its foreseeability holding in part on the federal regulations of the period, the 1972 OSHA asbestos standard among them; Louisiana courts have weighed the same 1972 provisions in finding a duty to a worker's spouse; and in the Alabama litigation the court pointed to the OSHA regulations together with the defendant's own workplace standards aimed at keeping fibers from leaving on employees' skin, hair, and clothing. The promulgation date of the provision, in other words, anchors the state-of-the-art timeline on which Section II's third variable turns, which is why the regulatory history belongs in the expert's file rather than in the brief alone.

The discoverable record follows the provisions. Whether change rooms and dual lockers existed, and when they were installed; the laundry contract, its invoices, and the hazard notification given to the launderer; the housekeeping procedures and the equipment purchased to perform them; the exposure monitoring the employer conducted and retained for thirty years under § 1910.1020, which is the only measured data anyone will ever have for the upstream end of Section III's chain; and the industrial hygiene reports, safety committee minutes, and internal correspondence in which someone raised — or did not raise — what the workers were carrying home. A program that provided change facilities, laundered on site, and documented the practice is the strongest defense available in this docket, and it is also, not incidentally, the program that prevented the disease. Its absence, in a period when the governing standard required it, is the plaintiff's case in a single exhibit.

V. Community and Emissions Claims

The same inversion operates at the fence line, with a different evidentiary spine. The community plaintiff alleges exposure from what the facility emitted rather than from what a worker carried, which removes the household relationship from the duty analysis and replaces the transport model of Section III with an atmospheric one.

The proof is built from records generated by the environmental regulatory system rather than the occupational one: the facility's operating permit and compliance history, the emissions inventories and toxics release reporting, stack tests and continuous monitoring data, and the dispersion modeling by which an emission rate at a stack becomes a concentration at a residence. The ethylene oxide docket is the current template, and its sequence deserves counsel's attention because it shows how quickly a regulatory science event becomes a mass tort. The Environmental Protection Agency finalized its evaluation of the inhalation carcinogenicity of ethylene oxide in December 2016, sharply raising the compound's estimated potency; the National Air Toxics Assessment released in 2018 applied that value and identified census tracts around sterilization facilities — Willowbrook, Illinois among them — as cancer-risk areas. The Willowbrook plant was closed under state order in 2019. In September 2022 a Cook County jury returned $363 million in *Kamuda v. Sterigenics U.S., LLC* — $38 million compensatory and $325 million punitive, in the first of the Illinois cases to reach trial — and in January 2023 the defendants resolved approximately 870 of the remaining Illinois claims for $408 million while denying liability. Chapter 28 traces where that frontier runs next.

Two features distinguish these claims technically, and the ethylene oxide sequence illustrates both. The exposure is chronic, low-level, and shared across a population, which makes the individual plaintiff's dose almost entirely a modeled quantity — the screening assessment that started the docket was itself a model, generating census-tract concentrations from an emissions inventory rather than from monitors — and puts the model's inputs, meteorology, and receptor placement at the center of the Rule 702 fight in a way no workplace sampling case ever is. Ambient background is the recurring second front: a compound present in the general airshed requires the plaintiff to separate this facility's contribution from everyone else's, which is Chapter 18's problem relocated outdoors. And the remedy sought is frequently medical monitoring for a population not yet sick, a claim whose availability varies by jurisdiction as sharply as the take-home duty does, and which converts the exposure assessment from one element of the case into the whole of it.

VI. Discovery Targets and Expert Assignments

The household plaintiff's discovery runs at the employer's compliance file and is framed by the standards of Section IV: the exposure monitoring and medical records retained under § 1910.1020, including the sampling that established what the worker was exposed to; the change-room and locker facilities, with the capital records and photographs that date them; the laundry contracts, invoices, and the notification given to the launderer; the housekeeping procedures, the equipment purchased, and the practices actually followed; the written programs required by the governing standard and the dates they were adopted; the industrial hygiene reports and safety committee minutes; the citation and inspection history, which

Chapter 28 notes is the most professional contemporaneous documentation of a workplace that will ever exist; and the worker's own personnel and job-assignment records, which establish where in the plant he actually worked and for how long — the point on which Schneider Electric turned, where the time the worker spent in or near the molding department was described as a matter of great contention.

The defense develops the same file affirmatively and adds three lines the plaintiff will not. The first is the state-of-the-art record for the specific decade, assembled from the published literature and the regulatory history rather than asserted, because in several jurisdictions it is dispositive of duty rather than merely relevant to breach. The second is the alternative-source inventory of Chapter 18, run against the household plaintiff's own life: her residences, renovations, avocations, and any employment of her own, characterized to the same standard rather than named. The third is the transfer step of Section III, which is where a household reconstruction is most often assumed. In the community case the corresponding targets are the permit and emissions record, the monitoring data, and the modeling files themselves — inputs, meteorological data, receptor grid, and model version — because a model that cannot be re-run by the opponent is a model that should not survive Rule 702.

The expert assignments follow. The industrial hygienist reconstructs the occupational exposure, the transfer, and the household or ambient exposure as one chain, with the uncertainty of each link stated and propagated. A second witness, or the same one with the qualifications to do it, must handle the state-of-the-art history: what was published, and what the standards required, in the years that matter. In the community case a dispersion modeler joins, and the hygienist's role shifts toward auditing the model's exposure assumptions. The physician's differential etiology receives all of it, and, as in every chapter of Part III, must dispose of each alternative source rather than pass over it.

VII. The Program That Holds on Both Sides

For the plaintiff, the discipline is that the household or community dose must be reconstructed rather than asserted, and that the employer's compliance record is the place to start rather than the place to finish. The missing change room proves that the pathway existed and that the obligation was recognized; it does not prove that this plaintiff inhaled a dose sufficient to cause this disease, and the opinion that treats the regulatory gap as the causation case will meet the same motion that ended *Pluck*.

For the defense, the discipline is that the exclusivity reflex is unavailable and the case must be defended on its merits: the state-of-the-art timeline, the transfer assumptions in the plaintiff's chain, the alternative sources Chapter 18 taught how to characterize, and the affirmative record of a program that provided the facilities the standards required.

And for the safety and industrial hygiene professional, this chapter carries the sharpest version of the question Chapter 25 makes explicit. The change rooms, the laundering program, the housekeeping procedure, and the exposure records are written today, for compliance, in a building where no one is thinking about a plaintiff who has not been born. They will be read decades from now by a lawyer representing that plaintiff, in a forum where the employer's usual immunity does not reach. The program that keeps the dust inside the fence is the same program that answers the claim when someone outside it gets sick, which is the closest thing to a complete argument for doing the work properly that this book contains.

References

29 C.F.R. § 1910.1001; § 1910.1020; § 1910.1025; § 1910.1053; § 1926.62; § 1926.1153.

Agency for Toxic Substances and Disease Registry. Libby asbestos site health consultations and exposure investigations. U.S. Department of Health and Human Services.

Kamuda v. Sterigenics U.S., LLC, No. 2018-L-010475 (Ill. Cir. Ct. Cook Cnty. 2022).

Kan. Stat. Ann. § 60-4905(a); Ohio Rev. Code Ann. § 2307.941(A)(1).

Kesner v. Superior Court, 1 Cal. 5th 1132, 384 P.3d 283 (2016) (decided with Haver v. BNSF Railway Co.).

In re N.Y.C. Asbestos Litig. (Holdampf v. A.C.&S., Inc.), 5 N.Y.3d 486, 840 N.E.2d 115 (2005).

Olivo v. Owens-Illinois, Inc., 186 N.J. 394, 895 A.2d 1143 (2006).

Pluck v. BP Oil Pipeline Co., 640 F.3d 671 (6th Cir. 2011).

Ramsey v. Ga. S. Univ. Advanced Dev. Ctr., 189 A.3d 1255 (Del. 2018).

Satterfield v. Breeding Insulation Co., 266 S.W.3d 347 (Tenn. 2008).

Schneider Electric USA, Inc. v. Williams, No. 2023-SC-0436-DG (Ky. Mar. 19, 2026).

Schwartz v. Accuratus Corp., 225 N.J. 517, 139 A.3d 84 (2016).

U.S. Environmental Protection Agency. (2016). Evaluation of the inhalation carcinogenicity of ethylene oxide (Final report) (EPA/635/R-16/350F).

U.S. Environmental Protection Agency. National Air Toxics Assessment (2018 release; now AirToxScreen).

CHAPTER 21

The Evidence Won't Wait

Spoliation and Evidence Preservation from the Technical Side

> *Every reconstruction of a workplace accident or exposure is built from evidence that began disappearing the moment the incident ended. The machine is repaired, the scene is swept, the camera loop overwrites itself in days, the laboratory discards the sampling media on a schedule no one thought to interrupt, and memory, the most perishable evidence of all, begins its decay within hours. The law of spoliation supplies the consequences; preservation itself is a technical discipline, and only the professionals in this book know what the systems generate and how fast it goes.*

I. Preservation as a Technical Discipline

A workplace accident or exposure event exists twice. It exists once as an occurrence — a press cycles with a hand in the die, a hose parts and a vapor cloud drifts across a unit, a worker spends a decade breathing an aerosol no one measured — and it exists a second time as a record: the machine in its post-incident condition, the scene as the first responders found it, the controller logs and camera files, the sampling media and calibration sheets, and the recollections of the people who were there. The occurrence cannot be litigated; it is gone in seconds or dispersed across years. Only the record can be litigated, and every claim, defense, causation opinion, and reconstruction that will ever be offered in the matter must be built from whatever portion of that record still exists when the experts arrive. The uncomfortable arithmetic of workplace litigation is that the record's half-life is measured in days while the litigation's timeline is measured in years, and the gap between those two clocks is where cases are won, lost, and sanctioned.

The legal system's answer to that gap is the doctrine of spoliation and its instrument, the litigation hold. But the doctrine is only a set of consequences; it does not preserve anything. What preserves evidence is a person who knows, on the first day, that the servo drive holds a fault log with a finite buffer, that the facility's cameras overwrite themselves on a fourteen-day loop, that the laboratory will discard the filter cassettes sixty days after reporting unless instructed otherwise in writing, that the interlock's bypass history lives in the safety controller and not the maintenance file, and that the memory of the operator who saw the whole thing will be materially degraded — and socially contaminated by breakroom consensus — before the week is out. That person is not the lawyer. It is the safety professional, the industrial hygienist, or the engineer who understands the systems, and the central argument of this chapter is that preservation succeeds only when that technical knowledge is placed in harness with counsel's legal authority from the first hours: counsel supplies the hold, the privilege, and the notice to adverse parties; the technical professional supplies the map of what exists, where it lives, and how fast it dies.

The stakes run in both directions, and it is worth saying so at the outset because preservation is too often treated as a plaintiff's project or a defendant's burden. The employer that scraps the failed sling has not merely exposed itself to sanctions; it has destroyed the one artifact that might have proven the sling was cut by the load rather than degraded by neglect. The plaintiff's family that authorizes repair of the vehicle before the defense inspects it may forfeit the claim outright. Preserved evidence acquits as often as it convicts, and the spoliation inference, once a court imposes it, is frequently worse than anything the destroyed evidence would actually have shown, because the jury is invited to imagine the worst. The record, as Chapter 6 puts it, is indifferent to who reads it. It cannot say anything at all, for anyone, if it no longer exists.

II. The Legal Framework for Preservation

The duty arises before the lawsuit does. The obligation to preserve evidence does not wait for a complaint, a preservation letter, or even a lawyer's involvement; it attaches when litigation is reasonably anticipated, and courts have found that anticipation in the circumstances that surround nearly every serious workplace event — a fatality or hospitalization, an amputation, an OSHA inspection, a catastrophic release, an internal escalation that treats the incident as a potential claim, or the retention of counsel or consultants for the event. The federal courts' formulation in *Zubulake v. UBS Warburg* — that once a party reasonably anticipates litigation it must suspend its routine document retention and destruction policy and put a litigation hold in place — has become the common-law baseline, and the Federal Circuit's decisions in *Micron Technology v. Rambus* confirmed that even destruction carried out under a facially neutral document-retention policy is spoliation if litigation was reasonably foreseeable when the shredder ran. For the safety professional, the practical translation is simple and severe: on the day of a serious incident, the organization should behave as though the duty has already attached, because a court examining the question years later, with the benefit of hindsight and a docket, will very likely conclude that it had.

The sanctions framework asks three questions and answers them on a sliding scale. Although the formulations vary by jurisdiction, spoliation analysis almost everywhere reduces to the same three elements: a duty to preserve at the time of the loss; a culpable state of mind, which in many states can be satisfied by ordinary negligence rather than bad faith; and the relevance of what was destroyed — meaning a showing that the lost evidence would actually have mattered to a claim or defense. The available remedies run the full length of the scale: an instruction permitting or requiring the jury to infer that the evidence was unfavorable to the destroyer, the exclusion of the destroyer's own testimony about the lost item, the shifting of costs, the striking of claims or defenses, and, at the extreme, outright dismissal or default. *Silvestri v. General Motors* remains the cautionary landmark: a plaintiff who allowed the vehicle at the center of an airbag claim to be repaired before the defendant could inspect it saw the entire case dismissed, the Fourth Circuit reasoning that the vehicle was the sole evidence from which the defense could evaluate the claim. *Flury v. Daimler Chrysler* reached the same terminus on similar facts. The lesson cuts identically for employers, manufacturers, and claimants: where the physical evidence is central and irreplaceable, its loss can be dispositive regardless of which party lost it.

The routine retention policy deserves its own paragraph, because it is the mechanism by which most of the records this chapter cares about actually disappear, and because it is not the safe harbor employers take it for. In *Stevenson v. Union Pacific Railroad Co.*, 354 F.3d 739 (8th Cir. 2004), a railroad recorded radio traffic between train crews and dispatchers and reused the tapes on a routine cycle; the tape covering a fatal grade-crossing collision was destroyed before suit, and track-maintenance records were destroyed under a one-year schedule both before and during the litigation. The Eighth Circuit sustained an adverse-inference instruction for the pre-litigation destruction of the tape, reasoning that the railroad knew such recordings mattered in the fatal-collision cases it litigated routinely, had preserved other records of this accident, and had kept comparable tapes when they helped it — and that the loss was prejudicial precisely because the tape was the only contemporaneous record of the conversations, whatever it might have contained. The court also fixed the circuit's threshold, holding that the instruction requires a finding of intentional destruction indicating a desire to suppress the truth, which is a higher bar than the ordinary negligence many states apply and one more reason the governing standard must be identified at intake. Both halves of the holding bear on the systems inventoried below. A camera loop that overwrites in fourteen days, a controller whose fault buffer holds a few dozen cycles, and a laboratory that discards media sixty days after reporting are all routine retention policies, and *Stevenson* is the reason a policy's routineness protects nobody who knew what the record was for.

Electronic data has its own rule, and it rewards documented diligence. The 2015 amendment to Fed. R. Civ. P. 37(e) governs the loss of electronically stored information in federal court and builds the analysis around a phrase that should organize every hold from the first day: whether the party took reasonable steps to preserve. Where ESI is lost because reasonable steps were not taken and it cannot be restored or replaced, the court may order measures no greater than necessary to cure the prejudice; the harshest remedies — the adverse-inference instruction, dismissal, default — are reserved for the party that acted with intent to deprive its opponent of the information's use. The rule's architecture is an instruction manual in reverse. The organization that can produce a dated hold notice, a list of the systems identified and the custodians notified, the tickets showing auto-deletion suspended and camera footage exported, and the chain of custody for what was collected has built its own defense to the motion before the motion exists. The organization that must reconstruct its efforts from memory, after the fact, has built its opponent's.

Centrality is a technical question the law has dressed as its own. Across jurisdictions, the variable that most often decides whether spoliation shrinks to a shrug or swells to a case-ending sanction is how central the lost evidence was: what a competent analysis of the scrapped machine, the overwritten video, or the discarded media could actually have established, and whether anything else in the record can substitute for it. That question is not answerable by counsel's assertion in either direction; it is answered by an expert affidavit that explains, concretely, what the examination would have measured and what the measurement would have resolved. The failure-analysis metallurgist who can describe the fracture-surface examination the discarded shaft would have permitted, or the industrial hygienist who can explain that the discarded cassettes were the only physical embodiment of the exposure ever collected, is the difference between a spoliation motion that reshapes the litigation and one that is denied as speculation. Preservation and reconstruction are, in this sense, the same discipline practiced at two different times, and the professional who understands what the reconstruction will need is the only one qualified to say, on day one, what must be kept.

III. Preserving the Physical Record

Quarantine the machine as found, and let the paper follow it. The single most consequential preservation decision in an equipment-involved incident is made in the first hours, usually by an operations supervisor under production pressure: whether the machine is repaired and returned to service or locked out, tagged, and quarantined in its post-incident condition. The technically disciplined answer is quarantine: de-energize the equipment under the facility's lockout program, leave guards, interlocks, selector switches, control settings, dies, tooling, and workpieces exactly where the incident left them, and move nothing until the condition is documented. The consensus standards that govern this work are not obscure: ASTM E1188 addresses the collection and preservation of information and physical items by a technical investigator, ASTM E860 governs the examination and preparation of items that are or may become involved in litigation — including the cardinal instruction that no examination that alters the item be undertaken without notice to the other interested parties — and ASTM E1459 and E1492 supply the labeling and chain-of-custody practices that let the artifact testify years later to being the same object in the same condition. An investigator who marks, bags, labels, logs, and secures the failed component under those standards has created evidence; an investigator who tosses it in a desk drawer has created a cross-examination.

If production genuinely cannot wait, document before you disturb, and keep what you remove. Real facilities face real constraints, and there are incidents in which the affected line must run again before any litigation timetable would permit. The preservation discipline does not evaporate in that circumstance; it compresses. Before any repair, the condition is exhaustively documented — photographs from fixed reference points, measurements, three-dimensional laser scans or photogrammetry where geometry matters, video of functional checks if the equipment can be safely exercised — and every component that is removed or replaced is preserved, labeled, and stored rather than discarded, along with the fasteners, fluids, and fragments that accompanied it. The maintenance work order for the repair becomes part of the record, and the technicians who performed it become witnesses whose observations should be captured while fresh. What a court will not forgive is the silent repair: the machine returned to service, the failed parts in the scrap bin, and the incident's physical record reduced to a supervisor's recollection that the guard "seemed fine."

Destructive testing is a negotiated act, never a unilateral one. Sooner or later the reconstruction may require examinations that consume or alter the evidence: sectioning a weld for metallography, tensile-testing an exemplar of remaining sling stock, disassembling a valve whose internal condition is the case. The governing principle, articulated in ASTM E860 and in the spoliation guidance long embedded in NFPA 921's treatment of fire scene evidence, is that any alteration of evidence proceeds only after notice to all known interested parties, an opportunity for them to attend or object, an agreed written protocol specifying what will be done, in what sequence, with what documentation, and — where feasible — the nondestructive work completed first so that the pre-test condition survives in scans, radiographs, and photographs even after the test article does not. Parties who follow that protocol convert a potentially case-ending act into an unassailable joint record; parties who cut first and notify later hand their opponent a spoliation motion with the protocol violation as Exhibit A.

The scene is more perishable than the machine, and it must be captured, not just kept. A machine can be quarantined; a scene mostly cannot. The housekeeping condition, the spill footprint, the position of barricades and signage, the staging of materials, the lighting, the weather, the location of every movable object: all of it is altered by rescue itself and erased by the next shift. The answer is capture at the highest fidelity available in the first hours: overlapping photographs from established positions, wide-to-close sequences with scale references, three-dimensional laser scanning or drone photogrammetry where the spatial relationships will matter to the reconstruction, measured sketches tied to permanent datums, and contemporaneous instrument readings where conditions are part of the story — illuminance where visibility is in issue, sound levels where audibility of an alarm is, air monitoring where a release is. Two additional records deserve deliberate attention because they are so often the only professional documentation that survives: the emergency responders' photographs and reports, and the OSHA compliance officer's file — the photographs, measurements, and interview notes gathered during the inspection — which should be pursued in full through discovery or records request rather than settling for the citation summary. And the decision to release a scene back to production should be made the way a destructive test is: deliberately, with counsel, after documentation is complete, and with a record of when, why, and by whom the release was authorized.

IV. The Electronic Record and the Litigation Hold

The plant is full of witnesses that never blink, and most of them are erasing themselves on a schedule. Modern facilities generate a continuous electronic record that is frequently more probative than any human account: the PLC and safety-controller event logs that show which interlock tripped and whether it had been bypassed; the drive and servo fault histories; the process historian trending temperatures, pressures, flows, and alarm states through the moment of the release; the machine's cycle counters and mode selections; the CMMS work-order history that documents what maintenance knew and deferred; the badge and access-control logs that place people in rooms; the forklift telematics recording speed and impact events; the building automation system trending the ventilation that an exposure reconstruction will one day need; and the video. Nearly every one of these systems is designed to forget. Camera systems overwrite on loops commonly measured in days or a few weeks; controller event buffers are finite and roll; historians compress or purge; telematics platforms age data out under subscription tiers no one in EHS has ever read. The first technical act of preservation is therefore an inventory: a walk through the facility's actual data-generating systems, made by someone who knows them, that lists what exists, where it resides, who administers it, and, for each source, the retention clock now running against the case.

The litigation hold is drafted by counsel, but it is scoped by the technical professional, and a hold scoped only to people will miss the systems. The conventional hold notice identifies human custodians and instructs them to preserve documents and communications. That model, inherited from commercial litigation, fails workplace cases in a characteristic way: the most important evidence is not in anyone's inbox. It is in machine controllers, camera servers, historians, and hosted platforms that no individual custodian 'holds' and that will destroy data automatically, without malice, unless someone with administrative rights intervenes. The technically competent hold therefore has two halves. The custodian half reaches the supervisors, operators, maintenance personnel, engineers, and EHS staff involved, and expressly covers the informal channels where the candid record actually lives: text messages, shift-turnover notes, radio logs, personal-device photographs. The systems half, built from the inventory described above, directs named administrators to suspend automatic deletion and overwriting for each identified source, export the perishable data now rather than merely promising to keep it, and document each step with dates and tickets. Exports should be made in native formats with metadata intact, hashed or otherwise verified where authenticity may be contested, and logged into the same chain-of-custody discipline as the physical evidence, because a spreadsheet screen-shotted from a historian is an anecdote, while a native export with an intact audit trail is a record.

A hold is a living instrument, not a memo. Rule 37(e)'s reasonable-steps standard is applied to conduct over time, and the holds that survive scrutiny share the same habits: they are issued promptly and in writing; acknowledgments are collected; the scope is revisited as the investigation teaches the parties what actually matters; reminders are reissued at intervals because employees leave, systems are upgraded, and vendors are replaced; departing custodians' data is captured before their accounts are deprovisioned; and the hold is released only deliberately, on counsel's written instruction, when the matter and its reasonably related successors are concluded. The safety professional's role in this maintenance is continuous, because the technical landscape keeps shifting beneath the legal instrument

— the camera server is replaced, the historian is migrated to the cloud, the machine at the center of the case is scheduled for decommissioning — and each of those ordinary business events is, for the held sources, a preservation decision that must be recognized as one before it happens rather than litigated after.

V. Sampling Media: What Cannot Be Re-Measured

An exposure, once past, exists only in its media and its paper. Accident reconstruction enjoys a luxury that exposure reconstruction does not: the machine can be examined next year. The air a worker breathed on the night of the release cannot. Whatever was captured on the filter, the sorbent tube, the passive badge, or the wipe — together with the pump calibrations, the field notes, and the laboratory's raw data — is the entire physical embodiment of the exposure that will ever exist, and everything not captured must later be rebuilt inferentially from surrogate data. That asymmetry should drive two behaviors. First, sample generously in the immediate aftermath of any event with exposure potential, because contemporaneous measurement, even imperfect, anchors a reconstruction in a way no model can; and document the conditions of sampling — tasks, controls, ventilation status, production rate — with the same care as the results, because a number without context is an invitation to competing narratives. Second, treat every physical sample as litigation evidence from the moment of collection: chain of custody initiated in the field, holding times and preservation conditions per the governing NIOSH or OSHA method observed and recorded, field blanks and duplicates logged, and the media's journey to the laboratory documented rather than assumed.

The laboratory is running a disposal clock, and only a written instruction stops it. Accredited laboratories retain analyzed media and digested samples for defined periods, often measured in weeks, and then discard them in the ordinary course, exactly as a document-retention policy discards documents. When litigation is reasonably anticipated, the same *Zubulake* logic that suspends the shredder must suspend the laboratory's disposal schedule: a written retention request for the media, the extracts, and the raw analytical data, acknowledged by the laboratory and renewed as its own retention milestones approach. The full analytical record deserves the same treatment, because reconstructions and admissibility challenges alike reach past the summary report to the underlying material: the chromatograms and spectra, the calibration and quality-control data, the analyst's bench notes, the method deviations. An industrial hygienist who can produce the cassette, the chain of custody, the pump's pre- and post-calibration, the field blank results, and the laboratory's raw data has an exposure measurement that will survive any gatekeeping challenge; one who can produce only a summary table has an exhibit waiting to be excluded.

The regulatory retention rules are a floor the litigation hold builds upon, and a discovery map for the other side. The exposure record is unusual among workplace records in that federal law already compels its long preservation: employee exposure records must be kept for thirty years under OSHA's access-to-records standard, medical records for the duration of employment plus thirty years, the OSHA 300 logs and incident reports for five, and process safety incident investigations for five under the PSM standard. Those obligations exist independent of any litigation, which carries two consequences. For the preserving party, the destruction of records the law independently required to be kept is close to indefensible in a spoliation analysis, because the duty and the culpability elements arrive pre-proven. For the requesting party, the retention rules are a map of what must still exist: decades of monitoring results, safety data sheets for the products actually in use in the relevant years — not the current revisions, which may describe a reformulated product — ventilation surveys, respirator program records, and the production and throughput records from which task frequency and duration can be rebuilt. In the long-latency disease case, where the exposure being reconstructed ended decades before the diagnosis, this archival record is the case, and the expert's first assignment is to demand all of it before concluding what can and cannot be reconstructed.

Post-incident 'improvements' change the workplace being measured, sequence the sampling accordingly. There is a final trap peculiar to exposure matters: the instinct to fix the condition immediately competes with the need to characterize it. Worker protection always wins that competition, no litigation interest justifies continued exposure, but the two can usually be sequenced rather than traded: interim controls and exclusion protect people while contemporaneous sampling, scene capture, and instrument data logs characterize the condition; the permanent abatement follows once the condition is documented. What cannot be undone is the reverse order. Sampling performed after the ventilation is upgraded, the process enclosed, or the housekeeping transformed measures a different workplace, and both sides' experts will spend the rest of the case arguing about what the first one was like. The parallel to the quarantined machine is exact, and so is the discipline: document the as-found condition first, preserve what you collect, and let the record — not recollection — carry the burden of describing what the workplace was.

VI. Witness Statements: When, How, Whose Privilege

Memory is the fastest-decaying evidence at the scene, and the schedule must respect that. Human recollection of a sudden event degrades rapidly and nonlinearly, the steepest losses come in the first hours and days, and what decay does not take, contamination does: witnesses who discuss the event with one another converge on a shared account that belongs to the group rather than to any observer, and each retelling quietly edits the original. The operational consequences are concrete. Statements should be taken as soon as the demands of medical care and hazard control permit: the same shift where possible, and within a day or two at the outside for the key observers. Witnesses should be identified immediately and interviewed separately, before the breakroom synthesis forms, with a request that they not compare accounts until the interviews are complete. And the roster should be broader than the eyewitnesses: the operator on the prior shift, the mechanic who last serviced the interlock, the coworker who filed the near-miss report a month earlier all hold pieces of the reconstruction, and their memories are running on the same clock.

The craft of the statement is to capture observation and exclude conclusion. A useful witness statement is a record of perception, and the interviewer's discipline is to keep it one. The interview opens with an uninterrupted narrative — what the witness saw, heard, smelled, and did, in the witness's own words — and only then proceeds to specifics: positions and distances marked on a diagram or photograph, timing anchored to fixed events, lines of sight, lighting, sound, what the witness was doing in the moments before. Questions are open-ended and neutral; the interviewer supplies no facts, suggests no sequence, and asks no question that presumes its answer, because a leading interview does not merely produce a weak statement — it can overwrite the memory itself. Conclusions, fault attributions, and speculation are gently redirected to observation: not whether the guard was 'inadequate' but where it was, not who was 'careless' but who was standing where, doing what. The statement is reduced to writing or recorded, reviewed by the witness, corrected in the witness's hand, signed, and dated, and the diagram the witness marked is preserved with it, because five years later that marked-up plan view may be the most reliable spatial evidence in the file.

Privilege is not a stamp; it is a structure, and it must be built before the interviews begin. The legal protections available for investigation materials are two, and they are narrower than the labels routinely applied to them. The attorney-client privilege, under *Upjohn v. United States*, protects communications between company counsel and company employees — at any level, not merely a control group — when made in confidence for the purpose of enabling counsel to provide legal advice to the organization. The work-product doctrine, rooted in *Hickman v. Taylor* and codified in Rule 26(b)(3), protects documents and tangible things prepared in anticipation of litigation, with near-absolute protection for the attorney's mental impressions and a qualified protection for the rest. Neither protection attaches because a document is marked privileged, and both are lost by the habits that plague post-incident files: routine distribution beyond those who need the material for the legal purpose, statements taken by the same personnel under the same procedure used for every ordinary incident, and reports written to serve operations and then re-labeled when the claim arrives. The recurring battleground is the dual-purpose investigation — every serious workplace event is investigated partly for safety and partly for the coming claim — and the In re Kellogg Brown & Root line of authority supplies the workable test: the privilege applies where obtaining or providing legal advice was one of the significant purposes of the communication, even if a regulatory or business purpose coexisted. But the significant legal purpose must be real and demonstrable, which is why the structure matters more than the label.

The practical architecture is a two-track investigation, and the tracks must be genuinely separate. The design that reconciles the safety function's need for shareable findings with the legal function's need for protected candor is deliberate separation. The safety track proceeds as it always has — the incident investigation the management system requires, conducted under the standard procedure, producing the corrective actions and the recordable determinations, and its output should be written on the assumption that it will be produced in discovery, because it almost certainly will be. The legal track is initiated by counsel in writing at the outset, with a memorandum establishing that the investigation is undertaken to provide legal advice in anticipation of litigation; the sensitive witness interviews are conducted by counsel or at counsel's express direction by consultants retained for the litigation purpose; employees receive the *Upjohn* admonition — that the interviewer represents the company and not the individual, that the privilege belongs to the company and may be waived by it alone, and that the conversation should be kept confidential; the resulting memoranda are marked, held by counsel, and distributed to no one who does not need them for the legal purpose. Two cautions complete the design. First, privilege protects communications, never the underlying facts: the witness can always be asked at deposition what she saw, and no interview structure changes that. Second, privilege is a shield for advice, not a vault for evidence: routing the only copy of a critical measurement or photograph into the privileged file does not exempt it from preservation or production, and attempting to use the doctrine that way is the fastest known method of losing it.

Regulatory interviews run on their own rules, and the two processes must not be confused. When OSHA investigates, compliance officers are entitled to interview non-managerial employees privately, and employees are entitled to speak to the agency without employer retaliation; management representatives, by contrast, may have counsel present, and their statements can bind the employer. The company's internal statement program must operate alongside that process without ever appearing to interfere with it: no instruction to employees about what to tell the agency, no debriefing that could be characterized as coaching or intimidation, no discipline that could be read as retaliation for candor. The internal interviews and the agency's interviews will eventually be compared line by line in the litigation, which is one more reason the internal program should be built on the disciplines above: early, separate, neutral, observational, and honestly structured. A statement program designed to find out what happened will survive that comparison. A statement program designed to build a defense will not, and will usually have failed at building the defense as well.

VII. The First Seventy-Two Hours

The doctrine, the systems, and the craft described above compress into a sequence that any organization — and any claimant's counsel — can execute, and the sequence is worth stating plainly because in the event it will be executed under stress. In the first hours: care for the injured, control the hazard, and lock down the perishables — quarantine the equipment as found, secure the scene, order the export of camera footage and controller logs before the loops overwrite, and identify the witnesses. In the first day: counsel engaged and the two-track structure established; the litigation hold issued with both its custodian half and its systems half, built on a real inventory of the facility's data sources and their retention clocks; the laboratory's disposal schedule suspended in writing for any media in play; the preservation letter sent, or, on the receiving end, answered with documented compliance; and the separate, neutral witness interviews under way while memory is still the witness's own. In the first weeks: the scene documented to reconstruction fidelity before release; any repair preceded by documentation and followed by retention of every removed part; destructive examination deferred until all interested parties are noticed and a protocol agreed; and the hold maintained as a living instrument as systems change and people leave. None of this requires knowing who will sue whom. All of it determines what the eventual experts, on both sides, will have to work with.

Everything above rests on a claim worth stating plainly, because it is usually assumed to be false. Preservation is routinely discussed as though it were a burden imposed on defendants for the benefit of plaintiffs, and litigation holds as though they were traps. The technical reality is the opposite: preservation is the only act in the entire lifecycle of a workplace claim that is guaranteed to serve whichever party the truth favors. The quarantined machine exonerates the employer whose guard was defeated by an unforeseeable act as surely as it condemns the one whose interlock had been bypassed for a year; the preserved sampling media protects the manufacturer whose product was never airborne at a meaningful concentration as surely as it proves the exposure of the worker who breathed it; the early, neutral witness statement is the best friend of whichever account is true. What spoliation destroys, in every case, is not one side's evidence but the possibility of an honest reconstruction: leaving both sides to litigate inferences about a record that no longer exists, before a jury invited to assume the worst about whoever let it die. The professionals who prevent that outcome are the ones reading this chapter: the safety manager who knows the camera loop is fourteen days, the industrial hygienist who knows the laboratory discards at sixty, the engineer who knows where the fault log lives, and the lawyer who puts their knowledge under privilege and behind a hold before the end of the first day. The evidence won't wait. Neither can you.

References

Access to employee exposure and medical records, 29 C.F.R. § 1910.1020 (thirty-year retention of exposure records); recording and reporting occupational injuries and illnesses, 29 C.F.R. pt. 1904 (five-year retention); process safety management — incident investigation, 29 C.F.R. § 1910.119(m) (five-year retention).

ASTM International. ASTM E1188, Standard Practice for Collection and Preservation of Information and Physical Items by a Technical Investigator; ASTM E860, Standard Practice for Examining and Preparing Items That Are or May Become Involved in Criminal or Civil Litigation; ASTM E1459, Standard Guide for Physical Evidence Labeling and Related Documentation; ASTM E1492, Standard Practice for Receiving, Documenting, Storing, and Retrieving Evidence in a Forensic Science Laboratory.

Fed. R. Civ. P. 26(b)(3); Fed. R. Civ. P. 37(e) (as amended Dec. 1, 2015).

Flury v. Daimler Chrysler Corp., 427 F.3d 939 (11th Cir. 2005).

Hickman v. Taylor, 329 U.S. 495 (1947).

In re Kellogg Brown & Root, Inc., 756 F.3d 754 (D.C. Cir. 2014).

Micron Tech., Inc. v. Rambus Inc., 645 F.3d 1311 (Fed. Cir. 2011).

National Fire Protection Association. NFPA 921, Guide for Fire and Explosion Investigations (evidence preservation and spoliation guidance).

National Institute for Occupational Safety and Health. NIOSH Manual of Analytical Methods (sampling, media handling, and holding-time specifications). Centers for Disease Control and Prevention.

Occupational Safety and Health Act § 8 (inspections, investigations, and recordkeeping); § 11(c) (protection from retaliation).

Occupational Safety and Health Administration. OSHA Technical Manual and Field Operations Manual (inspection documentation, sampling, and interview practices). U.S. Department of Labor.

The Sedona Conference. The Sedona Principles, Third Edition: Best Practices, Recommendations & Principles for Addressing Electronic Document Production (2018).

Silvestri v. Gen. Motors Corp., 271 F.3d 583 (4th Cir. 2001).

Stevenson v. Union Pacific R.R. Co., 354 F.3d 739 (8th Cir. 2004).

Upjohn Co. v. United States, 449 U.S. 383 (1981).

Zubulake v. UBS Warburg LLC, 220 F.R.D. 212 (S.D.N.Y. 2003).

CHAPTER 22

The Investigation Writes the Record

Conducting the Accident Investigation — When Counsel Comes In, and How the Evidence Is Collected Correctly and Completely

> *Every chapter in this Part reads a record someone else created. This chapter is about the hours in which that record is created: the investigation conducted in the first day and week after a serious injury, before anyone knows which side of the caption they will occupy. Between the preservation duties of Chapter 21 and the privilege doctrine of Chapter 26 sits the operational craft neither of them teaches.*

I. The Case Is Investigated Before It Exists

A serious workplace injury sets three clocks running at once, and none of them belongs to the litigation. The regulatory clock is the shortest: a work-related fatality must be reported to OSHA within eight hours, and an in-patient hospitalization, an amputation, or the loss of an eye within twenty-four (29 C.F.R. § 1904.39), which means the agency frequently knows about the incident, and may be en route to it, before the employer has finished securing the scene. The physical clock runs almost as fast: the scene will be altered by rescue and medical response, then by the pressure to resume production; surveillance systems overwrite on loops measured in days; the machine will be repaired, the trench backfilled, the damaged equipment discarded by someone who was never told not to. And the human clock is the least forgiving of all, because the witnesses' accounts are never again as accurate, as detailed, or as uncontaminated by conversation and consequence as they are in the first hours. Everything this book says about the record that decides — the reconstruction of Chapter 16, the incident chapters that follow, the trial chapters that precede — assumes that someone, in those first hours, did the collection right.

The uncomfortable truth is that the investigation is conducted by the party with the least perspective and the most at stake, at the moment of maximum confusion, usually by people who have never been deposed and do not know that every note they write is a future exhibit. The employer's safety manager who investigates candidly creates the document the plaintiff will enlarge for the jury; the one who investigates defensively creates the document that convicts the employer of investigating defensively, and the one who does not investigate at all creates the inference that there was something not worth finding. This chapter's thesis is that the tension is real but manageable: an investigation can be honest, complete, and structured for protection at the same time, but only if the structure is built in the first hours, deliberately, with counsel, because none of it can be retrofitted once the notes exist.

II. The First Call: The Reporting Clock and Counsel

The decision that shapes everything downstream is when the lawyer enters, and the case law has made the answer unusually concrete. An investigation initiated by management in the ordinary course — the routine safety inquiry every incident triggers, conducted by the safety department because that is what the safety department does — is a business record, discoverable in its entirety, however damaging its candor; that is the teaching of *Dowling*, where routine pre-accident safety-committee minutes drew no work-product protection because they would have been created the same way with or without litigation on the horizon. An investigation directed by counsel in anticipation of litigation, by contrast, can claim the work-product footing of *Hickman* and Rule 26(b)(3), and the communications within it the privilege footing of *Upjohn*, and *Martin v. Bally's* shows the structure working: a consultant's post-incident safety report, commissioned through counsel, protected even against an OSHA subpoena because the engagement was built for litigation from the outset. The difference between those outcomes is not the content of the investigation. It is the architecture of its first day — who commissioned it, for what stated purpose, reporting to whom — and the architecture cannot be improved after the fact.

The practical rule for the serious injury — the fatality, the amputation, the incident with obvious litigation gravity — is that the call to counsel belongs in the first hours, alongside the § 1904.39 report and the preservation steps of Chapter 21, for three reasons beyond privilege. First, the reporting decision itself has judgment in it: what qualifies as work-related, what counts as an in-patient hospitalization, and what is said in the report are questions with enforcement consequences, and the report is the first statement the employer makes on a record it cannot later amend casually. Second, OSHA's investigators may arrive while the scene is still controlled, and the employer's rights and obligations during that inspection — the walkaround, the interviews, the document requests of Chapter 7 — are managed better by a team that includes someone who has managed them before. Third, and least appreciated, early counsel protects the investigation from the client: the executive who wants the scene cleaned up, the manager who wants to interview the crew alone, the adjuster who wants the equipment shipped out — each is proposing, in good faith, a spoliation problem, and the instruction that stops them lands harder from a lawyer than from the safety manager they outrank.

III. The Two Tracks: Honest and Protected at Once

The structural answer to the candor-versus-discovery tension is not to choose between them but to run the two investigations the situation actually requires, separately and honestly labeled. The first track is the compliance and prevention investigation the safety function must conduct: the inquiry that feeds the 300 log determination of Chapter 23, satisfies the process-safety incident-investigation duty where it applies — 29 C.F.R. § 1910.119(m) requires an investigation of covered incidents initiated within forty-eight hours, by a team with process knowledge, with findings retained for five years — and produces the corrective actions that prevent the next injury. That track is discoverable, and it should be written knowing it: factual, precise, and free of the speculation, the self-flagellation, and the legal conclusions that convert a prevention document into an admissions exhibit. The second track is the litigation assessment counsel directs: the frank evaluation of exposure, fault, and defense, conducted under privilege, using consultants retained through counsel in the *Kovel* structure Chapter 26 describes. The tracks may examine the same facts; they must not share drafts, personnel confusion, or a file drawer, because the commingled investigation forfeits the protection of both, and the party that later places its investigation at issue, as *Harding* teaches, can waive even the protection it preserved.

Two disciplines keep the structure honest rather than cynical. The first is that the compliance track must be real: an employer that routes every incident inquiry through counsel and calls it all privileged is asserting a position courts see through and punish, because the routine safety investigation the company always conducts does not become work product by copying a lawyer. The dual-purpose document, created both because litigation loomed and because the safety program required it, is tested in most federal courts under *Adlman*'s because-of standard, and the honest answer for most root-cause work is that it would have been created anyway. The second discipline runs the other way: the prevention track's candor is not a liability to be minimized but the employer's best long-run defense, because the company whose files show incidents investigated seriously, causes found honestly, and corrections closed out verifiably has built the good-faith record that every enforcement chapter of this book shows paying dividends, and Rule 407 exists precisely so that the subsequent remedial measure, the fix itself, cannot be offered to prove the negligence it corrected. The employer that fixes the hazard and documents the fix has not confessed; it has complied.

IV. Collection: The Evidence That Cannot Be Re-Collected

A. The Scene, Controlled and Documented

Scene work begins with control and ends with a documented release, and everything between should be captured on the assumption that the scene will not exist next week. Control means a perimeter that survives the emergency response: nothing moved, removed, or cleaned beyond what rescue and safety require, with the incident commander's alterations themselves documented, because the defense will otherwise attribute every anomaly to the response. Documentation means photographs and video taken systematically rather than opportunistically: wide establishing views from fixed reference points, then mid-range, then close-up, with a scale in frame; the perishable conditions first — lighting, weather, housekeeping, the position of controls, guards, and warning devices — because those are the facts that change by morning. Measurements are taken to fixed features and recorded on a sketch or, increasingly, captured by laser scanning that preserves the geometry for the reconstruction and the demonstratives of Chapter 5. The release decision — when production resumes, when the trench is backfilled, when the scene stops being a scene — is a litigation decision as much as an operational one, and it is made after the documentation is complete and, where litigation is probable, after opposing interests have been offered the inspection Chapter 21 contemplates, not before.

B. Physical Evidence, Custody, Destructive Testing

The physical evidence — the failed component, the parted sling, the tripped breaker, the tool, the PPE the worker was or was not wearing — is collected into a chain of custody that begins at the scene, because provenance is half of admissibility and all of credibility. Each item is photographed in place before collection, tagged, logged, and stored in its as-found condition: the machine is not test-run, the guard is not refitted, the settings are not disturbed, because the as-found state is itself evidence and every undocumented manipulation is a cross-examination. The governing discipline for anything that may be examined further is the one the spoliation cases enforce with dismissals: no destructive or condition-altering testing without notice to the parties whose interests are known, a protocol agreed or ordered, and an opportunity to attend — the lesson of *Silvestri* and *Flury* applied prospectively, while the choice is still cheap. The employer that unilaterally tears down the press to get the line running has converted its best physical evidence into its worst litigation fact, whatever the teardown found.

C. The Electronic Record

The modern scene extends into systems that overwrite themselves, and the first-day preservation list should treat them as perishable evidence with expiration dates. Surveillance video is the most urgent: retention loops of days or weeks, and the footage of the incident and of the preceding shifts, which establish how the task was actually performed, are lost by default rather than by decision. Behind the video sit the machine and vehicle data: PLC fault and event logs, safety-system and interlock histories, engine control modules and telematics, access-control and badge records, equipment inspection apps, and the maintenance system's work-order history for the equipment involved. Each is a timestamped, employer-independent witness, and each is governed by the litigation-hold obligations Chapter 21 details, but the hold letter does not preserve anything by itself; a named person must actually pull the video, export the logs, and image the data before the systems do what they were configured to do. The investigator's checklist should say who, for each system, by when.

D. Witnesses, and the Form a Statement Takes

Witness evidence decays fastest and is mishandled most. The collection principles are settled: interview early, separately, and without an audience of coworkers or supervisors whose presence shapes the account; let the witness narrate before questioning; document contemporaneously; and identify everyone with knowledge, including the medical responders and the workers who left at shift change, before the roster scatters. The legal form of the statement matters as much as its content, and it is decided at creation, not later. A statement taken by counsel, or at counsel's direction for the litigation assessment, carries the work-product and *Upjohn* footing the first-call decision established, with the caution *Upjohn* itself requires, that the interviewer make clear to the employee who the client is and is not. A statement taken in the routine compliance inquiry is discoverable and should be taken with that understanding: accurate, signed, dated, and free of the interviewer's characterizations. Where a union is present, an employee facing an investigatory interview that could lead to discipline is entitled on request to representation, and the interview conducted in disregard of that right produces both a labor charge and a statement whose voluntariness will be attacked. And in every setting, the injured worker's own statement is approached with restraint — taken, if at all, with candor about purpose, never in the emergency room, and never as a condition of benefits — because the statement extracted from a medicated worker is worth less than nothing when the jury learns how it was obtained.

V. The Root-Cause Analysis as a Litigation Document

The root-cause analysis is where the investigation's findings take the form the litigation will inherit, and it should be written by someone who understands both of its audiences. The methodological point comes first: a named, recognized methodology — the causal-tree and barrier-analysis families, the structured why-analysis the process-safety literature codifies — applied as designed, distinguishes an investigation from an opinion, and the analysis that identifies contributing causes at the levels the method prescribes (the immediate deviation, the preconditions, the management-system factors) is both better prevention and better evidence than the single-cause narrative that stops at the injured worker. The drafting point follows: findings are stated as facts and causal relationships, not as legal conclusions or corporate self-indictments. There is a difference — visible to any juror — between the lockout procedure for the press did not address the pneumatic energy source, and we violated the standard and this was entirely preventable; the first is a finding, the second is a closing argument, and the investigator who writes the second has done the plaintiff's work without the plaintiff's burden of proof. Recommendations are phrased as improvements with owners and completion dates, tracked to verified closure, because the recommendation that was never implemented is the exhibit that proves knowledge without cure, the precise pattern the willfulness and punitive-damages chapters of this book turn on. The analysis that is honest about causes, disciplined about language, and closed out in verified correction is simultaneously the best prevention document and the most defensible one the employer can hold; the file that is none of those things should not exist in the form it is in, and by the time anyone realizes that, it is too late to fix. Chapter 25 extends these drafting disciplines from the investigation file to the whole of the program record.

VI. Rule 407: What the Fix Costs You, and What It Does Not

Every recommendation in this chapter, and in Chapters 24 and 25, tells the reader to find the hazard and correct it. The obvious anxiety follows immediately, and the book has so far left it unanswered: if we fix it, does the fix become the exhibit that convicts us? The doctrine that answers the question is Fed. R. Evid. 407 and its state analogues, and it deserves a proper treatment because the safety professional's entire relationship to correction runs through it, and because the rule protects considerably less than its reputation suggests.

A. What the Rule Says

Rule 407 provides that when measures are taken that would have made an earlier injury or harm less likely to occur, evidence of the subsequent measures is not admissible to prove negligence, culpable conduct, a defect in a product or its design, or a need for a warning or instruction. The rationale is partly evidentiary — the fix is weak proof of prior fault, since prudent parties improve things that were adequate — and mostly social: the law does not want to tax repair.

Four limits define the rule's actual reach, and each of them matters more in practice than the rule itself.

B. The Four Limits

The rule bars a purpose, not a document. Rule 407 excludes the evidence only when it is offered to prove negligence, culpable conduct, or defect. The same evidence comes in for *another* purpose, and the rule names three: impeachment, ownership or control, and feasibility of precautionary measures *if controverted*. That last exception is the one that swallows cases. A defendant who argues that guarding the nip point was impossible, that the ventilation could not have been improved, or that the sequencing could not have been changed has controverted feasibility, and has thereby admitted the retrofit it installed three weeks after the incident, which is the single most damaging exhibit available to the plaintiff. Counsel and the technical expert should agree, before the answer is filed and again before the expert is disclosed, whether an infeasibility position is worth what it costs. Frequently it is not.

Impeachment is broader than it looks. A witness who testifies that the machine was "as safe as it could be made" or that the procedure "was fully adequate" has opened the door, and the fix comes in to contradict him. This is a discipline problem for the expert as much as for the corporate witness: superlatives invite the exhibit that the qualified opinion would have excluded.

Many states decline to apply the rule in strict liability. The federal rule was amended in 1997 to cover product-liability claims expressly, but Rule 407 is a rule of evidence, and in state court the state's rule governs. A substantial number of states apply their version only to negligence and admit subsequent remedial measures in strict-liability design cases: on the reasoning that the question is the product's condition rather than the manufacturer's conduct. In a diversity case the *Erie* question is contested and jurisdiction-specific. A defendant that structured its post-incident conduct around federal Rule 407 and then finds itself in a state court that admits the evidence in a design case has planned against the wrong rule.

The rule does not restrain OSHA, and it does not create a privilege. Rule 407 is an admissibility rule in litigation. It has no application to an OSHA inspection, where abatement is not merely admissible but compulsory and certified, and it does not make the repair records privileged, undiscoverable, or confidential. They are produced; the fight is over what they may be used to prove. The employer who believes Rule 407 keeps the corrective-action file out of the case has misread it entirely.

C. Rule 407 Does Not Protect the Pre-Incident Finding

Here is the point the safety professional most needs and the one the rule's reputation most obscures. Rule 407 protects measures taken *after* the injury. It has nothing whatever to say about the audit finding, the near-miss report, the risk assessment, the internal recommendation, or the budget request that identified the hazard *before* the incident and was not acted on. That material is fully admissible, and it is where essentially all of the real exposure lives: the notice layer of Chapter 6 §IV, the leading-indicator stream of Chapter 19, the open corrective action of Chapter 24, the unfunded recommendation of Chapter 25.

The practical consequence reverses the intuition the anxiety produces. The employer's risk is not created by fixing things after an incident; Rule 407 covers most of that. The risk is created by finding things and not fixing them, and by keeping a record of having found them. Which means the correct response to a pre-incident finding is not to stop writing findings: a program of deliberate blindness reads exactly as such to a jury, and Chapter 25 shows why weasel-wording is not the remedy either. The correct response is to close the loop the finding opened, and to document the closure with verification rather than with a date. A finding closed out with evidence is an asset in the case. The same finding closed out with the word "completed" and nothing behind it is the plaintiff's exhibit with the employer's own letterhead on it.

Two operational rules follow for the investigation and the post-incident period. Correct the hazard immediately and without regard to the litigation consequence; the safety obligation is not contingent, the OSHA abatement duty is not waivable, and the delayed fix compounds the exposure with a second, worse fact. And describe the correction in the corrective-action record as what it is — a change to a procedure, a guard installed, a sequence revised — rather than as an admission of what preceded it. The distinction Section V draws between a finding and a closing argument applies with equal force to the fix.

VII. A Worked Investigation: The Confined-Space Fatality

The disciplines above are easier to state than to run, and one incident type exercises all of them at once under the worst possible conditions. Confined-space fatalities are worth a worked treatment here for three reasons: the governing standard is unusually prescriptive and therefore unusually productive of records; the most probative evidence is an atmosphere, which is to say the most perishable evidence in this book; and the incident has a recurring multiple-casualty pattern that changes the whole shape of the case. The facts below are composite. The sequence is not.

A maintenance technician enters a below-grade wet well to clear an obstruction. He collapses within a minute. A second worker, hearing nothing over the radio, goes down after him and collapses beside him. A supervisor arriving thirty seconds later starts down the ladder, stops at the third rung when he feels light-headed, and backs out. Emergency services arrive in eleven minutes with supplied air. Both entrants are pronounced at the scene. That would-be-rescuer pattern is not incidental to confined-space work; it accounts for a large share of the deaths, because the hazard is invisible, the collapse is immediate, and the instinct to help is faster than the training. It also means the investigation has two decedents, two entries, and two separate failures to analyze, and the second entry is frequently the one that decides the case.

The atmosphere is gone before anyone arrives, and it is the whole case. Nothing in Section IV's collection discipline matters more here than acting within hours. Four things can still be recovered and each degrades on a different clock. The first is the instrument: the gas monitor used for pre-entry testing, which must be secured immediately, not returned to the charger, because most instruments log readings, alarms, and calibration events internally and many overwrite on a rolling buffer. Download the datalog before the unit is bumped, cleared, or recalibrated, and preserve the calibration and bump-test history alongside it, because an instrument out of calibration converts a documented atmospheric test into no test at all. The second is the space itself, in the condition it was in: the ventilation equipment as it sat, the blank or blind that was or was not installed, the valve positions, the lockout devices actually hung. The third is the process record — what was flowing into that space, from where, and under whose control — which lives in the SCADA or historian data of Section IV.C and typically purges faster than anyone expects. The fourth is a reconstruction of the atmosphere itself, which is a modeling exercise the industrial hygienist can run from the process data, the space volume, and the ventilation rate, and which is available only if someone preserved the inputs in the first week.

The paper the standard generates is the second half of the investigation, and it is unusually rich. A permit-required confined space program under 29 C.F.R. § 1910.146 produces, for every entry, a document that records the hazards evaluated, the isolation performed, the pre-entry atmospheric test results with the instrument identified, the acceptable entry conditions, the entrants and the attendant by name, the ventilation in use, and the rescue arrangements. It also produces a canceled permit that the employer must retain, and a program review the standard requires whenever an entry reveals that the program is not working. That means the investigation asks a series of questions with documentary answers: Was a permit written for this entry, and does it exist? What did the pre-entry test show, on what instrument, calibrated when? Was the space isolated, and by whom? Was there an attendant, was he at the opening, and did he have a means of summoning rescue that did not require him to enter? What were the rescue arrangements, and were they arrangements or a phone number? Were the canceled permits from prior entries into this space retained, and do they show the same hazard recurring and the same control never installed?

That last question is where the confined-space case usually turns, and it is the notice structure this book has described in every other context. A file of canceled permits is a dated series of the employer's own atmospheric measurements in the same space. If those permits show oxygen deficiency or hydrogen sulfide on prior entries and the response each time was a respirator rather than isolation or ventilation, the employer has documented a recognized hazard and its own decision about it, in its own hand, over a period of years. If they show the space consistently tested clean and the entry that killed two people was the anomaly, that same file is the strongest defense document in the case. Neither reading is available to anyone who did not ask for the whole permit file rather than the permit for the day in question, which is the same instruction Appendix E gives about the inspection file and Chapter 6 gives about near-miss reports.

The second entry gets its own analysis, and both chairs get it wrong in predictable ways. The plaintiff's error is to treat the rescuer's death as a straightforward second count of the same violation. The defense error is to reach for comparative fault, on the theory that the rescuer violated a rule he had been trained on. The rescue-provision requirements exist precisely because the standard's drafters understood that untrained rescue attempts are the foreseeable consequence of an entrant collapse, and an employer whose rescue arrangement was a telephone number has not arranged rescue — it has arranged for exactly what happened. The technical questions are whether retrieval equipment was in place and would have worked, whether the attendant could summon rescue without entering, whether the designated rescue service had been evaluated for its ability to respond to this space in a timely manner, and whether anyone had practiced. Each has a documentary answer, and the safety professional who investigates without asking them has produced a report that will be superseded by the one the other side commissions.

The general lesson is the one this chapter has been making throughout. A confined-space fatality looks, in the first hours, like an incident about a person who made a bad decision. It is nearly always an incident about a document that was not written, an instrument that was not calibrated, an isolation that was not performed, or a rescue arrangement that existed on paper and nowhere else — and every one of those is recoverable in the first week and unrecoverable in the second. The same structure, with different nouns, governs the combustible-dust deflagration and the trench collapse: a prescriptive standard, a records trail the standard compels, a hazard that leaves the scene faster than the investigators arrive, and a prior-occurrence file that turns the case.

VIII. The Other Chair: The Plaintiff-Side Investigation

The injured worker's team investigates the same incident from outside the fence, and the disciplines are mirror images with one asymmetry: the plaintiff controls almost none of the evidence and must therefore move faster on paper. The preservation letter of Chapter 21 goes out immediately: specific to the systems Section IV lists, naming the video retention loops and the machine data by system where possible, because the letter that says preserve all evidence preserves nothing in particular. Scene access is requested formally and early, through the employer or by inspection demand once suit is filed, and the plaintiff's expert attends the joint inspection with the same protocol discipline the defense owes, because *Silvestri* binds the plaintiff's side too: the worker's team that obtains the failed component and then loses or alters it faces the same sanctions doctrine, and the plaintiff's counsel who lets the client's family discard the tool has spent the case's best evidence. The public record fills what access cannot: the OSHA investigation file obtained through the playbook of Appendix E, the 300 logs and prior-incident history of Chapter 23, the citation history, the 911 and EMS records that fix times and conditions. And the retained expert enters now, not at the report deadline, because the expert's first contribution, on either side, is the collection checklist itself: knowing, from the mechanism, which records will matter, which conditions are perishable, and which question the reconstruction will eventually have to answer. That early retention is the single practice this book recommends more often than any other, and this chapter is why.

IX. Practice Implications

For counsel on the defense side, the investigation is won in the first day's architecture: make the call early, separate the tracks explicitly and keep them separate, put the preservation instructions in writing with named owners, and treat the scene release, the teardown, and every interview as decisions with litigation consequences, because they are, whether or not anyone treats them that way. For plaintiff's counsel, speed on paper substitutes for control of the scene: the specific preservation letter, the early access demand, the OSHA file request, and the expert retained while the evidence still exists. For the safety and industrial hygiene professional — who will usually be the person actually holding the camera and the interview notes — the chapter's lesson is that investigative craft and litigation awareness are the same discipline: collect completely, document provenance, write findings as findings, and understand which track you are working on and for whom, because the question who directed this investigation, and for what purpose will be the first one you are asked under oath. The incident lasts a moment. The investigation of it becomes the record, and the record, as every chapter before and after this one shows, decides.

References

29 C.F.R. § 1910.146 (permit-required confined spaces; entry permits, atmospheric testing, attendant duties, rescue and emergency services, and program review); 29 C.F.R. pt. 1926 subpt. AA (confined spaces in construction).

Abatement verification, 29 C.F.R. § 1903.19.

ANSI/ASSP. (2019). Occupational health and safety management systems (ANSI/ASSP Z10.0). American Society of Safety Professionals.

Center for Chemical Process Safety. (2019). Guidelines for investigating process safety incidents (3rd ed.). CCPS / American Institute of Chemical Engineers.

Dowling v. Am. Hawaii Cruises, Inc., 971 F.2d 423 (9th Cir. 1992).

Fed. R. Civ. P. 26(b)(3).

Fed. R. Evid. 407 (subsequent remedial measures), as amended 1997; Advisory Committee Note to the 1997 amendment.

Flury v. Daimler Chrysler Corp., 427 F.3d 939 (11th Cir. 2005).

Harding v. Dana Transp., Inc., 914 F. Supp. 1084 (D.N.J. 1996).

Hickman v. Taylor, 329 U.S. 495 (1947).

Martin v. Bally's Park Place Hotel & Casino, 983 F.2d 1252 (3d Cir. 1993).

NLRB v. J. Weingarten, Inc., 420 U.S. 251 (1975).

Process safety management of highly hazardous chemicals — Incident investigation, 29 C.F.R. § 1910.119(m).

Reporting fatalities, hospitalizations, amputations, and losses of an eye, 29 C.F.R. § 1904.39.

Silvestri v. Gen. Motors Corp., 271 F.3d 583 (4th Cir. 2001).

United States v. Adlman, 134 F.3d 1194 (2d Cir. 1998).

Upjohn Co. v. United States, 449 U.S. 383 (1981).

CHAPTER 23

The OSHA 300 Log as Evidence

What Injury Records Reveal (and Conceal) in Discovery

> *The injury log is the most under-used document in workplace injury litigation. Read correctly, it establishes what the employer knew and when, and read against its own gaps, it can reveal what the employer chose not to record.*

I. The Document Everyone Requests and Few Interrogate

Almost every workplace injury and OSHA matter includes, somewhere in the document requests, a demand for the employer's OSHA 300 logs. Far fewer include a plan for reading them. The OSHA 300 Log, the Log of Work-Related Injuries and Illnesses, and its companion forms, the 301 Incident Report and the 300A Annual Summary, are required of most employers with more than ten employees under OSHA's recordkeeping regulation at 29 C.F.R. Part 1904 (OSHA, n.d.). They are, in effect, a contemporaneous log of the hazards an employer encountered, maintained by the employer itself, and admissible as a business record. For a litigator, that combination is unusually valuable, because the central question in most injury and enforcement cases — what did the employer know, and when did it know it — is precisely what a properly kept log documents.

The thesis of this chapter is a practical one aimed at the paralegals and associates who actually work these records: the 300 log is a knowledge document, and its evidentiary power runs in two directions — what it records can establish employer notice of a recurring hazard, and what it omits, measured against the records the employer could not suppress, can establish underreporting. Both require someone who knows how the recordkeeping rule actually works.

II. What the Rule Requires, and Why It Is Evidence

A working knowledge of Part 1904 turns a stack of logs into a timeline. A handful of provisions do most of the evidentiary work.

A. What Must Be Recorded

A case must be entered on the 300 Log if it is a new, work-related injury or illness that meets one of the recording criteria: most commonly, medical treatment beyond first aid, days away from work, restricted work or job transfer, loss of consciousness, or death (29 C.F.R. §§ 1904.4, 1904.7). Work-relatedness is presumed for injuries resulting from events in the work environment unless a specific exception applies (29 C.F.R. § 1904.5). The recording threshold is deliberately low, which is what makes a missing entry meaningful: if a worker received sutures, lost a day, or was put on light duty, the case was recordable, and its absence from the log is not a matter of judgment but of compliance.

B. Retention and Updating

Employers must retain the 300 Log, the privacy case list, the 300A summary, and the 301 reports for five years following the end of the calendar year the records cover (29 C.F.R. § 1904.33). Critically, the obligation is not static: during that five-year period the employer must update a stored log to add newly discovered recordable cases and to reflect changes in a case's classification, if an injury first logged as a restricted-duty case later results in days away or a fatality, the entry must be corrected (29 C.F.R. § 1904.33). This updating duty is an underappreciated discovery lever: a log that was never updated to reflect the known progression of an injury is itself a recordkeeping deficiency, and a comparison of successive log versions can show what the employer learned and when.

C. Posting, Access, and Electronic Submission

The 300A annual summary must be posted in the workplace from February 1 to April 30 of the following year (29 C.F.R. § 1904.32). Employees, former employees, and their representatives have a right of access to these records on request (29 C.F.R. § 1904.35), and certain establishments must submit injury data electronically to OSHA's Injury Tracking Application each year (29 C.F.R. § 1904.41). Each of these creates a parallel record — a posted summary an employee may have photographed, an access request that was or was not honored, an electronic submission that can be compared against the internal log — and each parallel record is a check on the integrity of the log itself.

D. Privacy Cases

For a defined set of sensitive injuries — including injuries to intimate body parts, those resulting from sexual assault, certain mental-illness cases, HIV or hepatitis cases, and needlestick contamination — the employer must omit the employee's name from the log and keep a separate confidential privacy case list (29 C.F.R. § 1904.29). Counsel handling these records must understand the privacy-case mechanism both to interpret a log with unnamed entries and to frame discovery that reaches the confidential list through appropriate protective order without running afoul of the privacy rule.

III. Reading the Log to Establish Knowledge

The affirmative use of the log is to prove notice. In a General Duty Clause case, the Secretary must prove the employer knew or should have known of the hazard; in a civil case alleging a recurring dangerous condition, the plaintiff must show the employer was on notice that the condition was injuring workers. A log that records three prior lacerations on the same unguarded machine, or a cluster of heat-related entries across successive summers, or a pattern of back injuries tied to the same manual-handling task, is contemporaneous, employer-authored evidence of exactly that notice.

The technique is comparative and pattern-based. One reads the logs across multiple years, sorts by the injury type, the body part, the department, and the described mechanism, and looks for recurrence. A single entry proves an accident; a pattern proves knowledge. When the same mechanism appears repeatedly and the abatement never changed, the log establishes both that the hazard was recognized and that feasible correction was foregone: the recognition and abatement elements of a General Duty Clause case, or the notice and breach elements of a negligence case, assembled from the defendant's own records.

The 301 Incident Reports deepen this analysis. Where the 300 Log gives a one-line classification, the 301 captures how the injury occurred, and reading the narratives in sequence often reveals a common causal thread the summary entries obscure. A set of 301s describing the same jam-clearing task, the same missing guard, or the same absent lockout across several incidents is a more powerful knowledge exhibit than any single citation.

IV. Reading the Log to Detect What Is Missing

The concealment side of the analysis is subtler and requires triangulation, because underreporting is proven not from the log alone but from the log measured against records the employer did not control. Recordkeeping violations are a persistent feature of OSHA enforcement, and deliberate or systematic under-recording is both a citable violation and, in litigation, powerful impeachment.

The method is to compare the 300 Log against independent injury indicators for the same period. Workers' compensation first-report-of-injury filings, group health and short-term disability claims, emergency-room and occupational-clinic records, the employer's own incident and near-miss reports, and even payroll records showing unexplained light-duty reassignments each capture injuries that should also appear on the log. A recordable injury that surfaces in the comp file or the clinic records but never made the 300 Log is a discrepancy that demands an explanation, and a systematic pattern of such discrepancies suggests the log was managed rather than merely kept.

Motive matters here, and counsel should understand it. Employers have concrete incentives to under-record: injury rates drive workers' compensation experience modifiers and premiums, qualify or disqualify contractors for bids, and — under the current National Emphasis Programs — help determine which establishments OSHA targets for inspection, because compliance officers are directed to mine 300 logs to identify hazards and expand inspections. A demonstrated pattern of underreporting is therefore not merely a technical foul; it supplies a narrative of an employer suppressing the very evidence of the hazard that injured the plaintiff, which is why the missing entry can be more valuable than the recorded one.

V. The Two Signatures That Bind the Employer

Everything in the preceding sections treats the log as a compilation. Two provisions of Part 1904 do something different: they attach a name and a date to it, and in doing so convert a clerical record into a signed statement by the company. Both are routinely missed in discovery, and both are worth more than the entries they certify.

A. The 300A Certification

The annual summary is not merely posted. It must be certified, and the rule is specific about who may sign: a company executive, defined as the owner, an officer of the corporation, the highest-ranking company official working at the establishment, or that official's supervisor (29 C.F.R. § 1904.32). The certification is not a formality of the "reviewed and approved" kind. The signer attests that he has examined the 300 Log and that he reasonably believes, *based on his knowledge of the process by which the information was recorded*, that the summary is correct and complete.

Read that language the way an examiner reads it. The certifying executive has personally represented, under a federal recordkeeping obligation, three things at once: that he looked at the log, that he knows how the entries got there, and that the total is right. Where the log understates the year — where the comp file, the clinic records, and the short-term disability claims of Section IV establish injuries the summary omits — the certification is the document that makes the omission a person's rather than a system's. And where the log is accurate, the certification is equally useful in the other direction: it is a senior officer's contemporaneous acknowledgment that the establishment recorded, that year, four lacerations on the same machine.

The practical consequences run in three directions. In discovery, request the signed 300A for every year in the retention period, not the log alone; the summaries are separately retained under § 1904.33 and are frequently produced without signatures unless the request specifies the executed copy. In deposition, the certifying executive is a fact witness on the recording *process*, who decided recordability, on what advice, and by what criteria — and that examination reaches the judgment calls behind every entry the employer classified as first aid rather than medical treatment. And for the safety professional, whose recommendation typically drives the recordability call the executive certifies: the process you designed is the process he swore he understood, and the two of you will be examined about it in sequence.

B. The Reporting Clock of § 1904.39

The second signature is a phone call, and it is time-stamped by the government rather than the employer. Separate from the duty to record, an employer must report a work-related fatality to OSHA within eight hours, and a work-related inpatient hospitalization, amputation, or loss of an eye within twenty-four hours of learning of it (29 C.F.R. § 1904.39). Chapter 21 treats the reporting clock as a preservation problem — the report is what brings a compliance officer to a scene that would otherwise be released — but the log chapter is where its evidentiary character belongs, because a reporting failure has a property no other recordkeeping defect has: it is a single, discrete, precisely datable act, and it is provable from the agency's own records rather than the employer's.

That property matters more than its modest penalty exposure suggests. The employer who did not report a qualifying event has left the government with no inspection, no compliance officer's photographs, no contemporaneous sampling, and no citation, which is to say, has caused the disappearance of the neutral professional documentation that Chapter 6 §III identifies as frequently the only such documentation a case will ever have. The technical consequence is that the plaintiff's expert in such a matter is reconstructing from a thinner record than the rule contemplated, and the reason the record is thin is an act of the defendant. Whether that supports an inference, an instruction, or nothing at all is a question for counsel and the forum. That it should be developed, dated, and put in front of the trier of fact is not.

The point is not hypothetical, and this book's own anchor case supplies the illustration. The *Valdivia* matter of Chapter 6, No. C-731101, arose from an August 2022 injury that produced a catastrophic spinal cord and brain injury; the incident was, on the public reporting, never reported to OSHA, and by the time the verdict drew attention to that fact the six-month citation window had long since closed. The consequence was not merely a penalty forgone. It was that the agency never opened a file, and a case worth four hundred eleven million dollars was tried without the one set of contemporaneous professional measurements that an inspection would have produced. Both sides litigated the scene from photographs and recollection because nobody made a phone call within twenty-four hours.

VI. Suppression at the Source

Section IV treats under-recording as something done to the log after the injury is known. The more consequential failure happens earlier, at the point where the worker decides whether to report at all, and Part 1904 regulates it directly. An employer must establish a reasonable procedure for employees to report work-related injuries promptly and accurately, and the rule states expressly that a procedure is not reasonable if it would deter or discourage a reasonable employee from reporting (29 C.F.R. § 1904.35). The same provision prohibits discharging or otherwise discriminating against an employee for reporting.

That regulation makes three ordinary management practices into discoverable evidence.

The first is the reporting procedure itself. A rule requiring a worker to report an injury "immediately" or before the end of shift, applied to an injury with delayed onset — a back strain that stiffens overnight, a hearing loss, an exposure symptom — is a procedure that punishes the biology of the injury rather than the conduct of the worker, and it is exactly what the rule contemplates as unreasonable. Request the written procedure, the disciplinary policy that enforces it, and every instance of its enforcement.

The second is post-injury discipline. Discipline administered to an injured worker for a safety-rule infraction discovered only because he reported the injury is the most common form the problem takes, and its evidentiary value is double: it supports the retaliation analysis of Chapter 13, and it explains, to a jury asked why a hazardous condition never appeared on the log, precisely why it never appeared on the log.

The third is the incentive program. A safety bonus, prize drawing, or crew-level award keyed to injury *rate* rather than to leading indicators pays the workforce, in cash, not to report — and does it in a document the employer created, announced, and can be made to produce. An expert asked whether a log understates the injury history should read the incentive plan before reading the log, because the plan predicts the answer.

The synthesis for both audiences is uncomfortable and worth stating plainly. A 300 Log is only as complete as the reporting culture that fed it, and the documents that establish the culture — the reporting procedure, the discipline file, the bonus plan — are not safety documents in the employer's mind and are therefore rarely produced unless specifically demanded. Demand them. For the safety professional, the same three documents are the ones to audit first, because a program that suppresses reporting has not reduced its injuries; it has only stopped counting them, and it has created, in the meantime, the exhibit that will explain the silence in its own log.

VII. From Discovery to Trial: The Log's Limits

A log that establishes knowledge in discovery still has to survive the evidence rules at trial, and the path is narrower than the discovery value suggests. The affirmative foundation is straightforward — the 300 Log and 301 reports are records of a regularly conducted activity under Rule 803(6), kept under a regulatory duty, authenticated by the custodian or a certification — but the fight is rarely over hearsay. It is over Rule 403 and the prior-incident doctrine: entries describing other workers' injuries are other-accident evidence, and most courts admit them to prove notice or the existence of a hazard only on a showing of substantial similarity — same machine or task, same mechanism, same relevant conditions — with the required degree of similarity relaxing when the evidence is offered for notice rather than for causation. The pattern analysis of Section III should therefore be built to the trial standard from the start: the entries worth fighting for are the ones the 301 narratives tie to the same task and mechanism, and the spreadsheet that sorts the log by machine, body part, and mechanism is not merely an investigative tool — it is the substantial-similarity proffer, drafted early.

The log's limits deserve equal respect, because the over-read log is a gift to the objecting party. The recordkeeping rule says expressly what defense counsel will repeat at every opportunity: recording a case does not mean the employer or the worker was at fault, that a rule was violated, or that the injury is compensable, recordability is a reporting trigger, not an admission. The *Volks* Constructors decision marks a second limit, and its history is worth knowing because it is settled in a way few OSHA questions are. OSHA had long treated the duty to keep accurate logs as continuing, so that every day an inaccurate log sat in the file was a fresh violation citable throughout the five-year retention period. In 2012 the D.C. Circuit rejected that construction, holding that the obligation to record is a discrete act, that the Act's six-month citation window runs from the failure to record rather than from the end of retention, and that nothing in the statute lets the agency hold a discrete record-making violation over an employer for years (*AKM LLC dba Volks Constructors v. Secretary of Labor*, 675 F.3d 752 (D.C. Cir. 2012)). OSHA answered in December 2016 with a rule reinstating the continuing obligation by regulation; Congress disapproved that rule under the Congressional Review Act in the spring of 2017, and OSHA removed it from the Code of Federal Regulations that May. The practical residue is a hard limit on enforcement exposure, stale recordkeeping failures are not citable, and a softer but real limit on litigation narrative, because a concealment theory that reaches back years must be built from the log's own updating duty under § 1904.33 and from the independent injury indicators of Section IV, not from an enforcement theory the courts have foreclosed. And the electronic-submission data now published from the Injury Tracking Application carries its own caution in both directions: it enables the establishment-level benchmarking that makes an outlier employer conspicuous, and it produces rate comparisons that are only as sound as the hours and classification data beneath them — a point the expert who presents, or attacks, an injury-rate exhibit must be prepared to defend at the level of the denominator.

The practical synthesis is that the log is a knowledge document with an evidentiary passport that must be stamped: authenticated as a business record, matched to the incident through substantial similarity, framed within what recordability does and does not concede, and presented by a witness fluent enough in Part 1904 to hold the exhibit's meaning steady under cross. Prepared that way, the employer's own contemporaneous injury history is among the most persuasive exhibits a workplace case can offer; prepared casually, it is excluded, limited, or turned, and the party that built its notice theory on an inadmissible pattern discovers the gap at the worst possible hearing.

VIII. Practice Implications

Several practical points follow for the litigation team.

Request the full recordkeeping set, not just the log. A request for the 300 Logs alone leaves the most useful documents on the table. The 301 Incident Reports carry the narratives; the 300A summaries carry the posted totals; the privacy case list (through an appropriate protective order) carries the sensitive cases; and the electronic ITA submissions provide an external comparison. Request all of them, for the full five-year retention window, and request every version of a log that was updated.

Build the comparison early. The independent injury indicators — comp filings, clinic records, disability claims — are the backbone of an underreporting theory, and they are obtained through separate discovery channels that take time. Identifying the discrepancy is only useful if the parallel records are in hand to prove it.

Use a reader who knows the rule. The difference between a recordable case and a non-recordable one, the significance of a case that was never reclassified, and the meaning of an unnamed privacy entry are technical judgments under Part 1904. A log reviewed by someone fluent in the recordkeeping regulation yields a knowledge timeline and, where the records diverge, an underreporting case. A log reviewed as an undifferentiated list yields neither. The document is less dramatic than an accident reconstruction, but in establishing what the employer knew, it is frequently the most important paper in the file.

References

AKM LLC d/b/a Volks Constructors v. Sec'y of Labor, 675 F.3d 752 (D.C. Cir. 2012).

Annual summary, 29 C.F.R. § 1904.32.

Electronic submission of injury and illness records, 29 C.F.R. § 1904.41.

Employee involvement, 29 C.F.R. § 1904.35.

Fed. R. Evid. 803(6); Fed. R. Evid. 403.

Forms and privacy concern cases, 29 C.F.R. § 1904.29.

Occupational Safety and Health Administration. (n.d.). Recording. U.S. Department of Labor. https://www.osha.gov/recordkeeping/recording Purpose — recording does not establish fault, 29 C.F.R. § 1904.0 (note).

Recording and reporting occupational injuries and illnesses, 29 C.F.R. pt. 1904 (2024).

Recording criteria, 29 C.F.R. § 1904.7.

Retention and updating, 29 C.F.R. § 1904.33.

PART V

THE PROFESSIONAL'S PRACTICE

CHAPTER 24

Mock Regulatory Audits

Finding the Citation Before the Compliance Officer Does

An OSHA inspection is a compliance audit conducted on the government's schedule, scored in citations, and preserved in a public database that opposing counsel reads. A mock regulatory audit is the same examination conducted on the employer's schedule, scored in findings, and closed out in corrective action. This chapter describes what a credible mock audit covers in general industry and construction, assembles the enforcement data and penalty structure that make the economics unambiguous, and traces the regulatory and litigation doctrines under which a documented program of self-audit and prompt correction is among the strongest evidence of good faith an employer can create, and an uncorrected finding among the most damaging.

I. The Inspection Is an Audit on the Government's Schedule

Every establishment covered by the Occupational Safety and Health Act operates under a standing appointment it did not make. A compliance safety and health officer can arrive at the gate without notice — routed there by an employee complaint, a reported hospitalization or amputation, a referral, a national or local emphasis program, or the simple misfortune of programmed selection — and when the inspection is over, the employer learns its compliance status in the form of citations, proposed penalties, and abatement deadlines (Occupational Safety and Health Act of 1970 [OSH Act], 29 U.S.C. §§ 651–678). The results are not private. Citation and penalty histories are searchable in the Occupational Safety and Health Administration's public enforcement database, where they are read by prospective clients, general contractors performing prequalification, insurers setting terms, journalists, and — with particular attention — attorneys evaluating whether an injury case can be built into something larger than a workers' compensation claim.

The financial terms of that appointment have hardened considerably. For citations issued in 2026, the maximum penalty is $16,550 for each serious or other-than-serious violation, $16,550 per day for each failure to abate beyond the abatement date, generally capped at thirty days, and $165,514 for each willful or repeated violation; the 2025 amounts were carried forward into 2026 without an inflation adjustment, but the statutory scheme adjusts them annually and the trajectory since 2016 has been steadily upward (Occupational Safety and Health Administration [OSHA], 2026). The arithmetic that matters is that penalties are assessed per violation, not per inspection: a single walkaround that documents ten serious violations approaches $166,000 in proposed penalties before any willful or repeat classification enters the calculation, and a repeat classification — available to the agency for substantially similar violations cited anywhere in a corporate family within the preceding five years — multiplies the exposure tenfold per item. Employers whose citation history shows willful, repeated, or failure-to-abate violations involving high-gravity hazards face placement in the Severe Violator Enforcement Program, which brings mandatory follow-up inspections, corporate-wide agreements, and public identification (OSHA, 2022). Behind the penalties stand the losses the penalties are meant to prevent: the 2024 Census of Fatal Occupational Injuries recorded 5,070 fatal work injuries at a rate of 3.3 per 100,000 full-time equivalent workers (U.S. Bureau of Labor Statistics [BLS], 2026), and the National Safety Council (2025) estimates the total economic cost of work injuries at well over $170 billion annually: a figure in which OSHA penalties are a rounding error.

What makes this exposure manageable is a fact that receives less attention than it deserves: the examination is not a mystery. OSHA publishes its most frequently cited standards every year, and the list barely moves. The ten standards atop the fiscal year 2025 list — led, for the fifteenth consecutive year, by the construction fall protection standard — are the same ten that appeared in fiscal years 2024 and 2023, with only modest reshuffling of rank (OSHA, 2025a). The failure modes within each standard are equally predictable: unprotected roof edges, missing machine-specific lockout procedures, respirator programs without medical evaluations, training records without training. Figure 24-1 displays the list. An agency that publishes its examination questions year after year has, in effect, invited employers to take the examination privately first. The mock regulatory audit is the formal acceptance of that invitation: a comprehensive compliance evaluation conducted with the same standards, the same methods, and the same rigor a compliance officer would apply — document review, walkaround, private employee interviews, records analysis — but on the employer's schedule, under the employer's control, with correction rather than citation as the endpoint. It is not a housekeeping walkthrough with a clipboard, and the distinction between the two determines whether the exercise produces protection or paperwork.

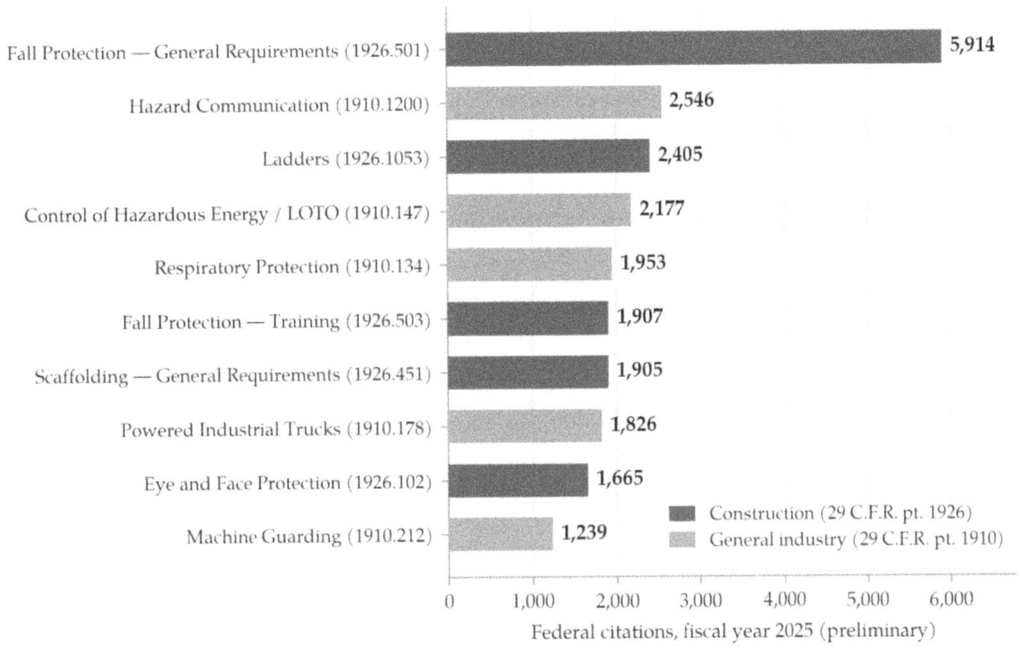

Figure 24-1 OSHA's Ten Most Frequently Cited Standards, Fiscal Year 2025 Note. Counts are preliminary federal citation totals for fiscal year 2025 (October 1, 2024, to September 30, 2025) as announced by OSHA (2025a); final figures are subject to revision. Standards beginning 1926 are construction standards; standards beginning 1910 are general industry standards. The same ten standards have appeared on the list, in varying order, for several consecutive fiscal years.

II. What a Mock Regulatory Audit Covers

The scope of a credible mock audit is not a generic checklist. It is derived, establishment by establishment, from three overlapping inventories: the standards that actually apply to the operations conducted there, the standards the agency actually cites — nationally, in the relevant industry, and under any active emphasis programs — and the hazards whose consequences are severe regardless of citation frequency. Table 24-1 summarizes the core scope elements common to most engagements. Within that structure, the general industry and construction portions of the audit examine different regulatory worlds, and the records portion examines the paper trail that binds them together.

Table 24-1. Core Scope Elements of a Mock Regulatory Audit

Audit element	What the auditor examines	Method mirrored
Written programs	Required written programs (e.g., hazard communication, lockout/tagout, respiratory protection, emergency action) against the standard's required elements and against actual site practice	Document request and program review
Physical conditions	Walkaround of all operations, including the areas and shifts a scheduled tour would miss; hazards classified as a compliance officer would classify them	Walkaround inspection
Records	OSHA 300/300A/301 accuracy and five-year retention; training, medical, exposure monitoring, equipment inspection, and certification records	Recordkeeping audit
People	Private interviews with employees and supervisors across shifts; what workers actually do and know versus what the written program says	Employee interviews
Prior history	Establishment and corporate citation history, prior audit findings, and the closure status of every previously identified hazard	History review
Corrective action	Findings ranked by severity and probable citation classification; owners, dates, interim controls, and verified abatement	Citation and abatement process

Note. Scope is tailored to the establishment's operations and regulatory applicability; the elements above are common to most comprehensive engagements. The "method mirrored" column identifies the corresponding element of an actual OSHA inspection (OSHA, 2025b).

A. The General Industry Scope (29 CFR 1910)

In general industry, the audit concentrates where the citations concentrate. Hazard communication — the second most cited standard in the nation for years running (OSHA, 2025a) — is audited not by confirming that a written program exists but by reconciling the program against the actual chemical inventory, pulling safety data sheets for chemicals selected at random from the floor, checking secondary container labeling in the areas where transfer actually happens, and asking workers what the pictograms on the drum beside them mean. Control of hazardous energy is audited at the machine: the standard requires machine-specific procedures, annual periodic inspections of those procedures, and training commensurate with each employee's role, and the most common finding in practice is a generic corporate procedure that names no equipment, verified by no one, covering machines that were modified years after the procedure was written. Machine guarding is audited by standing at the point of operation. Powered industrial trucks are audited through operator certifications, the triennial evaluations most programs quietly skip, and the daily inspection records that reveal whether the forklift with the dead horn has been reported every morning for a month. Respiratory protection is audited from the exposure assessment forward, and the single most frequent failure OSHA identifies under that standard is the absence of medical evaluation for the workers wearing the respirators (OSHA, 2025a). Depending on operations, the scope extends to walking-working surfaces and fall protection in general industry, electrical safe work practices, permit-required confined spaces, emergency action and fire prevention plans, occupational noise, air contaminant exposures requiring sampling, and, where threshold quantities of highly hazardous chemicals are present, the fourteen elements of process safety management.

B. The Construction Scope (29 CFR 1926)

Construction concentrates its risk, and OSHA concentrates its citations, in a shorter list, and the audit follows it. Fall protection is first among equals: the general requirements standard, 1926.501, has led the national citation list for fifteen consecutive years, and its training companion, 1926.503, sits on the same list, most often for the plainest of failures — no training at all (OSHA, 2025a). The audit examines the six-foot trigger discipline on every elevated surface, the condition and anchorage of personal fall arrest systems, guardrail integrity, hole covers, and the fall protection plan against the work actually underway. Ladders and scaffolds — third and seventh on the fiscal 2025 list — are audited for selection, condition, three-point access, and the competent person inspections the scaffold standard requires before each shift. Excavations are audited against Subpart P: protective systems in anything five feet or deeper, spoil pile setback, ladder access within twenty-five feet of lateral travel, and the daily competent person inspection that exists on paper at many sites and in practice at fewer. The remainder of the construction scope tracks the leading causes of death in the industry — struck-by and caught-between hazards around equipment and cranes, electrocution hazards including ground-fault protection and overhead line clearances — together with respirable crystalline silica exposure control plans, eye and face protection, and housekeeping. One structural feature distinguishes the construction audit from its general industry counterpart: on a multi-employer worksite, OSHA's citation policy reaches beyond the employer whose employees are exposed to the employer who created the hazard, the employer contractually responsible for correcting it, and the general contractor with supervisory authority over the site (OSHA, 1999). A construction mock audit that examines only the contractor's own crews has therefore audited a fraction of the contractor's exposure; the balance sits in subcontractor prequalification, site-specific safety plans, and the documented inspections by which a controlling employer demonstrates the reasonable care the citation policy demands.

C. The Records, the Programs, and the People

The third dimension of the audit is the one an inexperienced auditor undervalues and a compliance officer never does. The opening document request of a real inspection typically begins with the OSHA 300 logs, the 300A annual summaries, and the 301 incident reports for the preceding five years, and recordkeeping citations are among the easiest the agency writes: cases recorded as first aid that meet the recordability criteria, missing 300A certifications, summaries never posted. The mock audit reconciles the logs against workers' compensation claims and first aid registers, because the discrepancies between those three data sets are exactly what an investigator will find. Written programs are then read against practice rather than against the standard alone — a hazard communication program listing a chemical inventory last updated three years ago is a finding even if every required section is present — and training records are tested against demonstrable competence rather than attendance. Finally, and most decisively, the auditor interviews employees privately, because the compliance officer will: the OSH Act guarantees the agency that access, and the distance between what the program binder says and what the third-shift operator says is precisely where citations are written. An audit that omits the interviews has not simulated an inspection; it has simulated a tour.

III. How Proactivity Is Rewarded Under OSHA

The regulatory scheme does not merely tolerate self-audit; it is structured, at several points, to reward it. The first and most direct reward is the penalty reduction for good faith. Under the Field Operations Manual, the agency's internal penalty calculus grants reductions of up to 25 percent for employers with effective safety and health management systems, and a documented program of periodic self-inspection with tracked correction is central to what an area director accepts as evidence of such a system (OSHA, 2025b). The reduction structure those good-faith credits sit within was revised on May 21, 2026, and the revisions materially enlarge what a small or mid-sized employer can recover: the maximum size-based reduction was extended to employers with as few as one and as many as twenty-five employees, at 70 percent for those in the eleven-to-twenty-five band; the 80 percent reduction available on serious and willful items was extended to employers with twenty or fewer employees; and the history reduction — available to employers with no serious, willful, repeat, or failure-to-abate violations in the preceding five years — was raised from 10 percent to 20 percent. Because the reductions are multiplicative against the gravity-based figure rather than additive, the practical spread between an employer that can document a functioning system with a clean history and one that cannot is now considerably wider than the headline maximums suggest, and the mock audit is the instrument that produces the documentation on which two of the three reductions turn. The 2026 penalty structure those reductions operate on is set out at Table 8-1, in Chapter 8, where the classification and contest consequences are treated; state-plan penalty structures must be at least as effective as the federal scheme. The second reward is the one the repeat classification makes vivid: because a repeat violation carries a maximum penalty ten times the serious maximum and a five-year, potentially corporate-wide lookback, every hazard a mock audit finds and corrects today is a repeat multiplier that never attaches tomorrow. For an employer with one citation already in its history, the mock audit is not general prudence; it is the specific instrument that determines whether the next inspection produces a $16,550 problem or a $165,514 one (OSHA, 2026).

The third reward answers the objection that stops many audit programs before they start: the fear that the audit report becomes the government's road map. OSHA addressed that fear directly in its final policy on voluntary employer self-audits, and the policy has three operative commitments. The agency will not routinely request self-audit reports at the initiation of an inspection and will not use such reports as a means of identifying hazards on which to focus during an inspection; where a voluntary self-audit identifies a hazard and the employer promptly undertakes appropriate corrective measures, the agency will not treat the audit report as evidence of willfulness; and the agency will treat that same audit-and-correction sequence as evidence of good faith for penalty purposes (OSHA, 2000). Read carefully, the policy is both a shield and an instruction. The shield attaches to the audit that is acted upon. Nothing in the policy protects the audit whose findings sat unaddressed: a distinction Section IV returns to, because the plaintiffs' bar reads it the same way.

The remaining rewards are programmatic. OSHA's own recommended practices for safety and health programs — published in parallel general industry and construction editions — place periodic program evaluation, self-inspection, and verification of hazard controls at the center of what the agency defines as an effective program (OSHA, 2016a, 2016b); a mock regulatory audit is the most rigorous available implementation of that element. The consensus management system standards impose the same discipline in stronger terms: both ISO 45001 and ANSI/ASSP Z10.0 require planned internal audits at defined intervals, conducted with objectivity and impartiality, with results reported to management and nonconformities driven to corrective action (American National Standards Institute/American Society of Safety Professionals [ANSI/ASSP], 2019; International Organization for Standardization [ISO], 2018). And at the aspirational end of the distribution, the employers admitted to OSHA's Voluntary Protection Programs — admission that itself requires exactly this culture of self-evaluation — sustain days-away, restricted, and transfer case rates far below their industry averages (OSHA, n.d.), while small employers can obtain a cost-free, citation-free version of the same examination through the agency's On-Site Consultation Program. The direction of every incentive in the scheme is the same: the employer who finds its own violations pays less, in every currency, than the employer who waits to be found.

IV. Why the Audit Trail Wins Lawsuits

The OSH Act does not displace the common law. The statute says so expressly: nothing in it enlarges, diminishes, or affects the common law or statutory rights, duties, or liabilities of employers with respect to injuries to their employees (29 U.S.C. § 653(b)(4)). Workers' compensation exclusivity bars most employees' tort suits against their own employers, but the litigation perimeter around a serious injury extends well past that bar: the injured employee of a subcontractor suing the general contractor or site owner, the third-party action against an equipment manufacturer that draws the employer's conduct into the case, the premises liability claim, the wrongful death action with an exception to exclusivity in play, and, in a number of jurisdictions, intentional-tort theories that pierce exclusivity precisely by alleging the employer knew of the hazard and consciously disregarded it. In nearly all of these, OSHA standards enter the courtroom as evidence of the standard of care, and an OSHA citation — particularly one classified willful or repeat — arrives as the plaintiff's first exhibit, carrying with it the argument that punitive damages should be on the table.

Against that backdrop, the documented mock audit program is the defense's counter-narrative, and it is a powerful one because it is built from the same materials the plaintiff would otherwise use. Willfulness, conscious disregard, and punitive-damages theories all require a mental state: the employer who knew, or plainly should have known, and did nothing. A dated audit report that identified the hazard class, a corrective action record showing an owner, an interim control, a completion date, and a verification signature, and a re-audit confirming the fix held: that sequence is direct evidence of the opposite mental state. It shows an employer that looked for hazards it was not yet required to know about, found them, and fixed them. The same record defeats the more routine negligence theory that the employer's safety program was a paper program, and it does collateral work everywhere the employer's diligence is priced: in workers' compensation experience modification, in liability insurance underwriting, and in the prequalification files of every general contractor and owner who scores safety history before awarding work.

The candor the argument requires is this: the audit trail only defends the employer who closes it. A mock audit report with findings that were never corrected is the single best document a plaintiff's attorney can hope to find in discovery, because it converts an ordinary negligence case into a notice case — the employer's own expert told it about the hazard, in writing, before the injury — and OSHA's self-audit policy withdraws its protections on the same condition, shielding only the audit followed by prompt correction (OSHA, 2000). The operational rule that follows is absolute: no finding leaves the audit process without an owner, a date, an interim control where the hazard cannot be immediately eliminated, and documented verification of abatement. Counsel has a role in structuring the program — audits directed by an attorney in anticipation of litigation may attract privilege or work product protection in some circumstances, and decisions about privilege belong to counsel — but privilege is narrow, jurisdiction-specific, and easily lost, and an employer forced to choose should prefer the audit program built to be discoverable: one whose every page shows a finding met by a fix. The strongest legal posture is not the hidden audit. It is the audit the employer would be glad to produce.

V. Conducting the Audit

The design principles separating a protective audit from a ritual one are few and firm. The auditor must be qualified and independent: competent in the applicable standards and free of responsibility for the conditions being audited, whether that independence comes from a corporate auditor examining a site they do not manage, a qualified consultant, or a peer-facility exchange; a supervisor grading their own homework is the failure mode both consensus standards prohibit (ANSI/ASSP, 2019; ISO, 2018). The methods must mirror the inspection: an opening conference, a document review, a walkaround that includes off-shifts and the corners a scheduled tour avoids, private employee interviews, and a closing conference that previews findings. The findings must be classified as a compliance officer would classify them — serious, other-than-serious, and, critically, flagged wherever prior citation history or prior internal findings would support a repeat or willful theory — because that classification, multiplied by the Table 8-1 amounts, is what converts the audit report into a risk figure leadership can weigh. And every finding enters a corrective action system with the closure discipline Section IV describes, reported to management at a defined cadence. The cadence itself should be risk-based: a comprehensive audit at least annually for a fixed establishment, project-based audits in construction keyed to high-hazard phases — steel erection, excavation, roofing — and targeted reaudits after incidents, near misses, significant process or personnel changes, and any actual OSHA activity.

The first ninety days of the program follow the same diagnostic-then-design sequence any sound management initiative follows. In the first month, the safety professional assembles the exposure picture: an inventory of the standards that apply to each establishment or project, the organization's own citation history and that of its industry peers from the public enforcement database, the OSHA 300 logs and near-miss records re-read as a map of where the hazards live, and a crosswalk of operations against the current most-cited list (OSHA, 2025a) and any active emphasis programs in the region. In the second month, that picture becomes a pilot: the highest-exposure facility or project is audited first, in full, with findings classified and costed, because a pilot that surfaces — and then closes — a set of findings whose citation value would have exceeded the audit's cost by an order of magnitude makes the business case in the only language budget owners reliably hear. In the third month, the pilot's corrective actions are verified closed, the results are reported to leadership beside the lagging indicators they already watch, and the audit schedule, auditor assignments, and corrective action tracking system are formalized for the enterprise. The pilot-first structure is deliberate here for the same reason it is deliberate anywhere: an audit program imposed enterprise-wide by memo becomes a checklist ritual, while a pilot that demonstrably retired six figures of citation exposure and a season of litigation risk earns its expansion.

Read correctly, none of this is defensive. A mock regulatory audit is not an act of pessimism about the workforce or the operation; it is the ordinary discipline of an organization that prefers to learn its weaknesses from someone it hired rather than someone with a badge or a subpoena. The compliance officer's checklist is public. The penalty schedule is public. The citation history of every competitor is public. Nothing about the government's examination is a secret except its date, and the employer who audits first has made even the date irrelevant. The inspection is coming on somebody's schedule. The only decision available is whose.

References

American National Standards Institute/American Society of Safety Professionals. (2019). Occupational health and safety management systems (ANSI/ASSP Z10.0-2019). American Society of Safety Professionals.

International Organization for Standardization. (2018). Occupational health and safety management systems — Requirements with guidance for use (ISO 45001:2018). International Organization for Standardization.

National Safety Council. (2025). Injury facts: Work overview. National Safety Council. https://injuryfacts.nsc.org

Occupational Safety and Health Act of 1970, 29 U.S.C. §§ 651–678 (1970).

Occupational Safety and Health Administration. (1999). Multi-employer citation policy (CPL 02-00-124). U.S. Department of Labor.

Occupational Safety and Health Administration. (2000). Final policy concerning the Occupational Safety and Health Administration's treatment of voluntary employer safety and health self-audits. Federal Register, 65(145), 46498–46503.

Occupational Safety and Health Administration. (2016a). Recommended practices for safety and health programs. U.S. Department of Labor. https://www.osha.gov/safety-management

Occupational Safety and Health Administration. (2016b). Recommended practices for safety and health programs in construction. U.S. Department of Labor.

Occupational Safety and Health Administration. (2022). Severe Violator Enforcement Program (CPL 02-00-169). U.S. Department of Labor.

Occupational Safety and Health Administration. (2025a). Top 10 most frequently cited standards, fiscal year 2025. U.S. Department of Labor. https://www.osha.gov/top10citedstandards

Occupational Safety and Health Administration. (2025b). Field operations manual (rev. July 14, 2025). U.S. Department of Labor.

Occupational Safety and Health Administration. (2026). 2026 annual adjustments to OSHA civil penalties [Memorandum]. U.S. Department of Labor. https://www.osha.gov/memos/2026-05-21/2026-annual-adjustmentsosha-civil-penalties Occupational Safety and Health Administration. (n.d.). Voluntary Protection Programs. U.S. Department of Labor. https://www.osha.gov/vpp U.S. Bureau of Labor Statistics. (2026). National census of fatal occupational injuries in 2024 [News release]. U.S. Department of Labor. https://www.bls.gov/iif/

CHAPTER 25

Exhibit 14 Is Written Today

Drafting Safety and Health Documents — Wording, Completeness, Compliance, and the Over-Promise

> *This book has asked one question of every document it has examined — how will this read when it is Exhibit 14? — and this chapter answers it from the drafting side. The failures are symmetric, and both are drafting failures: the program that promises more than anyone intends to perform hands the examining lawyer the gap between the written program and the practiced one, while the program too frightened of discovery to document at all reads, in the same courtroom, as no program.*

I. The Second Audience

Every document a safety program produces is written twice. The first writing is the one the professional performs: the policy drafted for the workforce, the audit finding entered for the plant manager, the job-hazard analysis built for the crew that will run the task tomorrow morning. The second is performed years later, by a reader the author never met and never imagined — an examining lawyer with the complete file, unlimited time, and a theory — and it is performed aloud, one sentence at a time, in front of people deciding what the organization knew and what it cared about. Nothing about the second reading can be controlled when it arrives: the documents are in discovery (Chapter 26 explains why nearly all of them are), the author is under oath, and the sentence means what it says. The only point of control is the first writing, which is why the question this book has asked of every document it has examined, how will this read when it is Exhibit 14?, is not a litigation question at all. It is a drafting question, and it is answered at the keyboard, in peacetime, by professionals who have mostly never been told the second audience exists.

This chapter is the drafting manual the surrounding chapters presuppose. Chapter 22 taught the investigation report's language; Chapter 23 taught the injury log's entries; Chapter 24 taught the audit's closure discipline; Chapter 26 will teach what happens to all of it in discovery and how little of it privilege protects. What remains — and what this chapter supplies — is the craft that runs beneath every one of those documents and every other page the program generates: the calibration of the words, the completeness of the record, the fidelity to the governing text, and the discipline of promising only what the organization will actually perform. Four failures account for nearly every program document that ends up damaging its author, and all four are drafting failures: the over-promise, the miscalibrated word, the unfinished record, and the rule that drifted from the regulation and from the floor.

The failures come in symmetric pairs, and the symmetry is the chapter's first lesson. The program that says too much — the manual that pledges perfection, the finding that editorializes, the e-mail that prophesies — hands the examining lawyer the gap between what was written and what was done. The program that says too little — the audit never documented, the near-miss never recorded, the hazard flagged verbally so that nothing would be discoverable — reads, in the same courtroom, as the absence of a program, which is worse. The discipline this chapter teaches is therefore not defensive minimalism, and the professional who takes "write less" as its message has misread it. The discipline is accuracy: a record complete enough to prove the system worked, worded carefully enough to say exactly what happened and no more, promising exactly what the organization performs. Write what you do; do what you write; be able to prove both. Everything that follows is elaboration.

II. The Over-Promise: The Standard You Set for Yourself

Begin with the document that governs all the others, because it is also the one most often written by someone reaching for eloquence. The corporate safety manual is admissible in the injury case as evidence bearing on the standard the organization set for itself, and the examining lawyer will read its promises against the incident's facts. The law's formal position offers less comfort than employers hope and more than they fear. In *Wal-Mart Stores, Inc. v. Wright*, the Indiana Supreme Court held it error to instruct a jury that a company's own rules were evidence of the degree of care the company itself judged to be ordinary: precisely because organizations may, and should remain free to, set internal standards that exceed what ordinary care requires, and a rule that punished the higher standard by converting it into the legal one would teach employers to write worse manuals. That is the prevailing sensibility: internal rules do not define the standard of care. But the manual came into evidence in Wright, as it will in the next case, and the argument the defense is left with, that the written program exceeded the law and the practice satisfied it, is an argument that can be won only when the written program and the practice are close enough to be discussed in the same sentence.

That is the argument the aspirational pledge forfeits in advance. "Zero harm." "Safety is our highest priority." "We will never place production ahead of people." Each of these is written for a lobby wall and read, eventually, in a courtroom, where the arithmetic of cross-examination is unforgiving: the pledge on the poster, the denied capital request in the e-mail, the injury in between. The point is not that values have no place in a safety program, a program without stated values is a checklist, but that a value is a promise, and promises should be written in the currency the organization actually pays. "We investigate every recordable injury and near-miss and track corrective actions to verified completion" is a pledge an organization can perform, prove, and stand behind at deposition. "Nothing is more important to us than your safety" is a pledge the first budget cycle will impeach. The drafting rule: commitments phrased as conduct, in sentences the corrective-action log can corroborate.

The same calibration governs the small words. "Always," "never," "ensure," "guarantee," and "at all times" are absolute warranties, and a single contrary photograph defeats each of them; "all employees will be trained" is falsified by one expired certification. The honest modality — inspections conducted at defined intervals, training delivered before assignment and refreshed on a stated cycle, controls designed to prevent a stated failure — is not hedging; it is the difference between a specification and a slogan, and a specification can be met. The caution runs in one direction only. The remedy for the over-promise is never the weasel-word — the "as appropriate," the "where feasible," the "periodic" left undefined — because language written to be unenforceable reads exactly that way to a jury, and a program of deliberate vagueness is an admission with better grammar. Say precisely what will be done, by whom, how often, and mean it.

Beneath the language problem sits the deeper one: the gap between the written program and the practiced one, which Chapter 26 identifies as the deposition's destination. Every mature program drifts — procedures outlive the equipment they described, responsibilities outlive the organizational chart — and the drift is survivable when it is managed and fatal when it is discovered. The discipline is reconciliation on a schedule: the program reviewed against the floor at defined intervals, and one of two things done wherever they disagree — the practice corrected to the document, or the document revised, honestly and with a dated revision history, to the practice the organization actually performs and can defend. A revision made in peacetime is diligence; the identical revision made after the incident is consciousness of guilt; and the rule nobody follows is the worst artifact in the file, because it simultaneously proves the hazard was recognized, the control was known, and the organization's own requirements were negotiable — the combination that defeats the unpreventable-employee-misconduct defense (Valdak Corp. and Chapter 8) and invites the willful classification the settlement chapters describe.

The over-promise also escapes the manual and enters the contract, where it acquires consideration. The prequalification questionnaire that warrants an experience rating, a training regime, or "full compliance with all applicable regulations"; the bid that offers a dedicated full-time safety presence the project never staffs; the site-specific plan that assigns the contractor duties its crew size cannot perform — each is the program over-promising in a document the other side signed, and each will be read against the incident by a lawyer arguing that the duty was assumed in writing. Chapter 19 taught that retained-control liability follows the authority a hirer keeps and exercises; the drafting corollary is that safety obligations in contracts should describe the control structure the parties will actually operate, because the paragraph that promises supervision creates the duty to supervise, and the safest contract language is the language that matches the site.

The furthest reach of the over-promise is the undertaking doctrine, and every consultant and corporate parent should know it by section number. Under Restatement (Second) of Torts § 324A, one who undertakes to render services to another that the actor should recognize as necessary for the protection of a third person can be liable to that third person for failure to exercise reasonable care in the undertaking. In *Canipe v. National Loss Control Service Corp.*, the Fifth Circuit applied the principle to a consulting firm engaged to provide safety inspections and accident-prevention services at a plant where a worker was later injured by machine hazards the inspections allegedly should have found: the consultant owed the worker a duty to inspect with reasonable care, whether or not it held any authority to correct what it found. The doctrine reaches the retained consultant, the insurer's loss-control engineer, and the corporate parent whose headquarters audit program examines the subsidiary's plants, and its drafting lesson is scope. The engagement letter and the report should state what was undertaken and, where a reader might assume otherwise, what was not: the operations examined and the ones excluded, the standards applied, the limits of the sampling and the walkthrough. The audit that silently covered less than its title implies has promised, to a future reader, everything the title implies.

There is, finally, a flip side the profession must hear without flinching: the rule you wrote is proof of what you knew. An employer's own work rules, hazard analyses, and program records are evidence of hazard recognition under the General Duty Clause — in the Commission's Pepperidge Farm litigation, recognition of the ergonomic hazards was built in substantial part from the employer's side of the file, from what its own personnel, records, and responses showed it had already identified — and the internal document that names a hazard is the first exhibit in any willfulness or punitive-damages fight about it. None of this is a reason to stop writing rules; a program without written rules has conceded the recognition point anyway and abandoned its affirmative defenses with it. It is a reason to close the loop the rule opens. The document that names a hazard and the record that shows it controlled are, together, the strongest paper an employer can hold; the document that names a hazard and stands alone is the plaintiff's opening statement.

III. Wording: Verbs, Findings, and the Banned Vocabulary

Program language is built out of verbs, and the verbs carry legal weight the profession rarely assigns deliberately. "Shall" and "must" create requirements: they are the correct words for what the organization genuinely requires, audits, and enforces, and the wrong words everywhere else, because every "shall" is a self-imposed obligation the examining lawyer will test line by line — the procedure said the interlock shall be verified each shift; show me the verification records. "Should" recommends; "may" permits; and the difference between them is the difference between a violated requirement and an exercised judgment. Most programs were assembled over years by many hands, and their verbs were never calibrated by anyone: mandatory language decorating aspirations no one tracks, permissive language diluting requirements the organization actually enforces. The audit this chapter recommends before any other is a verb audit — every "shall" either backed by an enforcement and verification record or demoted honestly, every genuine requirement promoted out of "should" — because a document whose modality is accurate can be defended sentence by sentence, and a document whose modality is ornamental cannot be defended at all.

Findings — the working output of audits, inspections, observations, and investigations — have a formula, and Chapter 22 introduced it for the incident file: the condition observed, the standard or requirement it implicates, the correction recommended, the owner responsible, and the date. The formula deserves promotion from the investigation to the entire program, because it is simultaneously the best prevention writing and the best evidence writing the profession has: specific enough to act on, neutral enough to read aloud, structured enough that its closure can be proven. "Housekeeping needs improvement" is neither actionable nor defensible; "combustible accumulation inside the marked clearance of the battery-charging station, contrary to the facility's fire-prevention plan — removed today; clearance check added to the shift inspection; facilities supervisor; verification by photograph Friday" is both. The formula's discipline is also its protection: a writer reciting condition, standard, correction, owner, and date has no grammatical room left for the sentence that convicts.

Because certain words do convict, and they should be banned from program documents by name. The legal conclusions — "negligent," "reckless," "willful," "violation" used as a verdict rather than a finding, "liability" — are the lawyers' vocabulary; in a professional's document they are gifts to whichever side the professional's employer is not on, and they are almost never what the writer actually observed. The prophecy — "someone is going to get killed on that line" — feels like moral urgency when it is typed and reads like foreknowledge when it is projected; the urgency belongs in the risk ranking and the deadline, not the rhetoric. The hyperbole, the sarcasm, the self-flagellation — "a death trap," "an accident waiting to happen," "we got lucky again," "this was entirely preventable" — are closing arguments volunteered without the burden of proof. And the ban runs equally in the flattering direction, which safety writing discusses less: the minimizer that grades a missing guard a "minor housekeeping item," the euphemism that converts an amputation hazard into an "opportunity for improvement," and the agentless passive — "the guard was found to have been removed" — that launders responsibility out of the sentence read, one day, to a jury deciding whether anyone was responsible. The finding formula needs no adjectives in either direction. The condition, the standard, the correction: stated plainly, those are the whole truth, and they are enough.

Severity deserves language of its own, because the writer must be able to say that some hazards are grave without reaching for theatrics. The instrument is the defined scale: a risk matrix or ranking criteria adopted in the program and applied consistently, so that "serious" is a category with stated probability and consequence thresholds rather than an adjective chosen by mood. A hazard ranked at the program's highest severity under published criteria, with the correspondingly short abatement deadline, communicates everything the prophecy communicated — to management now and to the jury later — and it does so as a professional judgment applied through a method, which is exactly the posture this book has urged for every technical opinion in every other chapter. The ranking that triggered escalation and the deadline that was met are evidence of a system; the e-mail that screamed is evidence of a Cassandra, and juries punish the organizations that had one.

Timing disciplines the rest. The first documents after an incident should record facts — conditions, positions, readings, statements, photographs, custody — and should not speculate about cause, because the initial theory is usually wrong and always discoverable, and the file that opens with "operator error" or "the guard must have been off for weeks" has committed the organization to an analysis no one had yet performed. Causal analysis belongs in the structured, methodology-driven investigation of Chapter 22; legal evaluation belongs in counsel's privileged channel under the architecture of Chapter 26; and the same division answers e-mail, the ninety-second document where the record is most candid and least considered. The rule for e-mail is not silence — Chapter 26 explains why the professional who stops documenting hazards has made everything worse — but form: an e-mail that reports a hazard is a finding, and it should be written as one, in the formula, however fast the fingers are moving. Condition, standard, correction, owner, date. Send that, and the sentence the lawyer projects on the screen is the sentence the professional would write again.

IV. Completeness: The Record That Finishes Its Sentences

A finding is a sentence the program has started and must finish, because an unfinished finding is the most dangerous document a safety program can hold: it is knowledge, memorialized, without cure. The open audit item, the recommendation without an owner, the corrective action marked complete that no one verified: each is an instance of the pattern this book has traced through the willfulness and punitive-damages discussions, the pattern of the employer who knew and did not act. Completeness is therefore not clerical hygiene; it is the load-bearing element of the record's meaning. Every finding carries its correction, every correction its owner and date, every closure its verification — who confirmed the fix, how, and when — because "closed" is a claim, and claims in this file get tested.

Completeness does not require the impossible, and the program that understands this documents its judgment instead of hiding it. Corrections compete for capital and attention; a risk-based queue is legitimate and, in a mature program, inevitable. What converts the queue from evidence of neglect into evidence of management is writing: the ranking criteria applied, the interim control installed while the permanent fix awaits the outage or the appropriation — the guarding restored temporarily, the exposure held down with protective equipment and monitoring in the meantime, the schedule stated — and the deferral reasoned in a sentence someone is willing to read aloud. The organization that wrote "engineering correction scheduled for the spring shutdown; interim barrier and weekly verification until then" has documented diligence. The organization that let the finding age silently has documented the other thing, and the file does not record which of them meant well.

The temptation this chapter must confront directly is the one *Dowling* settled: there is no privilege for the routine program record, however candid, and professionals who learn that sometimes draw the wrong conclusion — that the safe program is the undocumented one. Chapter 26 dismantles the premise at length; here it is enough to state the drafting consequence. The discoverable record is not the risk; it is the defense. The audit that found, the correction that closed, the monitoring that confirmed — these are, in the defense chair, frequently the entire case: the proof that the organization ran a system rather than a filing cabinet. A program too frightened of discovery to write anything down has not avoided the second audience; it has simply arranged to meet it empty-handed, with no answer to the question of what the organization knew except the one the other side supplies.

Completeness also means coherence, because the program's documents testify as a group and contradictions impeach. The job-hazard analysis that lists a control the operating procedure omits; the training deck that teaches a rule the manual superseded two revisions ago; the permit that requires an attendant the staffing plan never assigned — each pair hands the examiner a choice of admissions, and the organization will be held to whichever reading is worse. The discipline is dependency: documents that share subject matter share a change process, so that a revision to the procedure propagates to the analysis, the training content, and the checklist that flow from it, with the linkage recorded. A program whose documents agree with each other can be defended as one thing; a program whose documents quarrel will be cross-examined against itself.

Finally, some completeness is not optional, and its absence is itself an entry. The recordkeeping regulation's log determinations (Chapter 23), the thirty-year exposure and medical records of § 1910.1020, the training and certification records the substantive standards require — these are records the law commands, kept to the law's content requirements, and a gap in a mandatory record is read the way Chapter 23 reads the silent log: as a decision. The mandatory records are the skeleton of the file the litigation will reconstruct; the program documents this chapter governs are the tissue around them; and both are judged, in the end, by the same reader.

V. Compliance: Writing Against the Text

A safety rule is a translation of a governing text, and translations drift. The procedure drafted from memory of the standard, then revised by committee, then localized by a site that never read the original, ends as a third thing — neither the regulation nor the practice — and the gap cuts in both directions at once: where the rule demands less than the standard, the rule documents noncompliance; where it demands more, the organization has volunteered a stricter requirement it will be held to under the logic of Wright. The discipline is drafting against the text: the work rule mapped to the specific provision it implements, cited by section and, where it matters, by edition; the standard's defined terms used as defined rather than paraphrased; the scope and exceptions carried over rather than rounded off. Chapter 8 made the litigation case for this practice — the unpreventable-misconduct defense requires work rules that implement the cited standard, and rules written in aspirational generalities fail it — and the drafting case is the same point in peacetime: a rule that tracks the text can be audited against the text, and a rule that tracks nothing can only be argued about.

Consensus standards deserve a deliberate adoption decision rather than an accidental one. When a program incorporates ANSI, NFPA, or ISO language — or simply requires "compliance with applicable consensus standards" — it has adopted, as its own requirement, provisions the law may not impose, and the organization will be measured against them precisely because it chose them. The choice is often right: the consensus document is frequently the current best practice, and the management-system standards a program may already claim (ANSI/ASSP Z10.0, ISO 45001) themselves require documented processes, document control, and internal audit. The discipline is to adopt on purpose: name the standard and the edition, record what was adopted and why, and put the adopted provisions into the reconciliation cycle of Section II, so that a revision to the consensus document is a decision for the program rather than a silent widening of the gap. The training file has its own consensus benchmark — ANSI/ASSP Z490.1's criteria for training development, delivery, and evaluation — and training materials drafted to it inherit a defensible method the way a sampling plan inherits one from a published protocol.

Document control is where compliance drafting is proven or lost. Every controlled document carries a revision number, an effective date, an approver, and a record of what changed; superseded versions are retained under the program's retention schedule, because the version in force on the day of the incident — not the current one — is the document the case will be about, and the organization that cannot produce it will litigate its contents from adverse memory. Retention itself must be a written schedule, uniformly applied, and applied before litigation is foreseeable: Chapter 21 teaches that destruction under a routine policy becomes spoliation once the duty to preserve attaches, and the undated policy, the unnumbered revision, and the drawer of drafts nobody can sequence are each, in their way, documents the other side will enjoy more than the author did.

Two review practices close the section. Legal review of program documents is worth having and easy to have badly: counsel's altitude is the promise, the conclusion, the warranty, and the retention trigger, not the technical content, and not the candor. A review that strips findings of their substance to manage discovery risk has converted a prevention document into a press release, and the waiver and privilege doctrines of Chapter 26 will not protect it anyway. And the borrowed program — the template purchased or downloaded and adopted wholesale, sometimes still bearing another company's name in paragraph four, describing equipment the facility does not own and duties no one on site performs — is the compliance failure and the over-promise fused into a single exhibit: proof, in the organization's own binder, that the written program was never about this workplace at all. Templates are starting points. The program the organization signs must describe the organization.

VI. The Documents, One by One

The disciplines above land differently on different documents, and a short tour fixes them. The policy manual sits at the top and should be the shortest document in the system: governance, responsibilities, and commitments phrased as conduct, with the operational detail delegated to procedures that can change without amending the constitution — everything in Section II bears on it hardest, because it is the document of promises. The job-hazard analysis is a specification of a task and its controls, and its credibility is specificity: this task, this equipment, these steps, these controls, dated, built with the workers who perform the work, and revised when the task changes; the generic analysis photocopied across a hundred dissimilar jobs proves the paperwork existed and the analysis did not. The operating and energy-control procedures inherit the same rule at machine level: Chapter 22's press, the lockout procedure that never addressed the pneumatic energy source, is what a procedure written for "equipment" rather than for the machine looks like in litigation.

The audit and inspection reports are the program's findings engine, and Sections III and IV are their entire law: the formula, the ranking, the closure, the verification. The safety-committee minutes are *Dowling*'s document — discoverable, always — and should record attendance, items, decisions, owners, and dates, in the finding formula, without the editorial debate; minutes that record "management again declined to fund the guarding proposal; discussion of prior incidents followed" have preserved an argument, not a program. The near-miss file is the program's early-warning system and its notice record in one, which is precisely why its reports must be written as facts and fixes: the near-miss narrative that assigns blame or predicts catastrophe converts the program's best prevention data into its worst exhibits, and the trend analysis that goes unread converts it into constructive knowledge. Training materials are program documents in every sense: the content must match the current rule, the delivery must be provable per employee — rosters, verifications, dates, cross-referenced to the rule taught — and the file must show, as one record, that the person on the machine was trained on the procedure in force. The injury log is Chapter 23's subject, and its entries are drafting decisions of exactly this chapter's kind. And e-mail is all of these documents at conversational speed, which is why it received its own rule in Section III and receives the last word in Chapter 26.

VII. Language, Literacy, and the Training Record

One document in the system fails more often, and more consequentially, than any other, and it fails for a reason none of the drafting disciplines above will catch: the signed training acknowledgment in a language the worker could not read. It is worth its own section because it is simultaneously the most common finding in this field, the easiest to prevent, and the one most likely to be treated as clerical by the person who could have prevented it.

The regulatory hook is not obscure. The hazard communication standard requires that employees be provided information and training on hazardous chemicals in their work area, and OSHA has taken the position for decades — in interpretation letters, in its training guidelines, and in the preamble to the standard itself — that training must be presented in a manner and language the employees can understand. The same requirement runs through the substance-specific standards, through the construction training provisions, and through the general duty to instruct employees in the recognition and avoidance of unsafe conditions. What the standards require is comprehension. What the file almost always documents is attendance.

The distinction is the whole of it. A sign-in sheet establishes that a body was in a room. It does not establish that the person understood the hazard, the control, or the refusal right, and it is exactly the document that will be enlarged and put on a screen. The problem is not hypothetical in this book's flagship docket: the engineered-stone fabrication workforce is overwhelmingly immigrant and substantially Spanish-speaking, a fact the case series behind Chapter 16 records in passing, and the same is true across residential construction, food processing, agriculture, warehousing, and janitorial services. A training file consisting of English-language materials, an English-language quiz, and a signature is not evidence of training in those workplaces. It is evidence of the opposite, and it is evidence the employer generated itself.

The symmetry this book insists on holds here as squarely as anywhere. For the plaintiff, the signed acknowledgment in a language the worker did not read is among the most efficient exhibits available: it converts a training defense into an admission, it needs no expert to explain, and the cross-examination writes itself from the document's own face. For the defense, the mirror is equally strong and much less often built. An employer who can produce the Spanish-language materials actually used, a bilingual trainer identified by name, a comprehension check answered in the worker's own hand, and a documented determination of the languages spoken by the crew has a training record that survives — and has it because someone made four inexpensive decisions years before the incident.

Those four decisions are the drafting instruction. First, determine and record the languages and literacy levels actually present in the workforce, by crew and by shift, and re-determine it when the crew changes; a program that has never asked cannot claim to have addressed it. Second, deliver in the language, using materials in the language rather than an English document read aloud through an ad hoc interpreter, and identify the interpreter by name and qualification where one is used. Third, document comprehension rather than attendance: a short check in the language of instruction, answered by the worker, retained with the acknowledgment, and, for a workforce with low print literacy in any language, verified by demonstration or teach-back rather than by a written quiz, with the demonstration recorded by the person who observed it. Fourth, write the acknowledgment itself in the language the worker reads, and have it say what actually happened: the topics covered, the language of instruction, the name of the trainer, the form the comprehension check took, and the worker's opportunity to ask questions. An acknowledgment that recites "I have received and understand the above training" in English, signed by a worker who does not read English, is worse than no acknowledgment, because it is a false statement the employer drafted and asked the worker to sign.

Three consequences reach beyond the training file itself, and they connect this section to the rest of Part V. The first is auditable: the language question belongs on the mock-audit protocol of Chapter 24 as a standing item, because it is invisible in a document review that only checks whether training records exist. The second is testimonial: the safety professional who signed off on a training program that was not comprehensible to the workforce is the witness of Chapter 26 §II, and the questions in that deposition are the four above, asked in order. The third is that the same reasoning governs every other communication the program depends on — the written hazard-communication program, the emergency action plan, the lockout procedure posted at the machine, the safety data sheets, the signage, and the mechanism by which a worker reports a hazard or refuses unsafe work. A refusal right the worker cannot read is a refusal right the employer has not provided, and Chapter 13 explains what that costs when the retaliation claim arrives.

VIII. What the Cases Teach

Wal-Mart (Ind. 2002), the manual in evidence, and its limits. A customer injured in a fall obtained a jury instruction that the company's own rules and policies could be considered as evidence of the degree of care the company itself judged to be ordinary; the Indiana Supreme Court reversed, holding that internal rules may reflect a standard higher than the law requires and that the instruction improperly invited the jury to substitute the company's standard for the objective one — a holding expressly protective of organizations that set demanding internal requirements. The comfort is real, and it is limited. The manual was admitted; the fight was over what it proved; and the decision's premise, that written programs may exceed ordinary care, is a defense only for the organization whose practice honors the program it wrote. For the drafter, Wright is both permission and warning: write the demanding rule the operation needs, and understand that from the moment it is written, the operation will be examined against it.

Grimshaw v. Ford Motor Co. (Cal. Ct. App. 1981) — the internal document, read aloud. The punitive award in the Pinto fuel-tank litigation rested on what the company's own files showed: crash-test results known to management, and internal analyses weighing the cost of design changes against the consequences of leaving the design as it was. The Court of Appeal affirmed punitive damages on a conscious-disregard theory built substantially from the defendant's own engineering and cost documents. The case is four decades old and remains the profession's clearest demonstration of the second audience: documents written as ordinary internal analysis, by authors who never imagined a jury, becoming the moral center of the trial. Its lesson for the drafter is not concealment — the analysis was the company's knowledge whether or not it was ever written down well — but the whole of this chapter: the organization that documents a hazard has assumed the obligation to document what it did about it, and the file that shows knowledge weighed against cost, with nothing done, is the punitive-damages case in the defendant's own hand.

Canipe (5th Cir. 1984), the undertaking that creates the duty. A food manufacturer engaged a loss-control firm to provide safety inspections and accident-prevention services at its plant; a worker injured on a machine sued the consultant, and the Fifth Circuit held that under Restatement § 324A the firm owed him a duty to perform its inspections with reasonable care — reversing summary judgment even though the consultant had no authority to implement corrections, because the duty follows the undertaking, not the authority. Every consultant, insurer loss-control engineer, and corporate parent running a headquarters audit program inherits the holding, and its drafting instruction is scope: the engagement and the report should state what was examined and what was not, which standards were applied, and what the review was and was not designed to find, because the undertaking a court later reconstructs will otherwise be the one the report's silence allowed a reader to assume.

Pepperidge Farm, Inc. (OSHRC 1997), the employer's own paper as recognition. In the Commission's landmark General Duty Clause litigation over lifting and repetitive-motion hazards, hazard recognition was established in substantial part from the employer's side of the file — from what its own personnel, records, and responses showed the company had already identified — and the decision stands for a proposition the profession must internalize: the documents a program generates about a hazard are proof the hazard was recognized, by the party whose recognition matters. The drafting response is not less paper. It is the closure discipline of Section IV, because recognition plus verified correction is a functioning program, the exact showing the self-audit incentives of Chapter 24 reward, while recognition standing alone is the Secretary's element, pre-written.

IX. The Program That Reads Well

The chapter reduces to a page of practice. Write the program against the governing text, in verbs that are audited before they are published, promising conduct the organization performs. Write findings in the formula — condition, standard, correction, owner, date — with the vocabulary bans enforced in both directions, severity carried by defined criteria rather than adjectives, and cause reserved for the analysis that has a method. Finish every sentence the program starts: closure verified, deferral reasoned and interim-controlled, the mandatory records complete, the document set internally consistent and version-controlled, the superseded revision retrievable. Reconcile the written program to the practiced one on a schedule, and amend whichever of the two is wrong. And when the temptation arrives — after the near-miss, after the close call the file describes too well — to write less, remember which failure is worse: the organization with a candid record of finding and fixing has a defense, and the organization with an empty file has an adverse inference with its name on it.

Counsel's reciprocal program is a single habit with a large return: read the client's program documents before the other side does. The drafting audit — the manual's promises against its practice, the verbs against the enforcement records, the open findings, the contradictions between documents, the contract's safety warranties against the site's actual structure — is loss prevention when performed in peacetime and case assessment when performed at intake, and it is an assignment for which the consulting safety professional of Chapter 1 is precisely built. Every weakness that audit finds is one the deposition would have found later, at a worse time, on the record.

And the professional should hear the chapter's point at its full size, because it is larger than litigation. Every discipline urged here — the accurate verb, the finished finding, the rule that matches the standard and the floor, the promise the organization keeps — is simply good safety practice wearing its courtroom clothes. The document that would survive cross-examination is the document that runs a program well: specific enough to act on, honest enough to trust, complete enough to manage. The second audience does not ask the professional to write for lawyers. It asks her to write exactly as well as the work always deserved, because the page she writes today is already Exhibit 14, and the only open question is which side it will help.

References

29 C.F.R. § 1910.1200(h) (hazard communication; employee information and training); 29 C.F.R. § 1926.21(b)(2) (instruction in the recognition and avoidance of unsafe conditions).

American National Standards Institute/American Society of Safety Professionals. ANSI/ASSP Z10.0-2019, Occupational Health and Safety Management Systems; International Organization for Standardization. ISO 45001:2018, Occupational Health and Safety Management Systems.

American National Standards Institute/American Society of Safety Professionals. ANSI/ASSP Z490.1, Criteria for Accepted Practices in Safety, Health, and Environmental Training.

Canipe v. Nat'l Loss Control Serv. Corp., 736 F.2d 1055 (5th Cir. 1984), cert. denied, 469 U.S. 1191 (1985).

Dowling v. Am. Hawaii Cruises, Inc., 971 F.2d 423 (9th Cir. 1992).

Grimshaw v. Ford Motor Co., 119 Cal. App. 3d 757, 174 Cal. Rptr. 348 (1981).

Occupational Safety and Health Administration. (2010, Apr. 28). Standard interpretation: Employer obligation to provide training in a language and vocabulary employees can understand. U.S. Department of Labor.

Occupational Safety and Health Administration. (2015). Training requirements in OSHA standards (OSHA 2254-09R). U.S. Department of Labor.

Pepperidge Farm, Inc., 17 BNA OSHC 1993 (No. 89-0265, 1997).

Recording and reporting occupational injuries and illnesses, 29 C.F.R. pt. 1904; access to employee exposure and medical records, 29 C.F.R. § 1910.1020.

Restatement (Second) of Torts § 324A (1965).

Valdak Corp., 17 BNA OSHC 1135 (1995).

Wal-Mart Stores, Inc. v. Wright, 774 N.E.2d 891 (Ind. 2002).

CHAPTER 26

Now the Record Calls You

The Safety Professional as Witness — You Wrote the Record; Now It Calls You to Testify

Every safety professional and industrial hygienist who audits a site, monitors an exposure, or sends a candid e-mail about an uncorrected hazard is writing testimony, and most do not learn this until a court reporter swears them in. When the injury becomes a lawsuit, the professional is summoned in one of three chairs — fact witness, corporate representative, retained expert, and each carries different obligations, different protections, and different ways to destroy a case or a career in a single answer. Meanwhile the professional's own paper moves through discovery under rules the safety curriculum never teaches.

I. The Three Chairs: Fact, Corporate, Retained

The safety professional enters litigation through three distinct doors, and the door determines everything that follows. The fact witness — the safety manager who conducted the orientation, walked the audit, or arrived at the scene — testifies to what she personally saw, said, did, and wrote, under a subpoena that compels attendance whether or not the employer is a party. The corporate representative, designated under Rule 30(b)(6) or its state analogues, testifies to something categorically different: not personal knowledge but the organization's knowledge, assembled from documents and interviews across the enterprise, on noticed topics, in answers that bind the corporation as its own admissions. And the retained expert, the consultant brought in after the incident, testifies to opinions, under disclosure rules that expose the basis of every one of them. The three roles carry different preparation duties, different privilege consequences, and different exposure, and the professional who does not know which chair she occupies is a danger to her employer, her client, and herself.

The chairs are not sealed compartments, and the seams between them are where witnesses are damaged. The fact witness who is asked, three hours into a deposition, whether a guard would have prevented the amputation is being invited to migrate from fact to opinion: to render, for free and without preparation, the expert testimony the examining lawyer could not otherwise obtain, and to render it against her own employer's record. The in-house professional whose observations are laced with professional judgment may be treated as a non-retained expert under Rule 26(a)(2)(C), disclosed and examined on opinions she formed on the job. The consulting expert retained to advise counsel quietly, protected from discovery absent exceptional circumstances under Rule 26(b)(4)(D), loses that protection the moment she is converted to a testifying role, and in many courts the conversion cannot be walked back once her opinions have been disclosed. Each seam has a discipline, and the disciplines are learnable; what is not survivable is discovering the seam mid-deposition.

One premise governs every chair, and it is the same premise that governs honest expert practice generally: the record was written before the case existed, by a professional who did not know which side of the v. she would someday serve. The audit says what it says. So does the e-mail. The witness's task in any chair is not to improve the record but to testify accurately about it, because the record will be in the examining lawyer's hands, tabbed and highlighted, and the witness who contradicts her own contemporaneous paper has handed the case to the other side more surely than any adverse finding in the paper itself. The sections that follow take the chairs in order, then turn to the paper, the protections, and the practices that determine how much of the paper the other side ever sees.

II. The Deposition: Testifying About Your Own Program

A. The Fact Witness: The Record Is the Testimony

The deposed safety professional is examined against a documentary record the examining lawyer knows better than the witness remembers it. The lawyer has the audit reports, the training rosters, the job-hazard analyses, the corrective-action log, the e-mail chains — obtained in discovery, read in sequence, and arranged to build toward the questions that matter: what did you know, when did you know it, what did you do, and what did you write about it. Preparation therefore begins with the record, not the memory. The witness must re-read every document she authored or received that touches the incident, in date order, before testifying, because the most damaging moment in a safety deposition is not the adverse finding but the confident answer the witness's own document contradicts, produced from the folder thirty seconds later. Memory reconstructed under oath and refuted on the spot is what impeachment looks like, and no privilege repairs it.

The answering discipline is the discipline of the honest technical witness everywhere: answer the question asked, completely and accurately, and stop. The safety professional's occupational instinct — to explain, to contextualize, to teach — is precisely the instinct the deposition punishes, because every volunteered sentence opens a door the question did not. The truthful "I don't know" and the truthful "I don't recall" are complete answers, and they are answers, not evasions, when they are true; the witness who guesses to appear competent has manufactured testimony the record will test. And the witness must recognize the opinion question when it arrives dressed as a factual one. Whether the guard was feasible, whether the exposure caused the disease, whether the program met the standard of care: these are expert questions, and the fact witness is generally entitled to decline them as beyond what she observed and did, while answering fully and without defensiveness what she did observe and do. Counsel defending the deposition will police the line, but the witness who understands the line defends it better than any objection.

The witness's own program will be used as the measuring stick, and this is both fair and survivable. The corporate safety manual is admissible as the standard the organization set for itself, and the examining lawyer will walk the witness through the provisions the incident implicates: the manual required monthly inspections, were they done? The manual assigned you stop-work authority, did you exercise it? Deviations between the written program and the practiced one are the examination's destination, and the witness cannot wish them away. What she can do is testify accurately about what the program required, what was actually done, and why — including the legitimate operational reasons practices evolve — without adopting the lawyer's characterizations, without agreeing that every gap was a violation, and without defending the indefensible. The witness who concedes the true gap plainly and holds the line on the false one is credible; the witness who fights everything is not, and juries reading deposition excerpts can tell the difference.

B. The Corporate Representative: Speaking for the Organization

The Rule 30(b)(6) designation transforms the assignment entirely. The organization, served with a notice listing topics — the safety program, the audit history, the training system, the incident investigation, the exposure monitoring — must designate one or more persons to testify about information known or reasonably available to the organization, and the designee's answers are the corporation's answers, binding in a way no individual witness's are. Personal knowledge is neither required nor sufficient: the designee must be educated, and the education is a duty, not a courtesy. Courts have sanctioned corporations whose designees appeared unprepared, treating the empty-headed designee as a nonappearance, and the sanctions decisions make clear that "no one ever told me" is an indictment of the preparation, not a defense of the answer. The safety director designated on safety topics must review the documents, interview the personnel, and absorb the institutional knowledge across departments, predecessors, and years, because "the corporation does not know" is itself binding testimony, and it forecloses the corporation from producing a knowledgeable witness at trial on the same topic.

The designated safety professional therefore prepares along two axes at once. The first is scope: the noticed topics define the examination, and counsel should have negotiated them to reasonable particularity, but within the topics, the designee answers for the organization, including for facts she personally never touched, decisions made before her tenure, and documents she did not write. The second is the distinction between the organization's knowledge and her own opinion, which matters more in this chair than any other: the designee asked what the corporation knew about the prior near-misses answers from the record; the designee asked whether the corporation's response was adequate is being asked to bind the corporation to an evaluative admission, and the preparation sessions must decide, before the deposition, what the organization's accurate position on those questions is. A corporate representative improvising the company's position under oath is drafting admissions in real time. The role, executed well, is among the most demanding assignments in civil litigation; executed casually, it is among the most expensive.

III. The Paper: Audits and E-mails in Discovery

The default rule of civil discovery is breadth: any nonprivileged matter relevant to a claim or defense and proportional to the needs of the case is discoverable, and the safety management system is a machine for generating exactly such matter. The audit reports, the inspection findings, the observation data, the near-miss reports, the safety committee minutes, the corrective-action tracking, the training records, the disciplinary file, the OSHA 300 logs, the industrial-hygiene monitoring results the exposure-records rule requires be kept for thirty years: all of it is created in the ordinary course of business, none of it is privileged by virtue of being about safety, and in the third-party injury case it will be demanded from the employer by subpoena even when the employer is not a party. The professional who imagines the audit file as internal is imagining a boundary the Federal Rules do not recognize. The file is written for the organization; it is read by the litigation.

E-mail deserves its own paragraph, because e-mail is where the safety record is at its most candid and least considered. The audit report is drafted, reviewed, and issued; the e-mail is typed in ninety seconds between meetings, and it is the ninety-second document — "this is the third time I've flagged this," "someone is going to get killed on that line," "we both know operations won't spend the money" — that becomes the exhibit blown up before the jury. The discipline this compels is not silence; a safety professional who stops documenting hazards to manage discovery risk has committed a worse error than any e-mail, both professionally and in the eventual litigation, where the absence of documentation reads as the absence of a program. The discipline is precision: the drafting rules of Chapter 25, applied at ninety-second speed. Write findings as findings: the condition observed, the standard implicated, the correction recommended, the date, the recipient responsible. Strip the adjectives, the sarcasm, the prophecy, and the legal conclusions — "negligent," "willful," "liability" are the lawyers' words, and in a professional's e-mail they are gifts to whichever side the professional's employer is not on. The document that states the hazard and the fix, plainly, serves the program when it is written and the truth when it is read; the document that editorializes serves neither.

The record also cuts the other way, and the professional should understand this before concluding that documentation is danger. The audit finding that was closed, the near-miss that produced the engineering change, the monitoring that confirmed the control worked — these are the affirmative record, and in the defense chair they are frequently the whole case: the demonstration that the organization generated safety data and acted on it rather than filing it. The corrected finding is not an admission that the hazard existed; it is proof that the system found and fixed it, and the distinction between a program that documents and corrects and a program that documents and ignores is the distinction juries actually draw. The professional's writing discipline, in other words, is not adverse-inference management. It is the construction, entry by entry, of the record the organization will someday stand on.

IV. The Shield: Privilege and Work Product

A. Attorney-Client Privilege: Communications, Not Facts

The attorney-client privilege protects confidential communications between client and counsel made for the purpose of obtaining or providing legal advice, and since *Upjohn* it has reached communications between corporate counsel and employees at every level — the plant safety manager and the line supervisor as much as the general counsel's direct reports — when the employee communicates at counsel's direction, about matters within the employee's duties, so that counsel can advise the corporation. Two boundaries define its real shape. First, the privilege protects the communication, never the underlying fact: the safety manager who described the missing guard to counsel cannot be asked what she told the lawyer, but she can and will be asked about the guard, and the guard's absence does not become privileged because it was discussed with an attorney. Second, the privilege attaches to legal advice, not business advice, and safety compliance work sits uncomfortably on that line. The audit conducted because the program requires audits is business; the investigation directed by counsel to assess legal exposure can be privileged even when compliance obligations also required an inquiry, provided obtaining legal advice was a significant purpose. The purpose is fixed at the commissioning, which is why Section VI's engagement discipline matters more than anything stamped on the report afterward.

B. Work Product: Anticipation, Documented

The work-product doctrine, born in *Hickman* and codified in Rule 26(b)(3), protects documents and tangible things prepared in anticipation of litigation or for trial, by or for a party or its representative: a category that expressly includes consultants. The operative question is why the document came into existence: under the prevailing "because of" formulation, a document prepared because of the prospect of litigation is protected even if it also serves other ends, but a document that would have been created in essentially similar form in the ordinary course of business is not, no matter how useful it later proves to the litigators. This is the doctrine that decides the fate of the post-incident investigation, and it decides it on facts established at the outset: who commissioned the work, what the commissioning documents say, whether litigation was reasonably anticipated, and whether the investigation differed from what routine procedure would have produced anyway. The consultant's report ordered by counsel after the incident, for the stated purpose of assisting counsel in the anticipated claims, is the classic protected document: the Third Circuit shielded exactly such an industrial hygienist's report from an agency subpoena in *Martin v. Bally's*. The same consultant's report, commissioned by the safety department under a standing purchase order as part of the annual assessment cycle, is ordinary business paper, and no legend converts it.

C. The Protections That Are Weaker Than They Sound

Two beliefs circulate in the safety profession that the case law does not support. The first is the self-critical analysis privilege: the intuition that candid internal safety evaluation must be protected, lest discovery chill it. Courts have overwhelmingly declined the invitation: outside a handful of statutory pockets, routine safety audits, committee minutes, and pre-incident hazard evaluations are discoverable, and the Ninth Circuit's reasoning in *Dowling* states the prevailing view — organizations have ample non-litigation reasons to review safety candidly, so the chilling premise fails, and the minutes come in. The second is the power of the label: the belief that stamping "Privileged & Confidential — Attorney Work Product" on a document protects it. The stamp is evidence of intent, nothing more; courts look to function, and a routine audit wearing a privilege legend is a routine audit, with the added cost that indiscriminate stamping teaches the court that the organization's legends mean nothing, which taxes the documents that deserved them. Protection is built into the commissioning of the work, or it does not exist.

D. Waiver: How the Shield Is Dropped

Every protection in this section is waivable, and waiver is where safety organizations actually lose them. Confidentiality is the privilege's predicate: the investigation report circulated beyond those who need it for the legal purpose, forwarded to the broader safety distribution list, shared with the insurer outside a common-interest arrangement, or handed to OSHA to show cooperation has been disclosed, and disclosure to an adversary or the world generally forfeits the protection — sometimes, under the subject-matter waiver principles Rule 502 now moderates but does not eliminate, for the entire subject and not just the document. Waiver also arrives by implication: the defendant who answers the negligence claim by affirmatively relying on its investigation — we inquired, we were thorough, we acted reasonably — has put the investigation at issue and cannot simultaneously shield it, as the *Harding* line of decisions holds. And the consulting expert's protection under Rule 26(b)(4)(D) dissolves on designation: the quiet consultant converted to testifying expert opens her considered materials to disclosure, and the dual-hat professional who consulted first and testifies later will litigate, expensively, which of her files crossed with her. The operational rule is simple to state and demanding to practice: decide what the protected work is, keep it in its channel, and never exhibit the shield's contents while asserting the shield.

V. When a Prosecutor Enters: The Criminal Overlay

Everything to this point has assumed a civil case and an administrative enforcement proceeding: two forums with different rules, both of which end in money. There is a third forum, it is rare enough that most safety professionals will never see it and common enough that a book about the record would be incomplete without it, and every discipline described in this chapter changes when it opens. This section is about what happens to the professional's file, her statements, and her testimony when the matter has become, or may become, a criminal one.

The reason it belongs in this chapter rather than in Part II is that the criminal overlay is felt least by the corporation and most by the individual. The company has counsel from the first hour. The safety director, the industrial hygienist, and the site supervisor frequently do not, and by the time they understand the distinction they have already given a recorded interview, written an incident report, and signed a certification.

A. The Federal Statute and Its Narrow Reach

Section 17(e) of the OSH Act makes it a criminal offense for an employer to willfully violate a standard, rule, or order where the violation causes the death of an employee. It is a Class B misdemeanor carrying up to six months of imprisonment, with the fine set not by the Act's own antique figure but by the general federal sentencing provision, which reaches $250,000 for an individual and $500,000 for an organization where the offense results in death. A second conviction doubles the imprisonment exposure. Two further OSH Act offenses matter more often than the fatality provision: knowingly making a false statement, representation, or certification in a required record or report, and giving advance notice of an inspection. Both are misdemeanors, and both reach conduct that occurs during the inspection this book has spent Part II describing.

The reach is narrow by design and narrower in practice. The prosecutor must prove beyond a reasonable doubt that the violated requirement was a standard, rule, or order — the General Duty Clause is not one — that the violation was willful in the criminal sense of a deliberate act taken with knowledge of both the hazard and the legal requirement, and that the violation was both the but-for cause and the legal cause of the death. Those elements, joined to a misdemeanor's sentencing exposure, explain the historical numbers: across the more than four decades and several hundred thousand workplace fatalities since the Act's passage, federal OSH Act criminal cases have been brought fewer than two per year, and convictions have been rarer still.

Counsel and professionals who stop reading at that statistic draw the wrong conclusion. The federal misdemeanor is not where the criminal exposure in this field actually lives.

B. The State Forum, Which Is Where It Lives

The OSH Act does not preempt state criminal law, and a state prosecutor confronting a workplace fatality has the ordinary homicide and endangerment statutes available: manslaughter, criminally negligent homicide, reckless endangerment, and, in the states that recognize it, corporate criminal liability. Those are felonies with felony sentencing, and they do not require the prosecutor to prove a violation of any OSHA standard at all: the OSHA file, when there is one, is evidence of the defendant's knowledge rather than an element.

The landmark is a trench. In April 2015 a twenty-two-year-old worker named Carlos Moncayo was killed when an unshored fourteen-foot trench collapsed on him at a Manhattan construction site, after months of documented warnings about the conditions. The Manhattan District Attorney prosecuted the general contractor, Harco Construction, and obtained convictions at trial for manslaughter, criminally negligent homicide, and reckless endangerment: establishing, as the District Attorney's office put it at the time, that a company could be held criminally responsible for the death of a worker at its site. Two individuals were convicted alongside the company, and they are the ones this chapter's readers should study. Wilmer Cueva, the excavation subcontractor's foreman — the man who ordered the worker into the trench after the warnings — was convicted of criminally negligent homicide and reckless endangerment and sentenced to one to three years in state prison. Alfonso Prestia, the general contractor's site supervisor, resolved his case by plea to probation and community service after his jury deadlocked. Neither man owned the company, set its budget, or wrote its safety program. Both were site-level supervisors of exactly the kind who read books like this one, and the sentence that ran was the foreman's. The sequel is the part practitioners should notice: the corporate fine available under then-existing New York law was ten thousand dollars. In December 2022 the legislature responded with Carlos' Law, which amended the Penal Law to make a corporation criminally liable where it negligently, recklessly, intentionally, or knowingly causes the death or serious physical injury of an employee, and replaced the old ceiling with a floor. For a felony conviction the corporate fine is now not less than five hundred thousand dollars and may reach one million; for a misdemeanor, not less than three hundred thousand and up to five hundred thousand. The direction of the change is the point. The defect the Harco sentence exposed was not that the maximum was too low in some abstract sense but that a court confronting a worker's death had no authority to impose a penalty a corporation would notice; the legislature's answer was to remove the court's discretion to go low rather than to extend its discretion to go high.

California reaches the same conduct through its labor code rather than its penal code: a willful violation of an occupational safety standard that causes death or permanent impairment is separately criminal under Labor Code § 6425, carrying up to three years' imprisonment and a fine of up to two hundred fifty thousand dollars for an individual and up to one and a half million for a corporation — three times the federal organizational maximum, and a hundred and fifty times the figure the OSH Act itself still carries on its face. Other states have followed with statutes of their own or with prosecutions under general homicide law, and district attorneys in construction-heavy jurisdictions increasingly work directly with OSHA area offices on fatality investigations. For the practitioner, the operative point is jurisdictional: the same trench collapse that produces a $165,514 federal willful citation may simultaneously produce a state felony indictment, and the second proceeding is the one that ends in a courtroom with a jury and a sentencing.

C. The Federal Overlay That Is Not the OSH Act

Since December 2015 the Departments of Justice and Labor have operated under a memorandum of understanding expanding what is called the Worker Endangerment Initiative, with the Deputy Attorney General directing all United States Attorneys to coordinate with the Environment and Natural Resources Division on worker safety matters. The strategy is straightforward and has nothing to do with Section 17(e): where a workplace event also involves an environmental release, a false submission, or the destruction of records, the government charges the felony available under the Clean Air Act, the Clean Water Act, RCRA, or Title 18 rather than the OSH Act misdemeanor.

Three Title 18 provisions do most of that work, and each is a document offense rather than a safety offense. Making a materially false statement to a federal agency is a felony carrying five years. Falsifying, concealing, or destroying a record with intent to impede an investigation of a matter within federal jurisdiction carries twenty, and — this is the provision safety professionals most need to understand — it does not require that any proceeding be pending, or even that the defendant know one is contemplated. Conspiracy supplies the third. The consequence is that the exposure of a safety department after a fatality frequently has less to do with the guard that was missing than with what happened to the audit finding, the inspection log, or the e-mail in the seventy-two hours afterward.

D. What Changes for the Professional

Every protection described in Section IV of this chapter behaves differently once a criminal referral is live or foreseeable, and the differences are not intuitive.

Privilege belongs to the company, not to you. The *Upjohn* doctrine that extends the corporate attorney-client privilege to communications with employees at every level also fixes who owns it. Company counsel investigating an incident represents the company. The safety manager she interviews is a source, not a client, and the privilege over that conversation is the company's to waive. Where the company later determines that cooperation is in its interest, it may waive, and the safety manager's candid account of what she recommended, when, and to whom goes to the government without her consent and frequently without much notice. Competent corporate counsel gives an *Upjohn* warning at the start of such an interview, explaining exactly this. The professional's obligation is to understand what she has just been told: that the lawyer in the room is not her lawyer, that what she says is not confidential as to her, and that she may retain her own counsel. A professional who hears that warning after a fatality and does not act on it has made a decision, whether or not she realized she was making one.

Your incident report may be a confession. Chapter 22 builds the post-incident investigation on two tracks precisely so that the operational inquiry can be candid without surrendering the legal analysis, and Chapter 25 teaches drafting on the assumption that every sentence will be read aloud. Under a criminal overlay both disciplines intensify. The root-cause narrative that concludes "supervisor knowingly permitted entry without shoring" is, in a state manslaughter prosecution, the government's mens rea evidence, authored by the defendant's own safety department, dated, and unimpeachable as to its author. This is not an argument for writing an evasive report; an evasive report is worse in every forum and is itself evidence. It is an argument for the discipline the whole book teaches: record findings of fact and record them precisely, state what the evidence supports and no more, and leave characterizations of state of mind to the people whose job it is to prove them.

Preservation stops being a civil question. Chapter 21 treats spoliation as a matter of adverse inferences and sanctions. Once a matter is potentially criminal, the same act of deleting a file, overwriting a log, or discarding sampling media is a twenty-year felony exposure under the federal obstruction provision, with no requirement that anyone had yet been served with anything. The practical instruction for the safety professional is absolute and simple: after a serious incident, nothing in the department's custody is destroyed, edited, or "cleaned up," on any schedule, for any reason, including reasons that were entirely routine the week before. Routine destruction under a retention policy is a defense in a civil spoliation motion. It is a substantially weaker one when the government is asking why the audit file was shredded on the Monday after the funeral.

The Fifth Amendment has a civil price. A witness who invokes the privilege against self-incrimination in a criminal proceeding may not have that invocation used against her there. In a parallel civil case, the rule is different: the trier of fact may draw an adverse inference from a party's or witness's invocation. That asymmetry sets up the central strategic problem of parallel proceedings. The safety professional who declines to answer at her deposition protects herself in the prosecution and hands the plaintiff a jury instruction in the civil case; the professional who answers fully to protect the civil defense creates sworn testimony the prosecutor will read line by line. There is no clean answer to that problem, and the professional facing it needs her own lawyer to work it, which is the practical reason the *Upjohn* warning above deserves to be taken seriously the first time it is given rather than the second.

The stay is not automatic. Counsel's usual response is to move to stay the civil case pending resolution of the criminal one. Parallel civil and criminal proceedings are constitutionally permissible, and courts weigh a familiar set of factors: the overlap of issues, the status of the criminal matter (a stay is much likelier after indictment than during an investigation), the civil plaintiff's interest in timely resolution, the burden on the defendants, and the interests of the court and the public. The practical translations are two. First, the expert and the safety professional should assume the civil case will proceed on its own schedule and prepare accordingly, because assuming otherwise is how depositions get taken without preparation. Second, the technical work does not pause: the site inspection, the exemplar testing, and the exposure reconstruction all remain perishable on exactly the timetable of Chapters 6 and 21, and the pendency of a criminal investigation is a reason to do them sooner, under counsel's direction, rather than a reason to wait.

E. Practice Implications

For counsel, the first question after a fatality or a catastrophic injury is not what OSHA will do. It is whether a prosecutor may become involved: a question answered by the jurisdiction, the presence of an environmental release, the existence of prior warnings in the employer's own records, and whether the event was reported to OSHA at all. If the answer is anything other than a confident no, the investigation is structured from hour one on the assumption of a criminal audience: counsel-directed, *Upjohn*-warned, with separate counsel offered to individuals whose conduct is squarely at issue, and with a litigation hold that is issued in writing, acknowledged in writing, and enforced.

For the safety professional, three things are worth carrying out of this section. The lawyer conducting the post-incident interview is probably not your lawyer, and you are entitled to ask the question directly. The report you write in the next seventy-two hours may be read by a jury deciding whether someone goes to prison, so write findings and leave conclusions about intent to others. And the file you are tempted to tidy is the one that will be reconstructed from the server backup, at which point the tidying, rather than the hazard, becomes the case.

None of this is a reason for the profession to practice defensively. It is a reason to practice the way the rest of this book describes: contemporaneously, precisely, against published methods, with the assumption that every document will be read by someone who was not there and did not wish the author well. The professional whose ordinary work already meets that standard has very little to change when a prosecutor appears. That is, in the end, the argument of the whole book, stated in the forum where the stakes are highest.

VI. The Practice: Engagement Letters and Two Files

The engagement letter is where protection is created, and it must be written as if a magistrate judge will read it, because one may. For post-incident consulting work intended for protection, the elements are settled. The retention should run from counsel, not from the safety department or the company directly: the lawyer engages the consultant to assist the lawyer in advising the client and preparing for anticipated litigation, and the letter says so in terms — identifying the incident, reciting that litigation is anticipated and why, and stating that the consultant's work is commissioned because of that prospect. It should designate the consultant as a non-testifying consulting expert under Rule 26(b)(4)(D), reserving any later testifying designation as a separate, deliberate decision. It should direct that the consultant communicate findings to counsel; that reports, data, photographs, and drafts be delivered to counsel and not distributed further; that the consultant's file be maintained confidentially, segregated from any routine work the consultant performs for the same client; and that the file's disposition at the engagement's end follow counsel's instruction. It should address compensation at the professional's regular rates, never contingent on outcome: a contingent technical fee is an ethics violation in most professional codes and an impeachment exhibit in all of them. And where the consultant's role is to help counsel understand technical matter, the letter should say that too, because the interpreter-of-facts rationale that extends privilege to accountants and engineers working under counsel is the rationale a court will test the engagement against.

Marking should be accurate, consistent, and reserved for documents that have earned it. Work genuinely commissioned by counsel in anticipation of litigation should carry a legend saying exactly that — "Privileged & Confidential — Attorney Work Product — Prepared at the Direction of Counsel in Anticipation of Litigation" — on every page, with the transmittal directed to counsel and the distribution list limited to those within the privileged circle. Drafts deserve the same legend and more caution, because drafts memorialize the evolution of findings, and although the rules now protect draft reports of retained testifying experts, consulting-side drafts are protected only as far as the underlying doctrine reaches. The complementary rule is the one the profession resists: routine work should not be marked. The annual audit, the monthly inspection, the committee minutes, and the monitoring reports are program documents; stamping them privileged protects nothing, undermines the legends that matter, and hands the examining lawyer a theme — this company marks everything secret — that plays badly in front of juries. The professional should ask, before affixing any legend, the question the court will ask afterward: who commissioned this, and why, and is that answer in writing somewhere I can point to.

The operating structure that makes all of this work is the two-file discipline. The safety program runs in one channel: audits, inspections, monitoring, training, corrective action — documented fully, written precisely, and treated as the discoverable record it is, built to demonstrate a functioning system because someday it will be asked to. Litigation-driven work runs in the other: commissioned by counsel, performed under an engagement letter that says so, communicated to counsel, marked accurately, and distributed to no one else. The channels must not touch. The consultant who copies routine findings into the privileged report, or privileged analysis into the routine file, has entangled what the doctrines require be separable, and the entanglement is resolved in a discovery motion the client pays for. In-house professionals live this discipline daily and hardest: the same person may audit the plant in March as program work and assist counsel's investigation in June as privileged work, and the person's files, e-mail habits, and reporting lines must keep the two identities distinct, because at the deposition, the examining lawyer's first project will be to merge them.

VII. What the Cases Teach

Upjohn Co. v. United States (U.S. 1981), the corporate privilege, and its precise limits. After *Upjohn*'s counsel investigated questionable foreign payments through questionnaires and interviews of mid-level employees, the government demanded the responses; the Supreme Court held them privileged, rejecting the narrow control-group test and extending the corporation's privilege to communications between counsel and employees at any level, made at the direction of corporate superiors, concerning matters within the employees' duties, for the purpose of legal advice. The decision is the charter for every counsel-directed safety investigation, and its qualification is as important as its holding: the privilege protects the communications, not the facts within them, and the government remained free to question the employees about what they knew. For the safety professional, *Upjohn* means the interview with counsel is protected and the hazard is not, and the investigation structured to fit *Upjohn*'s terms, with counsel's direction documented at the outset, is the one that holds.

Hickman v. Taylor and *United States v. Adlman* (U.S. 1947; 2d Cir. 1998), work product, and the dual-purpose document. *Hickman* shielded a lawyer's witness statements and memoranda from an adversary who demanded them without showing need, founding the doctrine that litigation preparation may proceed without the other side's discovery reaching into it; Rule 26(b)(3) codified the protection for parties and their consultants. *Adlman* answered the question safety work poses daily, what of the document serving both litigation and business ends?, by adopting the "because of" test: protection extends to the document prepared because of anticipated litigation even if it also informs a business decision, but not to the document that would have been prepared in essentially the same form regardless. The pairing defines the post-incident boundary. The investigation that routine procedure would have produced anyway is discoverable; the analysis that exists because the claim is coming is not; and the record establishing which is which — the engagement letter, the commissioning memorandum, the changed scope — is written at the beginning or it is never written at all.

Martin v. Bally's Park Place Hotel & Casino (3d Cir. 1993): the industrial hygienist's report the government could not reach. After employees complained of emissions from a glass-washing machine, the employer's general counsel ordered the machine tested for iodine — an exposure OSHA regulates under 29 C.F.R. § 1910.1000 — and directed that the consultant's results be reported to him. When the employees' union and then OSHA demanded a copy, the employer refused on work-product grounds; the Secretary of Labor served an administrative subpoena, pursued it no further, and instead cited the employer under the records-access rule then codified at 29 C.F.R. § 1910.20(e) (now § 1910.1020) for withholding the results. The route to the holding is worth knowing, because it was not a clean win. The administrative law judge upheld the citations, reasoning that sampling results are employee exposure records accessible whoever commissioned them. The Review Commission vacated them on work-product grounds. And the Third Circuit, on the Secretary's petition for review, held that the records-access rule incorporates the work-product doctrine and that the employer had properly invoked it —

the report having been prepared in anticipation of litigation at counsel's direction, and not stripped of protection by the agency's need. The decision is the profession's clearest proof that the protections in this chapter are real when the architecture is right: the same monitoring survey that would have been freely discoverable as a program document was shielded because counsel commissioned it, for a litigation purpose, after the dispute was in view — and in-house general counsel was enough to do it. It is equally a proof of the converse, and consultants should read it that way: the protection followed the engagement's structure, not the report's subject matter, and the identical report under a routine purchase order would have gone to the government with the rest of the file. The judge's contrary reasoning is the argument the other side will run today, and an employer leaning on this case should expect to meet it.

Dowling v. American Hawaii Cruises (9th Cir. 1992): the safety committee minutes, and the privilege that isn't. An injured seaman sought the shipowner's pre-accident safety committee minutes; the defendant resisted on self-critical analysis grounds, and the Ninth Circuit ordered production, reasoning that routine pre-accident safety reviews are conducted for the organization's own operational reasons — regulatory compliance, loss prevention, workforce protection — and that discovery poses no realistic chill to candor the organization already has every incentive to practice. The holding states what safety professionals most need to internalize: the ordinary output of the safety program enjoys no privilege at all, however candid, however self-critical, however internal it felt when written. The minutes, the audits, and the hazard evaluations will be read by opposing counsel, and the only question the professional controls is whether they will read as a system that found and fixed, or found and filed.

In re Kellogg Brown & Root (D.C. Cir. 2014) — the compliance investigation that kept its privilege. A False Claims Act relator won an order compelling KBR's internal investigation files on the theory that the inquiry was a routine compliance exercise, mandated by regulation and corporate policy and conducted largely by non-lawyers, rather than a privileged legal one; the D.C. Circuit granted mandamus and reversed, holding that the privilege applies so long as obtaining or providing legal advice was a significant purpose of the investigation, not the sole purpose, and that investigations conducted at counsel's direction through non-attorney personnel qualify. For safety organizations, whose investigations are almost always compelled in part by regulation and policy, the decision is the workable standard: the incident inquiry can serve OSHA compliance, corrective action, and legal advice at once and remain privileged, provided counsel's direction and the legal purpose are real, contemporaneous, and documented, and provided the organization does not later waive the protection by parading the investigation's diligence as a defense.

QBE Insurance Corp. v. Jorda Enterprises (S.D. Fla. 2012) — the corporate designee's duties, cataloged. In a subrogation dispute, the corporate plaintiff produced a 30(b)(6) designee unable to answer on noticed topics, and the court's opinion — now among the most-cited catalogs of the rule's obligations — spelled out the duties the designation imposes: the corporation must prepare the witness on all information known or reasonably available to it, from documents, past employees, and other sources beyond the designee's personal knowledge; the testimony binds the organization; an inadequately prepared designee is tantamount to a failure to appear; and the corporation that answers "we don't know" at deposition may be barred from knowing better at trial. For the safety director handed a designation letter, the decision is the job description: the assignment is to become the organization's knowledge on the noticed topics, however many files, predecessors, and departments that requires, because the alternative is sworn corporate ignorance — binding, sanctionable, and permanent.

VIII. The Program That Survives the Subpoena

The threads of this chapter braid into a single professional program, and every element of it operates before the case exists. Write the program record as if it will be read aloud, because it will be: findings stated as conditions, standards, and corrections; adjectives, prophecies, and legal conclusions left out; the corrective loop closed and documented, so the file testifies to a system that acts. When the incident arrives, sort the work into its channels on day one: the routine response through the program, fully documented; the litigation-driven analysis commissioned by counsel, under an engagement letter that recites the anticipation, the purpose, and the consulting designation, with communications to counsel, legends that are earned, and distribution that stays inside the circle. When the subpoena or the designation letter arrives, prepare from the record, not the memory: the fact witness re-reads everything she wrote; the corporate designee absorbs everything the organization knows; both answer the question asked, concede what the paper shows, decline the opinions that belong to another chair, and stop. And when the retained-expert engagement is offered, take it under a letter that would survive its own production — scope stated, rates regular, independence explicit — because the engagement letter of a testifying expert is frequently the first exhibit at her deposition.

For counsel, the reciprocal program is equally concrete. Treat the safety professional's preparation as a first-order investment, not an hour in the hallway: the fact witness prepared against her documents and the designee educated to the organization's knowledge are the difference between a record that holds and a transcript that becomes the other side's opening. Structure the post-incident investigation for the protection it needs on the day it is commissioned — counsel's direction in writing, the legal purpose recited, the consultant engaged by the lawyer — because no motion practice retrofits privilege onto work that was born as business. Police the channels against waiver with the same energy spent building them. And in selecting the consultant, whichever side you are on, apply the standard this series has urged throughout: retain the professional whose program record, engagement letters, and testimony would look the same whichever side produced them.

One last word about the witness herself, because the three chairs share one seat. The safety professional's authority in litigation — as fact witness, as corporate voice, as expert — derives from a single source: that she wrote the truth before she knew who would need it, and testifies to the same truth regardless of who asks. The record she built in the ordinary course is discoverable precisely because the law trusts contemporaneous, unlitigated candor more than anything constructed afterward, and that trust is the professional's asset, not her exposure. The documents will not change. The witness who says the same thing, in every chair, under any examination, is the one the record cannot impeach, and the only one worth calling.

References

Access to employee exposure and medical records, 29 C.F.R. § 1910.1020; recording and reporting occupational injuries and illnesses, 29 C.F.R. pt. 1904.

American Board of Industrial Hygiene / Board for Global EHS Credentialing, Code of Ethics; Board of Certified Safety Professionals, Code of Ethics.

Dowling v. Am. Hawaii Cruises, Inc., 971 F.2d 423 (9th Cir. 1992).

Fed. R. Civ. P. 26(b)(1), 26(b)(3), 26(b)(4)(B)–(D), 26(a)(2)(B)–(C), 30(b)(6), 45; Fed. R. Evid. 502.

Harding v. Dana Transp., Inc., 914 F. Supp. 1084 (D.N.J. 1996).

Hickman v. Taylor, 329 U.S. 495 (1947).

In re Kellogg Brown & Root, Inc., 756 F.3d 754 (D.C. Cir. 2014).

Martin v. Bally's Park Place Hotel & Casino, 983 F.2d 1252 (3d Cir. 1993).

QBE Ins. Corp. v. Jorda Enters., Inc., 277 F.R.D. 676 (S.D. Fla. 2012).

United States v. Adlman, 134 F.3d 1194 (2d Cir. 1998).

United States v. Kovel, 296 F.2d 918 (2d Cir. 1961).

Upjohn Co. v. United States, 449 U.S. 383 (1981).

CHAPTER 27

Building the Practice

The Safety or Industrial Hygiene Professional as Retained Expert — Engagement, Ethics, and the Long Career

Chapter 1 is written for the lawyer choosing an expert. This chapter is written for the professional deciding to become one, and for the lawyer who benefits from understanding the expert's side of the engagement. The qualities that make a witness worth retaining are built, not born, and most of them are decided before the first case: how the engagement is structured, which matters are declined, how the file is kept, and how a reputation for saying the same thing on either side of the v. is earned and protected across a career.

I. The Two Chairs, Chosen Deliberately

The distinction between the consulting and the testifying expert, introduced in the book's Introduction, is a choice the professional makes at the outset of each engagement, and it carries consequences the professional should understand before the retention letter is signed. The consulting expert educates counsel and, in most jurisdictions, works under a discovery shield that keeps the analysis privileged; the testifying expert performs comparable analysis knowing every note is discoverable and every opinion must satisfy Rule 702. A professional may occupy either chair, but occupying both in the same matter is a trap: the moment a consulting expert is designated to testify, the shield lifts, and the candid consulting file — the early doubts, the discarded theories, the frank memo about the client's exposure — becomes discoverable. The professional who wants to preserve the option of testifying keeps the consulting work in a form that could survive disclosure, or keeps the roles, and the files, genuinely separate.

II. The Engagement Letter, Conflicts, and the Cases to Decline

The engagement letter is the expert's first act of self-protection, and it should establish the retaining party, the assignment, the fee structure, the ownership and handling of materials, and — critically — that the expert's obligation runs to the analysis rather than to a predetermined conclusion. A letter that recites the opinion the client wants is a letter the opponent will read to the jury as proof the opinion was bought; a letter that recites the questions the expert was asked to answer is a letter that supports the independence the whole engagement depends on. Conflicts must be cleared before the first substantive conversation, because an expert who has learned a party's confidences, even in a fifteen-minute intake call, can be disqualified from the other side of that matter and can taint a firm.

The most valuable discipline of a long practice is declining the case the expert should not take. The matter outside the expert's genuine competence, the general safety expert asked to opine on a process he has never assessed, is declined because Chapter 1 explains exactly how it ends. The matter whose facts will not support the opinion the client needs is declined, or accepted only with the client's understanding that the analysis may not help, because the expert who trims the method to fit the retention has started down the road to exclusion and to a reputation the next opposing counsel will exploit. Saying no to the wrong case is how an expert stays useful in the right ones.

III. Fees, the Ratio, and the "Hired Gun" Problem

Chapter 1 teaches counsel to screen for a healthy ratio of consulting to testimonial work and a balance of plaintiff and defense retention. This chapter is the same lesson from the witness's side: those ratios are a professional asset to be cultivated, not an accident to be explained. The expert whose practice is mostly real consulting for operating companies and workers, and only partly testimony, arrives with an independent professional identity a jury can credit; the expert who testifies for a living, always for the same side, arrives with the cross-examination written. Fees should be reasonable, charged for time rather than outcome — a contingent expert fee is improper and, in most jurisdictions, disqualifying — and documented cleanly, because the compensation disclosure of Chapter 3 puts the arrangement on the record. The same scrutiny reaches how the engagement was obtained: a fee split with a referring lawyer or broker, a commission owed to whoever supplied the case, or an undisclosed payment for the introduction gives the cross-examiner a financial-bias exhibit no hourly rate ever creates, and the certification codes' conflict and compensation provisions reach these arrangements as surely as they prohibit the contingent fee. The workable rules are disclosure and severance: a directory or agency compensated by flat fee rather than by a percentage of billings, a referral acknowledged and unpaid, and an answer to the standard deposition question of how the witness came to be retained that the professional is content to give under oath. The vulnerability is never a fair hourly rate; it is the pattern the rate reveals over a career, which is why the pattern is worth managing deliberately from the first engagement.

IV. Ethics Codes as Testimony Constraints

The board certifications that make a witness credible — the CIH, the CSP, the CHMM — carry enforceable codes of ethics, and those codes are not decorative: they are testimony constraints the professional agrees to and the opponent can invoke. The Board for Global EHS Credentialing and the Board of Certified Safety Professionals each require members to practice only within their competence, to be objective and truthful in professional reports and testimony, to avoid conflicts of interest, and to represent qualifications accurately. An expert who overstates credentials, opines beyond competence, or shades a conclusion has not only handed the cross-examiner material: he has arguably breached the code that certifies him, and a certification held out to qualify the witness can be turned into the instrument of impeachment. The professional who treats the ethics code as the actual standard of the work is, not coincidentally, the professional whose testimony is hardest to shake.

The competence boundary the codes draw is also a scope-of-certification boundary, and it is tested most often in the witness chair. The credential qualifies the professional in a discipline, not in every question a case can raise: the industrial hygienist who measured the exposure is not thereby a toxicologist on causation, a physician on diagnosis, or a mechanical engineer on the failed interlock, and Chapter 14 has already drawn the line the causation cases enforce. The temptation arrives dressed as helpfulness — counsel needs one more opinion, the adjacent question looks answerable, the retention would be simpler with a single witness — and the cost arrives twice: the opinion beyond competence is the one Rule 702(a) excludes, and the testimony beyond certification scope is the conduct a board complaint recites. The disciplined answer is the referral the codes actually require — identify the adjacent specialty, tell counsel the case needs it, and confine the report to the questions the professional's training, experience, and certification genuinely reach — because the witness who names the limits of her own expertise on direct has purchased credibility no borrowed opinion could.

V. Pressure, Adverse Findings, and the Board Complaint

Every retained expert eventually receives the call in which counsel, courteously and in the vocabulary of clarity, asks for a different report. Some of the requested edits are legitimate and should be accepted with thanks — the corrected citation, the fixed record reference, the ambiguity a hostile reader would exploit — because Rule 26 contemplates counsel's participation in the drafting and the 2010 amendments protect the drafts precisely so that the collaboration can be candid. The line is substance: the opinions, the methodology, the assumptions, and the stated confidence are the expert's, and the request to strengthen a conclusion the analysis does not support, delete a limitation the method requires, or omit a finding that hurts the theory is a request to trade the opinion's admissibility and the professional's certification for the client's convenience. The refusal is easiest when the engagement letter of Section II already said the analysis controls; the practical script is to accept every edit that makes the report more accurate and decline every edit that makes it less true, in writing, with the reasons, and the expert who has done so holds the deposition's best answer: asked whether the lawyers wrote the report, she can describe, without flinching, exactly what changed at counsel's suggestion and why the opinions did not.

The harder problem is the adverse finding: the sample that came back clean when the theory needed contamination, the reconstruction that puts the fault on the client's side of the line. The testifying expert has no discretion to bury it: the report must disclose the facts and data considered, cross-examination will ask what the analysis showed, and the finding suppressed is the finding that, discovered, converts a losing issue into a lost witness. What counsel may do with the adverse finding is a different question with a legitimate answer this chapter has already drawn: the consulting expert's analysis is shielded, in most jurisdictions, from discovery absent exceptional circumstances, and a party that retains a consultant, learns the answer is bad, and quietly declines to designate has used the system as designed. The professional's obligations in that sequence are candor and timing — report the adverse result to counsel promptly, completely, and without editorializing it away, then respect the role: the consulting file stays a consulting file only if the two-chair discipline of Section I was honored, and the expert already designated to testify cannot be un-rung in every jurisdiction, which is why the honest early answer, delivered before the designation deadline rather than after it, is a service to the client and not a betrayal of it.

The certification boards enforce their codes through a complaint process the testifying professional should understand before, not after, an opponent invokes it. Anyone may file — and in practice the complaints arising from litigation are filed by opposing counsel, opposing experts, or the losing party, with the transcript attached — and the board's ethics function screens the complaint, notifies the certificant, and requires a written response, for which the professional should engage counsel and notify the professional-liability carrier, because the file assembled here follows the credential. A complaint with substance proceeds to investigation and review before an ethics committee or hearing panel, with outcomes running from dismissal through private admonition and public reprimand to suspension and revocation of the certification, under the notice, hearing, and appeal procedures each board publishes. Three features give the process its litigation weight. It is independent of the lawsuit: the case may settle, the verdict may be won, and the complaint proceeds anyway, judged against the code rather than the outcome. Its results follow the witness: the qualification screening of Chapter 1 and the deposition script of Chapter 4 both ask about discipline, and a sanction, or a credential surrendered while a complaint was pending, must be accounted for in every future engagement. And its best defense is retrospective: the disciplined report, the stated limitation, the refused edit, and the documented candor of this chapter are the record a review panel reads, which is one more reason the ethics code is treated throughout this book as the actual standard of the work rather than a frame on the office wall.

VI. The Expert's Own Legal Exposure

A chapter about building a practice owes the professional an account of what the practice exposes her to, and the sections above have been about obligations rather than risks. Four questions come up in every serious conversation with someone deciding whether to enter this work, and the honest answers are more reassuring on some points than on others.

The first is witness immunity. The traditional common-law rule granted absolute immunity to witnesses for testimony given in judicial proceedings, and for a long time that immunity was read to protect retained experts against suit by the party who retained them. That reading has eroded substantially. A number of jurisdictions now hold that immunity protects the witness against claims arising from the testimony's effect on the opposing party and on the proceeding, its historical purpose, but does not bar a claim by the retaining client for negligent performance of the professional services the client paid for, whether framed as malpractice, breach of contract, or breach of fiduciary duty. Others retain the broader protection. The rule is genuinely jurisdiction-specific, it has moved within living memory in several states, and the practical instruction is to determine the rule in the forums where the practice will operate rather than to assume the traditional one. The exposure, where it exists, is concentrated in the same failures this book has spent twenty-six chapters describing: the missed deadline, the opinion that was excluded, the analysis that did not consider the record.

The second is the consulting expert who is subpoenaed anyway. Fed. R. Civ. P. 26(b)(4)(D) protects facts known and opinions held by an expert retained in anticipation of litigation but not expected to testify, and it is a strong protection: discovery is available only on a showing of exceptional circumstances under which it is impracticable for the party to obtain facts or opinions on the same subject by other means. In practice that showing is made in a narrow band of cases, and the recurring one in this field is the destroyed or altered scene. Where the consulting expert examined a machine, a trench, or an atmosphere that no longer exists in that condition, the information is by definition unobtainable by other means, and the protection may not hold. The consequence for the professional is procedural rather than dramatic: the consulting file should be built as though it might one day be read, which is the same instruction Chapter 22 gives the investigator and Chapter 25 gives the drafter. A consulting engagement is not a license to keep a sloppier file.

The third is what happens when an opinion is excluded and the client loses. The professional consequences are real and mostly reputational: exclusion orders are published, they are searchable, and Chapter 4 §II explains how an opposing examiner finds them. The legal consequences depend on the immunity question above and on whether the exclusion followed from a professional failure or from a contested legal ruling, which are different things and should be documented as different things. The practical protection is built at the engagement stage. A scope letter that records what the expert was asked to do, a contemporaneous record of the materials requested and not provided, and a written record of any instruction that narrowed the analysis together distinguish an expert whose opinion was excluded from an expert who did the work badly. Chapter 3 §V and Chapter 27 §II describe both documents; this is the reason to keep them.

The fourth is insurance. Professional liability coverage written for consulting engineers, industrial hygienists, and safety professionals does not automatically respond to litigation support and expert testimony, and a policy that is silent is not a policy that covers. The question to ask the carrier, in writing and before the first engagement, is whether the policy covers professional services rendered in connection with litigation, including testimony; whether it excludes claims brought by a retaining client; whether defense costs erode the limits; and whether coverage is claims-made, which for a practice whose exposure surfaces years later makes the tail provision the important term. Where a policy will not respond, some professionals contract for indemnity from the retaining party in the engagement letter.

That is worth requesting and worth understanding for what it is: a promise from a party whose interests may later diverge, not a substitute for coverage.

None of this argues against entering the work. It argues for entering it with the same documentation discipline the rest of this chapter recommends for every other purpose, because the file that protects the opinion is the same file that protects the professional who signed it.

VII. What This Costs: Budgeting the Technical Work

Chapter 1 tells counsel to budget for the method rather than the conclusion, and Section III above treats fee structure from the expert's side. Neither says what any of it costs, and counsel deciding whether to fund a site inspection, an exemplar reconstruction, or a full retrospective exposure assessment deserves an order of magnitude. What follows is a planning framework rather than a price list; rates vary by region, discipline, and market, and the figures behind them will age faster than the framework does.

The rate structure. Testifying safety and industrial hygiene experts in this field commonly bill hourly, with a modest premium for deposition and trial testimony over analysis, and most require a retainer against which work is drawn. Consulting-only engagements are billed on the same basis. The rate is the least interesting variable in the budget, because the hours are where the money is and the hours are driven by the scope counsel sets.

The engagements, by magnitude. A *file review with a verbal opinion*, the engagement this book recommends most often and counsel commissions least, is typically ten to twenty-five hours, and it is the highest-yield expenditure available because it happens early enough to change the theory. A *site inspection with measurements and a photographic record* adds travel, a day on site, and the analysis behind it. A *sampling campaign* carries per-sample laboratory costs on top of professional time, and the sample count is driven by the statistical question rather than by the budget — which is why Chapter 15's discussion of what a handful of samples can support is a budgeting document as much as a technical one. A *retrospective exposure reconstruction* is the most variable item in the field, ranging from a bounded exercise where historical monitoring data exist to a many-month undertaking where the estimate must be assembled from job titles, task descriptions, and published analogues. An *exemplar acquisition and reconstruction* in a products case carries equipment cost, fabrication or instrumentation cost, and testing time; it is the item most often cut and most often decisive.

Which expenditures buy the most admissibility per dollar. The reported decisions answer this more usefully than any budgeting heuristic. Ranked roughly by admissibility purchased per dollar spent: first, early retention, which costs the least and prevents the most — every case in this book that failed on a missing record failed because someone was retained after it was gone. Second, the measurement or the calculation that converts an assertion into a number, because *Joiner*'s analytical gap is bridged by arithmetic and not by experience. Third, the literature search that surfaces the contrary studies before the deposition does, which is inexpensive and is the difference between the *Acetaminophen* experts who were excluded and the ones who were not. Fourth, the sensitivity analysis, which costs a few hours and converts a vulnerable assumption into a stated limitation. Fifth, physical testing, which is expensive, which *Sardis* shows cannot be skipped in a design case, and which should therefore be scoped early enough to be funded rather than discovered at the report deadline.

What is not worth buying. Two expenditures recur and rarely pay. Additional sampling collected to improve a conclusion the existing data already support is money spent on precision the decision does not need. And a second expert retained to say the same thing as the first, in a case that does not require two disciplines, buys the opponent a second deposition and a second set of transcripts to mine. Chapter 1's observation that the expert team is only as strong as its weakest opinion is a budgeting principle as well as a strategic one.

A word to the professional. Counsel who ask what this will cost are usually asking whether the analysis can be scoped to a budget, and the honest answer is that some of it can and some of it cannot. A reconstruction can be scoped; the arithmetic cannot be left half-finished, because *Amorgianos* is what a half-finished reconstruction is worth. The practice worth building is one that says clearly, at the outset, which parts of the proposed work are divisible and which are all-or-nothing, and that declines the engagement scoped so thinly that the resulting opinion could not survive the motion it will face.

VIII. The Testimony List, the File, and the Career Record

Every retained expert accumulates a record that follows them from case to case, and managing it is a career-long discipline. The prior-testimony list required by Chapter 3 must be maintained contemporaneously — reconstructing four years of depositions and trials under a report deadline is how omissions happen — and every prior report and transcript should be retained and, before each new engagement, consulted, because the expert's own words are the first exhibit the next opponent will reach for. Consistency across the career is the durable asset: the professional who has stated the same methodological commitments, applied the same standards, and drawn the same lines in matter after matter has built the record that makes the neutrality of Chapter 6 believable. The expert who has said opposite things for opposite clients has built the opposite.

File discipline within each engagement is the corollary. The testifying expert assumes discoverability and writes accordingly — not dishonestly, but with the understanding that a jury may read the file — while genuinely deliberative consulting work is kept, and marked, within whatever protection the engagement supports, as Chapter 26 §VI describes for the professional's own records. The habits that protect the file are the habits that protect the opinion, and both are built long before any particular case makes them matter.

IX. Entering the Work: The First Engagement

The chapter so far assumes an expert practice already in motion; this section is about starting one, because the first engagement is the hardest to obtain and the habits it sets compound for a career. Retaining counsel find experts through three channels, and the professional can cultivate all of them deliberately: the referral from another lawyer or another expert, which is earned by the work itself and by the reputation for candor Section II describes; the published record — the article, the conference talk, the standard-committee membership — through which counsel searching for a silica or machine-guarding expert first encounters a name; and the directories and prior-testimony databases where the professional's career record already sits, curated or not. The strategic point about the public record cuts both ways and should be understood before the first article is written: everything published is a prior statement, discoverable by any future opponent and readable to any future jury, and the professional who writes with that audience in mind — measured claims, stated limitations, positions that could be held for either party — is building qualification and armor in the same document. The professional who wrote the confident absolutism of a trade-magazine column will meet it again, enlarged on a screen, in a courtroom.

The transition itself has mechanics that deserve deliberate handling. The professional leaving or reducing an industry or consulting role must clear the accumulated conflicts of a working life — the former employer whose facilities, records, and confidences are off limits; the industry relationships that will be characterized as allegiance — and should expect the first cross-examinations to be built almost entirely from the resume, because there is not yet a testimony record to mine. The early-practice disciplines that answer them are unglamorous: take the first engagements small and within unquestionable competence, ideally as a consulting expert or a second chair supporting an established witness, where the craft of the report and the deposition can be learned without carrying a case; charge from the first hour at a rate that will not embarrass the tenth year, because the fee history is discoverable and the bargain rate reads later as eagerness; carry professional liability coverage that includes expert work, and read the policy's cooperation and notice clauses before, not after, the first subpoena for the file arrives. And calendar honestly: trial work is scheduled by courts, not consultants, and the professional who accepts a retention while unable to survive a continuance landing on a committed week has created the withdrawal that opposing counsel will characterize forever after.

The compounding logic ties this section to everything above it. Each engagement generates the four assets the practice runs on — the work product that becomes the next engagement's exemplar, the transcript that becomes the consistency record, the relationship that becomes the referral, and the outcome that becomes the reputation — or it generates their opposites, and the direction is chosen by the disciplines of this chapter, one matter at a time. There is no marketing strategy for an expert practice that outperforms a decade of files any lawyer could read without flinching. The practice is the record, accumulating; the professional who builds it as if every page will someday be Exhibit 14 is the one for whom, when the page finally is, it helps.

X. Practice Implications

For the professional building an expert practice, the through-line of this chapter is that credibility is an accumulation, not a performance: it is built engagement by engagement through structured retentions, declined conflicts, honest fees, an ethics code treated as the real standard, and a career record kept consistent enough that the neutrality the courts reward is simply true. For counsel, understanding the expert's side of these choices sharpens the screening of Chapter 1: the questions in the First-Call Checklist of Appendix A are answered best by the professional who has managed the practice deliberately, and the expert worth retaining in your case is the one who built, across every prior case, the record that lets them say the same thing in either chair. That expert supplies the theory of the case. This book has been, from the first page, about becoming or recognizing them.

References

American Industrial Hygiene Association. (2015). A strategy for assessing and managing occupational exposures (4th ed.). AIHA Press.

Board for Global EHS Credentialing. Code of Ethics (Certified Industrial Hygienist).

Board of Certified Safety Professionals. Code of Ethics (Certified Safety Professional).

Fed. R. Civ. P. 26(a)(2), (b)(4); Fed. R. Civ. P. 26(b)(4)(D) (facts known and opinions held by a non-testifying consulting expert; the exceptional-circumstances showing).

Institute of Hazardous Materials Management. Code of Ethics (Certified Hazardous Materials Manager).

CHAPTER 28

The Rule Stalled. The Docket Didn't.

The Enforcement Frontier and the Rising Case Landscape

Current as of July 2026. This chapter tracks directives, dockets, and rulemakings that change faster than a book can. Every directive number, docket status, and rulemaking stage below was verified as of that date and should be re-verified before it is relied on. Corrections and updates are posted at www.lfmartinllc.com.

> *The federal rulebook and the civil docket are moving in opposite directions. As the heat standard idles in the post-hearing file, as an emphasis program substitutes for an ergonomics rule OSHA abandoned a generation ago, and as a deregulatory agenda and a wave of constitutional challenges narrow the agency's reach, the claims those rules were written to prevent are arriving in civil courtrooms in growing numbers. This chapter maps where enforcement is actually concentrating and what each rising case type demands of the technical expert.*

I. The Frontier Has Moved

For most of the last decade the question that organized occupational safety and health practice was regulatory: what would OSHA require next, and when? That question has inverted. The heat injury and illness prevention standard, the most ambitious health rulemaking the agency has attempted since silica, completed its public hearing in July 2025 and its post-hearing comment period that October, and has sat since then with no target date for final action and no indication that finalization is an administration priority. The ergonomics standard remains where Congress left it in 2001, rescinded and effectively barred from resurrection in similar form. The July 2025 deregulatory package proposed more than two dozen rescissions and revisions, including a reinterpretation of the General Duty Clause itself. And the Supreme Court's administrative-law turn — *Loper Bright* ending Chevron deference, *Jarkesy* reviving the jury-trial right against agency penalty adjudication — has put the enforcement machinery that would apply whatever rules survive under a constitutional cloud of its own.

None of this has made workplaces safer or the underlying hazards smaller, and that is precisely the point for the professions this chapter addresses. Where the regulatory floor recedes, the civil docket rises to occupy the space: a substitution the last three years have made visible in real time. The engineered-stone silicosis docket has produced the largest silica verdicts in American history while the federal silica standard, fully in force, failed to prevent the epidemic underneath them. Heat-death litigation is accelerating through the very seasons in which the federal heat rule idles, with a half-dozen state standards and the General Duty Clause supplying the standard-of-care scaffolding a final rule would have provided. Warehouse musculoskeletal injury, the hazard the rescinded ergonomics rule was written for, is being regulated instead by state quota-disclosure statutes whose most important effect, for litigators, is the discoverable record they force into existence.

For the safety professional and the industrial hygienist, the shift changes the nature of the work: the opinions that matter increasingly issue not from area offices but from expert reports, and the standards matrix in any given case is now a layered composite of federal rules, state-plan standards, consensus standards, and the defendant's own program — read under an amended Rule 702 that polices the technical work harder than any compliance officer ever did. For counsel on both sides, and for the jury consultants downstream of them, the shift changes where the cases come from, what they turn on, and who decides them. The sections that follow map the enforcement frontier as it actually stands in mid-2026, then the dockets rising fastest, then the technical program each demands, because in this landscape, as in the third-party case generally, the side that does the technical work earlier, more completely, and more honestly is the side the record ends up serving.

II. The Enforcement Frontier

Federal enforcement has not disappeared; it has concentrated. Three instruments define its current shape — the revived heat emphasis program, the warehousing emphasis program, whose three-year term has now run without a posted successor, and the General Duty Clause operating as the residual standard for everything the rulebook does not reach — and each sits under two countervailing pressures: a deregulatory rulemaking agenda that would narrow the agency's tools, and a constitutional litigation campaign that would narrow its forum. Counsel on both sides should understand all five layers, because each generates evidence, standards, and inspection files that will surface in civil discovery long after the citation itself is resolved.

A. Heat: The Standard Stalled, the Program Did Not

The proposed Heat Injury and Illness Prevention in Outdoor and Indoor Work Settings standard was published on August 30, 2024: the first federal heat rule the agency had ever proposed, built around a two-tier trigger structure keyed to heat index, with an initial threshold requiring drinking water, break areas, and paid rest as needed, a high threshold adding mandatory rest breaks and monitoring, and acclimatization protocols for new and returning workers. The comment period closed January 14, 2025. Informal public hearings ran from June into July 2025, and the post-hearing comment period closed that October. And there the rule has sat. No final action has been scheduled, no target date published, and nothing in the current regulatory agenda suggests finalization is a priority; the more likely outcomes, on the present posture, are indefinite deferral or a substantially narrowed re-proposal. Employers and counsel who have been waiting for a federal heat standard to supply the standard of care should stop waiting.

What the administration did instead is the more consequential fact for the next several summers. The original heat National Emphasis Program, issued in April 2022 and extended once, expired on April 8, 2026, and two days later, on April 10, 2026, OSHA issued a revised NEP, CPL 03-00-024, effective immediately and in place for five years. The revision narrows and sharpens rather than retreats: it targets a data-driven list of fifty-five high-hazard industries — construction, agriculture, warehousing, foundries, freight trucking, delivery and logistics, manufacturing, landscaping, waste collection, and even high-exposure retail — drops the prior directive's numeric inspection-growth quota, and directs compliance officers to expand any inspection, on any subject, where evidence of heat hazards appears on heat priority days. The program's enforcement instrument remains the General Duty Clause, informed by the NIOSH criteria document: roughly seven thousand heat inspections since 2022 have produced only a modest number of 5(a)(1) citations, a ratio that testifies to how demanding the recognized-hazard and feasible-abatement elements are, and to how much of the enforcement action is really occurring in the civil courts instead.

The operative standards, meanwhile, are increasingly state standards. California enforces both its long-standing outdoor rule and, since 2024, an indoor heat standard; Oregon, Washington, Colorado, Minnesota, Maryland, and Nevada enforce heat rules of their own, several more prescriptive than the stalled federal proposal. Eighteen states saw heat-safety legislation introduced in 2025 alone, and the 2026 sessions have produced proposals from Arizona's written-mitigation-program bill to New Hampshire's combined extreme-temperature act; California's legislature passed, and the governor vetoed, a workers'-compensation presumption linking agricultural heat injury to noncompliant employers. For litigators the map matters twice over: in state-plan states the specific rule supplies the standard of care and sometimes negligence per se, while in federal-enforcement states the expert must build the standard of care from the General Duty Clause line, the NIOSH criteria, the ACGIH TLV for heat stress, and — increasingly — the stalled proposed rule itself, which both sides will characterize and which neither can ignore.

B. Warehouse Ergonomics: Enforcement Without a Standard

The warehousing National Emphasis Program, CPL 03-00-026, is the closest thing federal OSHA has had to an ergonomics enforcement instrument since the 2001 rescission. Issued July 13, 2023, with inspections beginning that October, it targets warehousing and distribution centers, mail and parcel processing, courier and last-mile delivery, and high-injury-rate retail establishments: a sector whose employment nearly tripled from 2011 to 2021 and whose serious-injury rate led all nineteen private sectors. Programmed inspections are comprehensive, covering powered industrial trucks, material handling and storage, walking-working surfaces, egress, and fire protection, and, critically for the health side, the directive requires every inspection to consider and document heat and ergonomic hazards, with a health inspection opened whenever the compliance officer learns they are present. The directive's three-year term ran through July 13, 2026. As of the date at the head of this chapter, that term has expired and OSHA has neither announced a renewal nor issued a successor directive: CPL 03-00-026 remains posted on the agency's emphasis-program list under its original 2023 date, with no extension, revision, or replacement of the kind the heat program received in April 2026. Three readings of that silence are available, and the practitioner should not choose among them without checking. The program may be renewed late, as emphasis programs sometimes are. It may be revised and reissued under a new number, which is how the heat directive was handled. Or it may be allowed to lapse, which under the current deregulatory posture would be the most consequential of the three, because it would leave warehouse ergonomics to the General Duty Clause and to the state statutes described in Section III.C. What is certain is what the expired directive already produced: three years of programmed comprehensive inspections whose files, targeting lists, and heat and ergonomic documentation exist and are discoverable regardless of whether the program continues. Verify the directive's current status on OSHA's emphasis-program page before relying on it either way; a lapsed NEP cited as live is the sort of error Chapter 17 catalogues.

The program's ergonomics record explains why the civil docket, not the citation docket, is where warehouse musculoskeletal injury is being litigated. A 2024 Government Accountability Office study found that warehouse and last-mile delivery employers drew more than 2,500 violations from fiscal 2018 through 2023, of which eleven related to ergonomic hazards. Without a specific standard, an ergonomics citation must travel through the General Duty Clause, with expert proof of a recognized hazard and feasible abatement; the agency brings few and wins fewer. The GAO recommended OSHA sharpen how it identifies and addresses ergonomic hazards and evaluate whether the NEP protects workers from them; OSHA agreed and has folded ergonomic screening into the inspection program. But the enforcement gap is structural, and the states have moved into it.

Seven states — California, New York, Washington, Minnesota, Oregon, Connecticut, and Rhode Island — now regulate warehouse productivity quotas, requiring written disclosure of quotas and work-speed data, prohibiting quotas that prevent meal, rest, and bathroom breaks, and protecting workers who complain. New York went furthest: its Warehouse Worker Injury Reduction Act, effective June 1, 2025, requires covered employers to implement written injury-reduction programs, train employees annually, and — a first in American law — retain a qualified ergonomist to evaluate musculoskeletal risk factors, with board-certified ergonomist review when workers raise concerns. Federal legislation on the same model, the Warehouse Worker Protection Act, is pending in both chambers and would direct OSHA to propose an ergonomic program-management standard for the sector within three years. Whatever their regulatory fate, these statutes have already transformed litigation: they compel the creation, retention, and disclosure of the exact records — quota definitions, individual work-speed data, ergonomic evaluations, injury-reduction program documents — that a plaintiff's expert previously had to reconstruct and a defendant could previously decline to generate. Section III.C takes up what both sides should do with that record.

C. The Emphasis-Program Map

Beyond heat and warehousing, the active emphasis-program map tells counsel where inspection files — and the photographs, measurements, and sampling data inside them — are being generated: the respirable crystalline silica NEP, whose targeting now reaches the engineered-stone fabrication shops at the center of the silicosis docket; the long-running national programs on amputations in manufacturing, trenching and excavation, combustible dust, process safety management in chemical facilities, and falls in construction, which remain the largest single cause of construction fatalities; and dozens of regional and local emphasis programs —

grain handling, oil and gas, food manufacturing — that vary by area office. Every one of these programs produces, in each inspection, the most professional contemporaneous documentation of a workplace that will ever exist, obtainable by records request; the expert who does not pull the complete inspection file, rather than the citation summary, in any case where an inspection occurred is leaving the best evidence in the drawer.

Running against the enforcement current is the deregulatory one. On July 1, 2025, OSHA published more than two dozen proposed rules in a single day — rescissions and revisions the agency frames as modernization, ranging across construction illumination, respiratory-protection provisions, medical-evaluation requirements, and consultation procedures — followed in April 2026 by a proposed revision to the walking-working-surfaces rule. The centerpiece for litigators is the proposal to reinterpret the General Duty Clause itself: under the proposed interpretation, OSHA would not cite hazards that are "inherent and inseparable from the core nature of a professional activity or performance" — the position then-Judge Kavanaugh took in dissent in the SeaWorld case, now offered as agency doctrine. The proposal is aimed at entertainment, sports, animal handling, and similar occupations, and its direct reach is narrow; its indirect significance is not. Every boundary drawn around the General Duty Clause is a boundary drawn around the instrument currently carrying heat, ergonomics, workplace violence, and every other hazard without a standard, and every rescission argued as "unnecessary" will be quoted back in civil litigation by one side or the other, either as proof that the obligation never reflected the true standard of care or as proof that the hazard was so well recognized the agency once wrote a rule about it. The deregulatory record, in other words, is itself discovery.

D. The Constitutional Overlay: Loper Bright and Jarkesy

The third pressure on the enforcement frontier is structural. *Loper Bright Enterprises v. Raimondo* ended judicial deference to agency statutory interpretation, inviting fresh challenges to OSHA's readings of the OSH Act, including the multi-employer citation doctrine and the scope of Section 5(a)(1), that courts once resolved by deferring. *SEC v. Jarkesy* held that when an agency seeks civil penalties for claims analogous to common-law causes of action, the Seventh Amendment may require an Article III court and a jury; although *Atlas Roofing* sustained OSHA's administrative adjudication in 1977, *Jarkesy*'s reasoning has invited direct attacks, employers have sued to enjoin Review Commission proceedings on exactly this theory, and commentators on both sides expect the boundary between public and private rights to be relitigated. Meanwhile the Occupational Safety and Health Review Commission has lacked a quorum since April 2023, its contested-case backlog frozen; challenges to administrative law judges' removal protections, which the Department of Justice has declined to defend, add a further layer; and the Supreme Court's denial of certiorari in *Allstates Refractory* left the OSH Act's delegation intact for now while the question circulates below.

For the professions this chapter serves, the practical consequences are three. First, contested citations will take longer, settle differently, and carry less collateral weight: counsel should be careful about treating a citation, or its absence, as more than what it is, and experts should ground opinions in the underlying standards and facts rather than the enforcement outcome. Second, if penalty adjudication migrates toward Article III courts and juries, safety and IH experts will find themselves testifying in enforcement matters that look procedurally like civil trials, with everything that implies about Rule 702, cross-examination, and jury communication. Third, and most broadly: as the agency's interpretive and adjudicatory authority narrows, the civil courts' role in defining the operative standard of care widens, which is the door through which the rising dockets of Section III walk.

III. The Rising Dockets

Four case areas are generating enough filings, verdicts, and appellate activity to deserve their own sections, and a fifth — the widening chemical frontier — supplies the compounds most likely to join them. What unites them is the structure described in the enforcement sections above: each involves a hazard the regulatory system recognized but did not, or could not, control; each turns on an exposure reconstruction, a control analysis, or both; and each is being decided, on both sides of the v., by the quality of the technical work. Plaintiff and defense counsel should read each subsection as a map of the other side's case as much as their own.

A. Engineered-Stone Silicosis: The Fastest-Moving Docket

The numbers first, because they explain everything downstream. Engineered stone — the quartz-resin composite that now dominates the countertop market — runs ninety percent or more crystalline silica by weight, several times the content of granite. Cutting, grinding, and polishing it dry produces exposures that have compressed a disease with a decades-long latency into one that presents in workers in their twenties and thirties as accelerated and complicated silicosis, progressive massive fibrosis, and end-stage disease requiring transplant. California, whose Los Angeles fabrication-shop cluster sits at the epicenter, had recorded 560 confirmed cases among countertop workers as of mid-May 2026 — up from 447 the previous December, and from thirteen when surveillance began in 2019 — with at least twenty-nine deaths and more than fifty lung transplants. Median age at diagnosis is forty-six; median age at death is forty-eight. Those are the statewide surveillance medians, and they are not in conflict with the median of forty-five reported at Chapter 16 §IV.A, which comes from the published fifty-two-worker case series rather than from the surveillance count. Because the figure is updated weekly on the state's public dashboard, any brief or report that cites it should cite it as of a date.

The regulatory chronology matters as much as the count, and counsel should have it straight, because each step created a different set of documents. Cal/OSHA adopted an emergency temporary standard effective December 29, 2023, prohibiting dry cutting and dry sweeping in fabrication shops; the Standards Board approved permanent rules in December 2024, effective February 5, 2025, now codified at Cal. Code Regs. tit. 8, § 5204. Then the legislature went further than the regulators. The Silicosis Training, Outreach, and Prevention Act — Senate Bill 20, the STOP Act — was signed October 13, 2025 and took effect January 1, 2026. It writes the dry-cutting ban into the Labor Code, reclassifies silicosis and silica-related cancer as a serious injury or illness, requires employers to report confirmed cases to both Cal/OSHA and the Department of Public Health, and builds a licensing regime on a staged clock: an annual training attestation from July 1, 2026, an application process by January 1, 2027, and, from July 1, 2027, a requirement that every fabrication shop hold a valid three-year certification to operate lawfully at all.

For the technical practitioner that last provision is the consequential one, and it is not primarily a compliance story. A certification regime manufactures a dated, adjudicated record of each shop's controls, training, and monitoring — which is to say it manufactures the exposure history that every silicosis case filed after 2027 will be reconstructed from, and the notice evidence that every one of them will turn on. An employer who cannot produce a certification, or whose certification was granted on representations the sampling data do not support, has created the Chapter 25 problem in its purest form. Enforcement, meanwhile, has not kept pace with the rules: a January 2026 review of Cal/OSHA inspections at more than a hundred fabrication shops found not one worker using the level of respiratory protection the high-exposure tasks require, and roughly a quarter of shops still dry-cutting in the third year of a prohibition. Australia — the comparison every jury will eventually hear about — banned engineered stone outright in July 2024.

The litigation has moved with corresponding speed, and the results are running in both directions, which is the fact each side's counsel most needs to absorb. In

August 2024 a Los Angeles jury returned a $52,437,366 verdict for a thirty-four-year-old fabricator who had already undergone a double lung transplant and faced the prospect of a second within three to five years, against manufacturers including Caesarstone, Cambria, and Color Marble — the largest silicosis verdict in American history — in a case where dozens of upstream defendants settled before or during trial. Post-trial motions have been resolved and the judgment is on appeal, which is the posture to state whenever the number is cited — and the posture is not the verdict. On October 9, 2024 the trial court granted judgment notwithstanding the verdict as to Color Marble, finding that the plaintiff had not introduced substantial evidence on all the essentials required of a distributor and that a negligence finding could not stand beside the jury's rejection of the strict-liability and failure-to-warn theories against that same defendant; an amended judgment followed on March 18, 2025, and cross-appeals were filed the month before it. The allocation deserves the same attention here that the Denver verdict below receives, and for the same reason: the jury assigned fifteen percent of the fault to Caesarstone, ten percent to Cambria, and two and a half percent to Color Marble — twenty-seven and a half percent in total to the three defendants who actually stood trial — with two and a half percent to the plaintiff and the remaining seventy percent to parties not named on the verdict form. The jury also split on the theories, finding all three negligent while rejecting design defect and failure to warn as to Color Marble — the inconsistency on which the judgment notwithstanding the verdict turned. The gross figure is what the docket is quoted at; the operative figure after that ruling is twenty-five percent, carried by the two manufacturers still in the case, and it is that number, not the gross, on which a slab manufacturer's reserve actually turns. Both figures, and the distance between them, are the first thing opposing counsel will say about the case. Early 2025 brought a settlement in excess of $26 million, resolved during jury selection, for an oxygen-dependent fabricator with progressive massive fibrosis. Mid-2025 brought the counterpoint: a Torrance jury returned a defense verdict, finding the product not defective and attributing the disease to fabrication-shop handling — the empty-chair defense, run through the employer's dry-cutting practices, working exactly as designed. And on April 30, 2026 a Denver jury awarded $17,450,000 to a thirty-one-year-old fabricator with silicosis and silica-associated chronic kidney disease, diagnosed at twenty-eight after ten years in his family's shop (Jordan v. Cambria Co., No. 2024CV31180 (Denver Cnty. Dist. Ct.)). The verdict's internal logic is worth more than its total. The jury found no design defect and found the lead slab manufacturer not negligent, but held that manufacturer liable for misrepresenting the character of the product, and held a second manufacturer liable for both negligence and misrepresentation — and found that the employer had followed the wet-cutting, ventilation, and respirator protocols it understood natural stone to require, having been told the material was pure natural quartz. That combination converts the warnings and downstream-communication file into the whole case, and it is the reason the misrepresentation finding is now on appeal. The allocation in that verdict is the more instructive half of it, and it is the empty chair of Chapter 19 §IV operating in the flagship docket of this chapter: the jury assigned sixty-three percent of the fault to the fabrication shop that employed the plaintiff, thirty-two percent to the manufacturer found liable on misrepresentation, three percent to a second manufacturer, and two percent to the plaintiff himself, while a third major slab manufacturer and two other defendants were assigned none at all. A plaintiff's verdict of that shape is also a partial defense success, and counsel on either side should read it for what the apportionment rewarded rather than for the headline number. Dozens of new filings are landing in Los Angeles Superior Court, a coordinated California proceeding is managing the state docket, federal coordination is underway, and the defendant class is widening from slab manufacturers to distributors, big-box retailers, and respirator suppliers, with 3M facing allegations that its masks failed to protect against the exposures at issue.

The technical case is an industrial-hygiene case before it is anything else. The plaintiff's reconstruction runs from the slabs actually fabricated — identified through supplier invoices and job records — through task-level exposure estimation for dry versus wet methods, ventilation, and respiratory protection, against the silica standard's permissible exposure limit and action level and the published exposure-response literature for accelerated disease; the warnings analysis runs through the safety data sheet revision history and the downstream-communication file, asking what the manufacturer told the fabrication trade and when. The defense case audits every one of those inputs — duration, frequency, task mix, respirator fit and use, the employer's own obligations under the silica standard — and builds the allocation the Torrance verdict rewarded: a shop-practices case, made with documents rather than adjectives. Both sides should treat the OSHA silica NEP inspection files for the shops at issue, the employer's thirty-year exposure records under 29 C.F.R. § 1910.1020, and the state surveillance data as the neutral spine of the exposure record. And both should be candid about the apportionment problem — multiple shops, multiple slab brands, mixed natural and engineered stone — because the amended Rule 702, as the exclusion decisions in other exposure dockets keep demonstrating, has little patience for the hygienist who compresses a career of mixed exposures into a single defendant's product.

B. Heat Illness and Death: The Docket the Stalled Rule Is Feeding

Heat kills more American workers than any other weather hazard — roughly 34,000 recorded heat injuries and 479 deaths from 2011 through 2022, figures everyone who works in the field understands to be undercounts — and heat-related workers'-compensation claims are rising across every sector, led by construction, agriculture, landscaping, and delivery. The litigation is rising with them, and its structure tracks the third-party architecture familiar from every workplace-injury docket: the deceased worker's employer is usually shielded by the exclusive remedy, so the suits run against the general contractor that set the schedule and controlled the site, the owner whose premises lacked water and shade, the staffing agency that supplied an unacclimatized worker on day two, and — in the gross-negligence and intentional-conduct exceptions some states allow — the employer itself. Recent filings include wrongful-death actions against contractors whose crews died without water breaks and suits against warehouse operators over triple-digit indoor temperatures; the pattern will intensify as summers do.

The absence of a final federal standard shapes both sides' cases without deciding either. The plaintiff builds the standard of care from the layered record the enforcement sections described: the applicable state standard where one exists; the General Duty Clause citation history and the revised NEP's heat-priority-day framework; the NIOSH criteria document and ACGIH TLV; industry and consensus guidance; and, handled carefully, the proposed federal rule, whose 80°F and 90°F trigger architecture maps cleanly onto what a reasonable program would have contained. The defense tests the physiological case as rigorously as the programmatic one: heat-stroke causation runs through core temperature, time course, medical response, and individual risk factors, and the reconstruction of wet-bulb globe temperature at the actual time and place — not the airport weather station — is the industrial hygienist's work whichever side asks for it. Acclimatization is the recurring battleground: a large fraction of occupational heat deaths occur in the first days on the job, which makes hiring records, task assignment, and the staffing chain — who knew the worker was new — the discovery that decides allocation. And in every heat case both sides should preserve and demand the perishable record immediately: weather data, workload and production records, hydration and break logs, body-camera and site video, and the emergency-response timeline, because the interval between collapse and effective cooling is frequently the causation case.

C. Warehouse Musculoskeletal Injury: The Statutes

Musculoskeletal disorder claims have historically been hard tort cases: gradual onset, contested causation, no single incident, and workers'-compensation exclusivity closing the courthouse against the employer. Three developments are changing the calculus. First, the state quota statutes described in Section II.B have created a documentary record that did not previously exist: written quota definitions, individual work-speed data, the employee's own annotations of non-productive time, ergonomic evaluations by qualified — in New York, board-certified — ergonomists, and formal injury-reduction programs. Second, that record is generated before injury and without litigation in mind, which gives it exactly the evidentiary character the exemplar third-party case prizes: it reports what it reports, for whichever side it favors. Third, the third-party structure of the modern warehouse — staffing agencies supplying labor, third-party logistics operators running the building, equipment vendors supplying the automation whose pace and reach define the ergonomic exposure — multiplies the defendants the exclusive remedy does not shield.

For the plaintiff, the ergonomist's case now writes itself from the defendant's file when the file is bad: a quota the employer's own risk evaluation flagged, an injury-reduction program that identified the lifting task and changed nothing, work-speed data showing the pace the program said was hazardous. For the defense, the same statutes supply the affirmative program evidence Chapter 24 §IV describes — the evaluation conducted, the finding closed, the workstation redesigned — and the comparative record of the worker's own data. For both, the expert discipline is the same: MSD causation opinions must engage the epidemiology honestly, quantify the exposure — force, frequency, posture, duration, against the recognized assessment tools — and address personal and non-occupational contributors, because an ergonomics opinion that characterizes rather than measures is precisely the opinion the amended Rule 702 exists to exclude. Counsel should also watch the enforcement horizon: the warehousing NEP's renewal decision and the pending federal Warehouse Worker Protection Act, which would order an ergonomics rulemaking for the sector, will determine whether the next phase of this docket is built on state statutes alone or on a federal standard with negligence-per-se consequences.

D. Ethylene Oxide: The Sterilization-Plant Docket at Maturity

The ethylene oxide docket is the most instructive of the mature chemical dockets, because it shows the full life cycle the newer compounds are entering. The scientific predicate was regulatory: EPA's 2016 IRIS assessment reclassified EtO as carcinogenic to humans by inhalation, and that assessment — contested by industry, adopted by plaintiffs — became the hinge of the litigation. The early results were seismic: a $363 million Illinois verdict for a woman who lived near the Sterigenics Willowbrook plant, followed by a $408 million settlement of the remaining Willowbrook docket, with a further $31 million resolving ninety-seven additional cases in April 2025 and $34 million resolving one hundred twenty-nine more that July (Sotera Health Co., 2025a, 2025b). Then the correction: on March 14, 2025 Terumo BCT won a complete defense verdict in Colorado in *Isaacks v. Terumo BCT Sterilization Services, Inc.*, No. 2022CV031124 (Jefferson Cnty. Dist. Ct.), a six-week bellwether brought by four neighbors of its Lakewood plant who sought more than $217 million for physical impairment, $7.5 million in medical expenses, and punitive damages; the jury found the company not negligent and so never reached causation or damages, crediting the facility's regulatory compliance and emissions practices — the program-as-it-actually-operated defense, run successfully in a docket that had looked one-sided. The aftermath is worth knowing before a bellwether is volunteered: in September 2025 the trial court taxed roughly $5.18 million in costs against the losing plaintiffs. Georgia has since produced a $20 million compensatory verdict in *Walker v. C.R. Bard, Inc.* (Gwinnett Cnty. State Ct. May 2, 2025), brought by a retired truck driver who made regular pickups at a Covington sterilization plant and lived a mile and a half from it; the jury found negligence and public nuisance and that punitive damages should be awarded, the punitive phase then ended in a mistrial when a juror dissented on polling, and the compensatory award and the scope of any retrial are on appeal; Pennsylvania has produced both halves of the picture: in December 2024 a Philadelphia Court of Common Pleas jury returned a defense verdict on all counts in Glass v. B. Braun Medical Inc., after a five-week trial, on a claim that seven years of work at the company's Allentown-area sterilization plant caused the decedent's leukemia — the first ethylene oxide case tried anywhere in the country on behalf of a plant employee rather than a neighbor, and therefore the one occupational data point the docket has so far produced; and a Lehigh County judge subsequently denied class certification for a community medical-monitoring class near the same plant, questioning the reliability of several of the plaintiffs' expert opinions while leaving individual claims open. Counsel should read the two outcomes together. The occupational plaintiff had the better exposure record — a payroll history, a job title, a plant with a monitoring obligation under the ethylene oxide standard — and still lost, which is the clearest available evidence that in this docket the exposure reconstruction is necessary but nowhere near sufficient, and that specific causation is where these cases are actually decided; and the Fourth Circuit's 2025 *Sommerville* decision revived an EtO medical-monitoring claim on standing grounds, signaling that exposure-based monitoring theories, the leading edge of every chemical docket, remain live in at least some circuits.

The occupational dimension deserves counsel's particular attention, because it is where the next wave of filings sits. Sterilization-plant workers, hospital central-sterile technicians, and spice-industry employees carry the highest documented exposures, and OSHA's EtO standard — with its permissible exposure limit, excursion limit, monitoring, and medical-surveillance provisions — supplies both the exposure records the hygienist reconstructs from and the compliance framework the defense builds on. The lessons of the docket generalize: the agency risk assessment that reclassifies a compound is the event that creates the litigation; the emissions and monitoring record decides it; juries reward documented compliance and punish documented indifference; and the medical-monitoring theory, which converts exposure without present disease into a compensable claim, is the doctrinal frontier on which the plaintiff and defense bars are now fully mobilized.

E. The Widening Chemical Frontier: PFAS and the Next Compounds

PFAS is the largest of the coming occupational dockets and the least mature. The aqueous film-forming foam multidistrict litigation, already among the largest in history, has resolved major water-provider claims and is now working through the personal-injury tier, in which firefighters are the paradigmatic occupational plaintiffs: exposure through decades of foam use and, in the fastest-growing theory, through the fluoropolymer-treated turnout gear itself. Downstream, occupational claims by chemical-plant, textile, plating, and paper workers are forming behind the community-exposure cases, and medical monitoring — recognized in some states, rejected in others, revived on standing grounds by decisions like *Sommerville* — will be the vehicle that determines the docket's ultimate size. That decision is the same Fourth Circuit opinion Chapter 2 §I discusses as the Rule 702 counter-trend: *Sommerville v. Union Carbide Corp.*, 149 F.4th 408 (4th Cir. 2025), reversed on both the Article III standing question and the exclusion of the plaintiff's dispersion-modeling expert, and the two holdings compound. The standing holding enlarges the class of exposure-without-disease plaintiffs; the admissibility holding lowers the barrier to the reconstruction each of them will need. A defense practitioner who reads only the medical-monitoring half of the opinion will miss the reason the docket it opens is a docket about expert testimony. The industrial hygienist's problem in PFAS is the field's hardest: no OSHA standard, few occupational exposure limits, serum-level biomonitoring that integrates lifetime exposure from all sources, and an attribution question, occupational versus consumer versus environmental, that will make apportionment the whole case.

Behind PFAS, the compounds and case types to watch share a signature: a regulatory or scientific event creating notice, a defined occupational cohort, and an exposure susceptible to reconstruction. Benzene claims continue against solvent and fuel suppliers for hematologic disease, sustained by the low-dose literature. Paraquat's Parkinson's-disease docket, having survived its gatekeeping battles in part, holds thousands of agricultural claims. Welding fume, diesel exhaust, and hexavalent chromium persist in the trades. Lithium-ion battery fires are generating a new thermal-runaway docket — product claims against cell and pack manufacturers, premises and storage claims in warehouses, and first-responder exposure claims for the combustion products — that sits squarely at the intersection of process safety, fire science, and industrial hygiene. And the respirator and PPE-supplier theory that has attached to the silicosis docket, the protective equipment as a defendant in its own right, is a template plaintiffs will carry into every exposure case in which a mask, a cartridge, or a program stood between the worker and the dose. In each of these, the side that begins the exposure reconstruction first, with the employer's thirty-year records, the purchasing file, and the task history, will own the causation question the way the completed arithmetic always does.

IV. The Technical Work the Rising Dockets Demand

The rising dockets described in the preceding section do not reward general expertise; each demands a specific technical showing, built early, from records that are still recoverable. What follows is the work those cases actually require of the expert, in the order the docket asks for it. The 2023 amendment to Rule 702 governs throughout, and Chapter 2 treats it in full.

The record is perishable, and the rising dockets add new perishables. To the familiar list — the machine, the scene, the camera footage, the daily logs — the new dockets add the slab inventory and shop dust before the fabrication shop closes or cleans; the real-time environmental and physiological data of the heat case before the weather becomes a historical average; the work-speed data and quota files the state statutes require be produced to current and former employees on request, a demand the injured worker can make before any suit is filed; the emissions, stack-test, and monitoring records of the chemical facility; and the serum and biomonitoring evidence whose collection timing is itself a scientific decision. Preservation letters in these dockets should be written by counsel with the expert at the elbow, because only the expert knows which records each monitoring program, surveillance system, and statute generates, and which absences are themselves findings.

And the employer's file remains the spine of every exposure case. OSHA's exposure-and-medical-records rule requires employee exposure records be preserved for thirty years, making the industrial-hygiene file one of the few durable records in any workplace, and its completeness, or its documented absence, is frequently the first finding in the case. The non-party employer should be subpoenaed early and its safety and IH personnel deposed while institutional memory is intact; the OSHA inspection file, in any matter where an emphasis-program or complaint inspection occurred, should be obtained in full by records request; and both sides should remember that in the dockets above, the employer's comparative share — the fabrication shop's dry cutting, the staffing chain's acclimatization failure, the warehouse operator's quota — is usually the largest number on the verdict form, and the side that quantifies it with documents rather than blame is the side whose allocation the jury adopts.

V. The Venire and the Verdict: Jury Selection

The rising dockets carry attitudinal structure that trial teams and jury-selection consultants should treat as seriously as the science. Three axes recur. The first is regulatory trust: these are cases about hazards the government knew of and did not — or could not — regulate to prevention, and venire members split sharply on whether that fact indicts the companies, the agencies, or the litigation itself. Jurors who believe compliance with existing rules discharges a company's full obligation are the defense's foundation in the EtO and silicosis dockets, where documented compliance has produced defense verdicts; jurors who believe the rules are a floor written by the regulated are the plaintiff's. The stalled heat rule and the rescinded ergonomics standard give both narratives a factual hook, and voir dire should find where each juror locates responsibility when the rulebook is silent.

The second axis is the worker and the workforce. The silicosis docket's injured population is heavily immigrant, largely Spanish-speaking, and employed in small shops; heat's decedents are disproportionately new hires, outdoor laborers, and agricultural workers; the warehouse docket's claimants worked under algorithmic pace management that some jurors regard as ordinary modern employment and others as the hazard itself. Attitudes toward those workforces — toward immigration, toward manual labor, toward the gig-and-quota economy — predict verdict orientation in these dockets at least as strongly as generic tort attitudes, and they interact with the empty chair: the same juror who assigns fault generously to an absent employer may withhold it when the employer's failures read as the predictable product of the defendant's own pricing, pace, or product design. The third axis is comfort with quantification. Every one of these cases is decided on a reconstruction — a dose, a wet-bulb temperature, a force-frequency-posture profile — and jurors vary widely in whether arithmetic performed years after the fact strikes them as science or as advocacy. Both sides should test their reconstruction's presentation, not merely its admissibility, because the expert who survives the Rule 702 gate and loses the jury has won the wrong contest. None of this substitutes for the technical record; it determines who will be persuaded by it.

VI. Mass Tort Mechanics: One Case of Nine Hundred

Each of the rising dockets above — engineered stone, ethylene oxide, PFAS — is arriving as aggregate litigation rather than as a series of individual cases, and aggregation changes the technical work in ways the individual-case chapters of this book do not describe. Four features matter to the safety professional and the industrial hygienist who will be retained into one.

A single Rule 702 ruling propagates. In an individual case an exclusion loses one case. In a consolidated proceeding, a general-causation ruling made once at the leadership level governs every case in the inventory, and the practical effect on the expert's work is that the general-causation opinion is prepared to a standard the individual case would never justify economically: systematic literature review with stated inclusion criteria, every null study addressed, the weighting of the Bradford Hill considerations of Chapter 14 §II made explicit rather than left to judgment. The *Acetaminophen* and *Onglyza* rulings discussed throughout this book are aggregate rulings, which is precisely why they are so widely cited: they were made once, with full briefing, and they killed inventories.

Bellwethers select for atypicality. The cases tried first are selected — by the parties, by the court, or by a mix — and each side's selection criteria pull in opposite directions, with the result that the tried cases are systematically unrepresentative of the inventory whose value they are supposed to inform. For the exposure expert this produces a specific hazard: a reconstruction methodology developed on a bellwether plaintiff with unusually complete records may be unworkable across an inventory where most claimants have none, and an expert who commits to a data-hungry method in the bellwether has committed the inventory to it. The methodological question to ask before the first report is not can I reconstruct this plaintiff's dose but can I reconstruct dose for the median claimant in this inventory, on the records that actually exist.

Common-issue and case-specific work must be architecturally separate. The general-causation and state-of-the-art opinions are common; the exposure reconstruction and specific-causation opinions are individual. An expert who blends them, offering a common opinion that quietly assumes facts true only of the named plaintiff, has created a vulnerability that will be exploited once and applied everywhere. The discipline is to write the common report so that it states its assumptions as a class of assumptions, and to write each case-specific report so that it identifies which common findings it relies on and which individual facts it adds.

The economics change the expert's exposure. An expert anchoring hundreds of cases will testify repeatedly, over years, on a record that accumulates. Every prior transcript is available to every subsequent examiner; an opinion adjusted between the fortieth and the sixtieth deposition, however defensibly, becomes an impeachment exhibit in the eightieth. The corresponding fee volume is itself a cross-examination subject of a kind the individual case never generates, and Chapter 27's guidance about the retention balance and the answer given under oath applies with multiplied force. The protective practice is the one this book has recommended throughout, executed with unusual rigor: write down the method, state the limits at the outset, and change an opinion only for a stated reason that is itself in writing.

VII. Reading the Next Five Years

Forward-looking judgments should be labeled as such; these are the author's, offered with the confidence intervals the record supports. The federal heat standard is unlikely to be finalized in its proposed form in the near term, and if finalized in any form will draw an immediate *Loper Bright*-inflected challenge; the operative growth will be in state standards, in the revised NEP's five-year enforcement run, and, fastest of all, in the civil docket, where each summer's fatalities become the next term's filings. The warehousing NEP's lapse, if it proves to be one rather than a delayed renewal, will signal that federal ergonomics attention has receded to the General Duty Clause; either way, the state quota-and-ergonomics statutes will spread, the records they generate will become standard discovery, and a federal ergonomics standard, if it ever returns, will arrive sector-by-sector through legislation rather than by general rulemaking. The engineered-stone docket will consolidate, produce bellwether structures, and press the substitution question — whether a product that cannot be fabricated safely at prevailing practice should be sold at all — toward legislatures as Australia's ban did; retailer and PPE-supplier theories will mature or fail on the warnings and fit-testing records now in discovery. The chemical frontier will widen along the path EtO mapped: agency risk assessment, community docket, occupational docket, medical-monitoring battle, with PFAS the largest instance and the attribution science its decisive terrain. And the constitutional pressure on the enforcement machinery will continue regardless of administration, because its drivers sit in the courts; safety and IH professionals should expect more of their work to be tested before juries, in both enforcement and civil postures, than at any time in the OSH Act's history.

VIII. The Program for the Decade Ahead

The practical program follows from the map. Track the enforcement frontier as a source of evidence, not merely of obligation: the emphasis-program inspection files, the state-plan standards, the deregulatory docket's admissions and arguments, and the statutes that force records into existence are all discovery waiting to be demanded. Retain the technical expert — and, in every exposure matter, the industrial hygienist — at the front of the case, before the perishable record perishes and before the theory hardens against whatever the reconstruction will actually show. Build every opinion for the amended Rule 702, whoever is asking: quantified, sourced, limited, and honest about the records that point the other way. And in the rising dockets above all, remember the symmetry that governs the whole field: the exposure records, the quota data, the emissions files, and the shop invoices were written by people who did not know which side of the v. they were serving, and they will testify, through whichever expert reads them, to exactly what they contain. The rule may have stalled. The record didn't, and the side that reads it first, completely, and honestly is the side it will end up serving.

References

Access to employee exposure and medical records, 29 C.F.R. § 1910.1020; respirable crystalline silica, 29 C.F.R. § 1910.1053; ethylene oxide, 29 C.F.R. § 1910.1047; Fed. R. Evid. 702 (as amended Dec. 1, 2023).

Allstates Refractory Contractors, LLC v. Su, 79 F.4th 755 (6th Cir. 2023), cert. denied, 144 S. Ct. 2490 (2024).

Atlas Roofing Co. v. OSHRC, 430 U.S. 442 (1977).

Cal. Code Regs. tit. 8, §§ 3395 (outdoor heat), 3396 (indoor heat, 2024); heat illness prevention standards of Oregon, Washington, Colorado, Minnesota, Maryland, and Nevada.

California Department of Public Health, Occupational Health Branch. Silicosis in engineered stone countertop workers: case surveillance data (2019– 2026).

Glass v. B. Braun Medical Inc. (Pa. Ct. Com. Pl., Phila. Cnty. Dec. 2024) (defense verdict on all counts after a five-week trial; the first ethylene oxide case tried on behalf of a sterilization-plant employee rather than a neighbor).

Reyes-Gonzalez v. Caesarstone USA, Inc., No. 22STCV31907 (Cal. Super. Ct., L.A. Cnty.) (verdict Aug. 2024; $52,437,366 gross; fault apportioned fifteen percent to Caesarstone, ten percent to Cambria, two and a half percent to Color Marble, two and a half percent to the plaintiff, and seventy percent to parties not named on the verdict form; judgment notwithstanding the verdict granted as to Color Marble, Oct. 9, 2024; amended judgment, Mar. 18, 2025; on appeal); related coordinated engineered-stone silicosis proceedings, Los Angeles Superior Court.

Heat Injury and Illness Prevention in Outdoor and Indoor Work Settings, 89 Fed. Reg. 70698 (proposed Aug. 30, 2024) (to be codified at 29 C.F.R. pt. 1910); informal public hearing concluded July 2, 2025; post-hearing comment period closed Oct. 30, 2025.

Isaacks v. Terumo BCT Sterilization Servs., Inc., No. 2022CV031124 (Colo. Dist. Ct., Jefferson Cnty. Mar. 14, 2025) (defense verdict; no negligence found; costs of approximately $5.18 million taxed against plaintiffs, Sept. 2025).

Jordan v. Cambria Co., No. 2024CV31180 (Colo. Dist. Ct., Denver Cnty. Apr. 30, 2026) ($17,450,000 verdict; no design defect found; lead manufacturer not negligent but liable on misrepresentation, second manufacturer liable for negligence and misrepresentation; fault apportioned sixty-three percent to the employing fabrication shop; on appeal).

Kamuda v. Sterigenics U.S., LLC (Ill. Cir. Ct., Cook Cnty. 2022), and related Willowbrook resolutions (2023, 2025).

Loper Bright Enters. v. Raimondo, 603 U.S. 369 (2024).

National Institute for Occupational Safety and Health. Criteria for a recommended standard: occupational exposure to heat and hot environments (2016). Centers for Disease Control and Prevention.

N.Y. Warehouse Worker Protection Act (2022) and Warehouse Worker Injury Reduction Act (eff. June 1, 2025); Cal. Lab. Code §§ 2100–2112 (AB 701); warehouse quota statutes of Washington, Minnesota, Oregon, Connecticut, and Rhode Island; Warehouse Worker Protection Act, H.R. 4896 / S. 2613, 119th Cong. (2025).

Occupational Safety and Health Standards; Interpretation of the General Duty Clause: Limitation for Inherently Risky Professional Activities, 90 Fed. Reg. 28370 (proposed July 1, 2025) (Docket No. OSHA-2025-0041), and companion deregulatory proposals of July 1, 2025, and April 6, 2026.

OSHA, National Emphasis Program on Warehousing and Distribution Center Operations, CPL 03-00-026 (July 13, 2023) (three-year term expired July 13, 2026; no successor directive posted as of the date given at the head of Chapter 28).

OSHA, National Emphasis Program — Outdoor and Indoor Heat-Related Hazards, CPL 03-00-024 (Apr. 10, 2026) (superseding the 2022 directive; effective for five years).

SEC v. Jarkesy, 603 U.S. 109 (2024).

SeaWorld of Fla., LLC v. Perez, 748 F.3d 1202 (D.C. Cir. 2014) (Kavanaugh, J., dissenting) (source of the "inherent and inseparable" limitation OSHA's 2025 General Duty Clause proposal adopts).

Sommerville v. Union Carbide Corp., No. 24-1491, 149 F.4th 408, 2025 WL 2383496 (4th Cir. Aug. 18, 2025), reh'g en banc denied, cert. denied sub nom. Union Carbide Corp. v. Sommerville, No. 25-919 (U.S. May 26, 2026).

Sotera Health Company. (2025a, April 3). Form 8-K: Sterigenics settlement of 97 ethylene oxide claims [SEC filing]. U.S. Securities and Exchange Commission.

Sotera Health Company. (2025b, July 23). Form 8-K: Sterigenics settlement of 129 ethylene oxide claims [SEC filing]. U.S. Securities and Exchange Commission.

U.S. Environmental Protection Agency. Integrated Risk Information System (IRIS) assessment: evaluation of the inhalation carcinogenicity of ethylene oxide (2016).

U.S. Gov't Accountability Off., GAO-24-106413, Workplace Safety and Health: OSHA Should Take Steps to Better Identify and Address Ergonomic Hazards at Warehouses and Delivery Companies (2024).

Walker v. C.R. Bard, Inc. (Ga. State Ct., Gwinnett Cnty. May 2, 2025) ($20 million compensatory verdict on negligence and public nuisance; punitive phase ended in mistrial; on appeal 2026).

CHAPTER 29

The Machine in the Method

Artificial Intelligence, the Expert Record, and the Systems That Now Watch the Workforce

> *Every generation of technical practice acquires an instrument that arrives before the rules governing it, and this one has arrived on four surfaces at once: the model that produces the exposure estimate, the report that carries it, the demonstrative that teaches it to a jury, and the employer's own systems, which now watch the workforce continuously and generate a record no one has yet learned to demand. The obligations have not changed; the amended Rule 702 is indifferent to what produced a number so long as another qualified professional can check it.*

I. The Standard Has Not Changed

It is tempting to treat these tools as a new category of evidentiary problem requiring a new body of law, and the temptation should be resisted, because the rule that governs them is the rule that governed the slide rule and the spreadsheet. Rule 702 does not ask what produced a number. It asks whether the number rests on sufficient facts or data, whether the method that generated it is reliable, and whether that method was reliably applied to the facts of this case — and each of those questions is answered by showing the work, not by naming the instrument. An exposure estimate produced by hand from a NIOSH method and an exposure estimate produced by a machine-learned model stand or fall on the same property: whether a second qualified professional, given the same inputs, can arrive at the same result and say why.

What has changed is not the standard but the difficulty of satisfying it, and the difficulty runs in one direction. Every one of these tools makes it easier to produce output and harder to reconstruct the reasoning that produced it. That asymmetry is the whole of the problem. A hand calculation carries its own audit trail because the steps are the artifact; a model output carries none, and the expert who cannot supply one has not committed a technological sin but an evidentiary one — she has offered a conclusion whose basis exists nowhere the court can reach. After the

2023 amendment that is an admissibility defect rather than a matter for cross-examination, which is the practical consequence this chapter is organized around.

Three cautions follow, and they structure the four sections below. The first is that opacity and complexity are different things, and only one of them is fatal: courts have admitted difficult methods for a century and will continue to, but a method whose internals are unavailable to the opponent is not difficult, it is unreviewable. The second is that the failure modes are not exotic. They are the failure modes this book has already catalogued — an unvalidated input, an unstated assumption, an extrapolation beyond the data, a demonstrative that improves on the record — arriving faster and with better production values. The third is that the expert bears the burden of describing her own workflow, because the question is now asked, and an answer assembled in the deposition chair is not an answer.

II. The Model

Rule 702(c) and (d) presuppose a method a court and an opponent can follow. A machine-learned exposure model, a vendor tool that returns a concentration from task inputs, or a statistical engine whose parameters are not disclosed does not fail that test because it is a computer; it fails if it cannot be reproduced. The distinction the courts will draw is between a model that is merely complex and a model that is opaque, and the second is fatal where the first is not. The audit of such an opinion asks for six things: the identity and version of the tool; the provenance, period, and industry of the data it was trained or calibrated on, and whether that data resembles the operation at issue; every input variable and the value supplied for it; the validation performed against measured data, with the error characterized rather than asserted; a sensitivity run across the plausible range of the contested inputs; and re-runnability — the opponent's ability to execute the same model on the same inputs and obtain the same number. That last is the one that decides motions. It is Chapter 20 §VI's point about dispersion modeling generalized: a model the other side cannot re-run is a model the court cannot check, and after 2023 that is an admissibility problem rather than a matter for cross-examination.

III. The Report

Generative drafting has arrived in expert practice faster than the disclosure rules have adapted to it. Rule 26(a)(2)(B) requires a complete statement of the opinions and their bases; Rule 26(b)(4)(B) protects drafts of a testifying expert's report and (C) protects most attorney-expert communications while leaving discoverable the facts, data, and assumptions counsel supplied. None of that answers the question these tools actually raise, which is whether the expert can describe the provenance of her own sentences. A generative tool that supplies a factual assertion, a citation, or an inferential step has introduced material the witness did not derive and cannot source, and the failure mode is no longer hypothetical: the sanctions decisions against lawyers who filed briefs citing authorities that do not exist are now numerous enough to be a genre, and an expert report is a sworn document held to no lesser standard. The working discipline is a boundary rather than a prohibition. Such a tool may organize material the expert has assembled, summarize documents she has read, and improve prose she has written. It may not supply a fact, an authority, a measurement, or an inference. Every citation is verified against the source document before the report is served, not the tool's description of the source, the source. And the expert prepares for the question that is now asked in deposition and will shortly be asked at trial: did you write this report, and what part of it did something else write?

Disclosure deserves its own sentence, because the trend is one-directional. Standing orders in a growing number of federal and state courts require counsel to certify whether generative tools were used in a filing and whether the output was verified, and the practice is migrating toward expert submissions. The prudent course for the professional is to keep a contemporaneous record of what any such tool was used for in each engagement and to assume the question has already been asked. A witness who can describe her own workflow plainly is unremarkable; a witness who cannot is the story.

IV. The Demonstrative

The foundation requirement for a demonstrative has always been that it fairly and accurately represents what it depicts, and the current generation of image tools sits squarely across that line. Upscaling, denoising, interpolation, and generative fill do not recover detail that was in the original; they synthesize detail that is plausible. Applied to a scene photograph, a security-camera still, or a fracture surface, they produce a reconstruction offered under the authority of a photograph: an authenticity problem under Rule 901, a prejudice problem under Rule 403, and, if it is discovered on cross rather than disclosed on direct, a credibility loss no redirect repairs. The rule is simple: retain the original file, log every processing step and its parameters, disclose any processing performed, and never let generated or enhanced imagery be presented as depiction. Animation and simulation are governed by the older and more forgiving law of illustrative aids, but only where the underlying physics is the expert's and the animation merely renders it.

V. The Employer's Systems and the Second Audience

Computer-vision observation platforms that score behaviors from fixed cameras, wearable sensors that log exposure and ergonomic load in real time, forklift and vehicle telematics, and algorithmic pace-management systems have moved from pilot to fleet across exactly the sectors this chapter's dockets come from. What they generate is a dated, machine-recorded leading-indicator stream, which is to say, precisely the notice record Chapter 6 §IV builds a substantial-certainty case from, with two differences that matter. It is denser than any near-miss binder by orders of magnitude. And it is far more perishable: retention windows are configured by vendors, frequently in days, under contracts the safety department did not negotiate and often has not read. The preservation duty of Chapter 21 attaches to these systems on the day of the incident like any other record, and the litigation hold must name the system, the vendor, and the retention setting rather than gesturing at electronic records generally.

The mirror is the whole lesson, and it runs in both chairs exactly as the rest of this book does. A vision system that flagged the same unguarded nip point ninety times across four months and produced no work order is the most efficient notice exhibit in the modern record, and no adjective a plaintiff's lawyer could write would improve on the timestamp. A system that flagged it and produced a dated correction, a verification, and a closure entry is the strongest defense document the employer will ever hold, for the same reason and by the same mechanism. The discovery accordingly runs to the system rather than to its summaries: the platform and vendor, the retention schedule, the complete alert log rather than the dashboard, the routing — who received each alert and when — the closure records, and the configuration history, because a detection threshold quietly raised after an incident is itself a finding, and it is a finding the configuration log will carry long after everyone involved has forgotten making the change.

The perishability is the part the profession keeps underestimating, and it is worth being concrete about why. A near-miss binder sits in a file cabinet until someone throws it away. A computer-vision platform overwrites. The alert stream lives in vendor infrastructure under a retention parameter set during implementation, usually by the vendor's deployment engineer, usually at the product default, and usually with an eye to storage cost rather than evidentiary value. Thirty days is common. Seven is not unusual for full-motion video, with only extracted event metadata surviving longer. Wearable-sensor platforms and vehicle telematics follow the same pattern at different intervals. The consequence is that the most probative record in a modern notice case can be gone before the employer's counsel has finished reading the incident report — gone not because anyone destroyed it, but because nobody told the system to stop.

That is a spoliation problem of the worst kind, because the destruction is automatic, documented, and attributable to a party who had the ability to stop it. Fed. R. Civ. P. 37(e) asks whether electronically stored information that should have been preserved was lost because a party failed to take reasonable steps to preserve it. A retention window running at its default while a hold notice sits unsent in the general counsel's outbox is close to the paradigm case for that finding. Chapter 21 sets out the duty and its trigger. What follows is what the duty actually requires when the custodian of the record is a vendor rather than the employer.

A hold letter addressed to a computer-vision, wearable, or telematics vendor has to do something an internal hold notice does not. It has to override a running automated process operated by a third party who is not a defendant, has no stake in the litigation, and will do exactly what the contract and the letter tell it to do and nothing more. A letter demanding preservation of "all documents and electronically stored information relating to the incident" will produce nothing, because nothing in the vendor's system is indexed to an incident. Six elements make such a letter effective, and the first is identification: the product name, the deployment or tenant identifier, the site, and, where the incident location is known, the specific camera, sensor, or vehicle identifiers. A vendor operating a hundred installations cannot act on a reference to the cameras at the plant.

The second is a date range stated wide. The temptation is to ask for the day of the incident. The notice case lives in the months before it, in the alert history showing the same hazard flagged repeatedly and never closed, so the range should begin well before the earliest plausible date of notice and run through the date of the letter. The third is naming the data layers separately, because they are stored separately and they expire separately: raw and full-motion video, extracted clips, still frames, event and alert records, detection-confidence scores, the model or firmware version in force during the period, alert routing and notification records, user acknowledgments and closure entries, and the configuration and threshold history. A request for the footage gets the footage and loses everything that makes the footage mean anything.

The fourth element is a direction to suspend automated deletion, stated in the vendor's own vocabulary — retention, purge, overwrite, rolling window, archival tiering — together with a request that the vendor confirm in writing the date and time the suspension took effect and the earliest data still available at that moment. That confirmation is the document that later establishes what was recoverable when the duty attached, and it is worth more than any assurance given by telephone. The fifth is a demand for the retention configuration itself and its history: what the window was set to, when, by whom, and whether it changed after the incident. A threshold or retention setting quietly altered after an event is a finding in its own right, and the configuration log will carry it long after everyone involved has forgotten making the change.

The sixth is the contract. Most vendor agreements contain a data-ownership clause, a retrieval or export provision, a fee schedule for out-of-cycle exports, and a termination clause under which the vendor purges customer data on a schedule of its own. The letter should invoke the retrieval right expressly and should state that the account must not be terminated, downgraded, or migrated while the hold is in place. Counsel should read the master services agreement before drafting the letter rather than after the vendor answers it. Two practical additions: send the letter both to the contracting entity and to the account's named technical contact, because the first will route it to legal and the second is the person who can actually flip the switch; and follow within days with a request for export in native format with metadata intact, rather than a dashboard report or a summary in PDF. The summary is the vendor's interpretation of the record. The export is the record.

The second question is the one any preservation program should be tested against first, and at most employers it has an uncomfortable answer: the person who knows the record exists is not the person authorized to preserve it. The safety manager knows the cameras are there and knows what they flag. Hold authority sits with the general counsel or outside counsel. The contractual relationship sits with procurement or information technology, who negotiated it. The administrative credentials sit with a systems administrator or a site engineer. In many organizations no single person holds more than one of those four things, and the retention window is the only one of the four that is running on a clock.

The remedy is structural, and it belongs in the preservation protocol written before there is an incident rather than in the response to one. Three provisions do most of the work. The first is an inventory — a current list of every observational, sensing, and telematics system in use, with the vendor, the contracting entity, the account identifier, the retention setting for each data layer, and the name of the internal administrator — maintained by the safety function, because the safety function is the only one that knows what the systems are for. The second is standing authority: a written delegation permitting the safety director or the incident commander to issue a preservation demand to a named vendor immediately upon a recordable event, without waiting for a legal determination that litigation is reasonably anticipated. Suspending a purge is cheap and reversible; losing the record is neither, and the delegation converts a decision that takes days into one that takes an hour. The third is contract language negotiated at signature rather than discovered under deadline: a defined litigation-hold obligation, a maximum response time, an export right in native format with metadata preserved, and a stated fee schedule for out-of-cycle retrieval. A hold provision costs nothing when the agreement is signed and is unobtainable at the moment it is needed.

For the safety professional this is the chapter's practical point, and it is why this section speaks principally to the second audience. These systems were bought to reduce injuries, and they do. They also generate the most detailed contemporaneous record of hazard recognition that has ever existed in this field, and they generate it whether or not anyone in the organization has decided what to do with it. An employer that can produce a complete, dated alert log showing recognition, routing, correction, and verification holds a defense document no reconstruction can match. An employer that can produce only the last thirty days holds an admission that the rest existed and was allowed to expire. The distance between those two positions is a retention setting and a paragraph in a services agreement, and both are decided long before anyone gets hurt.

VI. What the Tools Changed, and What They Did Not

None of this is a new body of law, and the professional who waits for one will be waiting while the cases are decided. What the tools have changed is the volume of the record, the precision of its timestamps, and the number of steps between a measurement and an opinion at which something can happen that the expert did not do and cannot explain. The obligations are the ones this book has argued for twenty-nine chapters: work from the primary record, state the method, disclose the assumptions, quantify the uncertainty, and be able to show another qualified professional how the answer was produced. An expert who cannot explain how her own number was arrived at has failed the oldest test in this field. That the failure now has a faster instrument behind it changes nothing except how quickly it happens.

References

Committee Note to the 2023 amendment of Fed. R. Evid. 702 (reliable application of the methodology to the facts of the case; the proponent's burden as to each step of the analysis).

Fed. R. Civ. P. 26(a)(2)(B) (complete statement of all opinions and the basis and reasons for them); Fed. R. Civ. P. 26(b)(4)(B)–(C) (draft-report and attorney-expert communication protections, and the facts, data, and assumptions that remain discoverable).

Fed. R. Evid. 403; Fed. R. Evid. 702 (as amended Dec. 1, 2023); Fed. R. Evid. 901 (authentication and identification of photographic and machine-generated evidence).

National Institute for Occupational Safety and Health. (2024). Center for Direct Reading and Sensor Technologies: Direct-reading instruments and wearable sensor systems in occupational exposure assessment. Centers for Disease Control and Prevention. https://www.cdc.gov/niosh/cdrst/ Standing Order, In re: Artificial Intelligence ("AI") in Cases Assigned to Judge Baylson (E.D. Pa. June 6, 2023) (disclosure of any AI use in the preparation of a filing, and certification that every citation to law or record has been verified as accurate).

Judge Brantley Starr, Judge-Specific Requirements, Mandatory Certification Regarding Generative Artificial Intelligence (N.D. Tex. May 30, 2023) (certificate that no portion of a filing was drafted by generative AI, or that any such language was checked for accuracy by a human being using print reporters or traditional legal databases).

Order on Artificial Intelligence, Chambers Procedures of Judge Stephen A. Vaden (Ct. Int'l Trade June 8, 2023) (disclosure identifying the program used and the specific portions of text drafted with it, plus certification that its use disclosed no confidential or business proprietary information to an unauthorized party); accord Order on Artificial Intelligence, Chambers Procedures of Judge Claire R. Kelly (Ct. Int'l Trade Mar. 21, 2024).

Proposed amendment to 5th Cir. R. 32.3 (published for comment Nov. 2023) (certification as to generative-AI use and human review of AI-generated citations and analysis), not adopted (5th Cir. June 12, 2024) — cited here as evidence that the trend is contested at the appellate level even as it advances in the district courts.

PART VI

APPENDICES

APPENDIX A

The First-Call Checklist

Ten Questions to Ask Before Retaining a Safety or Industrial Hygiene Expert

For plaintiff and defense counsel alike: the answers matter, and so does the candor with which they are given. You are auditioning the witness's response to hard questions. These are the first ten.

1. How would you build the opinion in this case?

Listen for: named standards, specific records, planned measurements, and alternatives to rule out, not credentials and conclusions. An expert who can sketch the methodology in this call can defend it in a Rule 702 hearing.

2. Which published standards and methods govern this work?

Listen for: OSHA standards, the ANSI/ASSP and NFPA consensus series, ACGIH TLVs, NIOSH criteria documents, the AIHA exposure assessment strategy — cited from memory and matched to your facts.

3. Have you personally done this work, in this industry?

Listen for: hands-on practice with the process and equipment at issue, and command of the right regulatory overlay (general industry vs. construction vs. maritime vs. MSHA).

4. What share of your practice is consulting versus testifying?

Listen for: a working professional who also testifies. A career-only witness gives the jury no independent identity to trust and hands opposing counsel the "hired gun" theme for free.

5. How is your retention history split between plaintiff and defense?

Listen for: reasonable balance, stated without hesitation. A witness who has only ever testified for one side arrives with a credibility discount already applied.

6. Have you ever been challenged or excluded under Rule 702?

Listen for: candor. An exclusion is not disqualifying; concealing one is. Ask for the opinions and prior transcripts: opposing counsel will read them, so you should first.

7. What evidence in this case is perishable — and what should we preserve today?

Listen for: site conditions, equipment before repair or scrap, records on retention schedules, witness memories, the OSHA file (submit the FOIA request now). The basis of a reliable opinion has a shelf life.

8. What will you need to inspect or measure yourself?

Listen for: a site inspection, the actual or exemplar equipment, calibrated instruments, reconstruction where feasible. Facts the expert gathered personally anchor the opinion; inherited facts merely decorate it.

9. What alternative explanations will you have to rule out — and how?

Listen for: named competing causes and a method for addressing them on the record. Failure to account for the obvious alternative is one of the most common grounds for exclusion.

10. What facts would make the opinion we need unsupportable?

Listen for: a specific answer. It proves independence, honesty, and command of the subject in one response, and shows you exactly where the case is vulnerable before you spend a dollar.

Verify, don't accept. Confirm certifications with the issuing board, not the CV: CIH → Board for Global EHS Credentialing (gobgc.org) · CSP → Board of Certified Safety Professionals (bcsp.org) · CHMM → Institute of Hazardous Materials Management (ihmm.org). Check active status, recertification currency, and the absence of ethics actions.

APPENDIX B

Glossary of Terms

Definitions are practical rather than exhaustive, oriented to how each term arises in the chapters.

ACGIH / Threshold Limit Value (TLV). The American Conference of Governmental Industrial Hygienists publishes Threshold Limit Values: health-based exposure guidelines updated regularly, unlike most OSHA PELs. TLVs are not law, but they are widely treated as the professional benchmark of acceptable exposure and routinely anchor standard-of-care testimony where the PEL is outdated.

Action Level. A trigger concentration, typically one-half of the permissible exposure limit, at which substance-specific standards impose obligations — periodic monitoring, medical surveillance, training — even though no overexposure has occurred. In discovery, exceeding an action level without follow-up is often more probative of knowledge than the exposure numbers themselves.

AIHA Exposure Assessment Strategy. The American Industrial Hygiene Association's published framework for judging exposures: defining similar exposure groups, characterizing the exposure profile, and rating acceptability against the limit. It is the professional method an industrial hygienist is expected to follow when deciding what a handful of samples can and cannot support.

Analogous Data. Exposure measurements imported from another worker, task, time, or facility and asserted to describe the case at hand. Analogous data are indispensable to most retrospective reconstructions and are also their most common point of failure: the inference is only as good as the documented similarity of tasks, materials, and controls, and an ill-fitting analogy is attacked as a reliable-application defect, not merely as weight.

Arc Flash. An explosive release of thermal energy from an arcing electrical fault, capable of causing severe burns at a distance without contact. NFPA 70E requires a risk assessment that calculates the incident energy at the working distance and matches labeling and PPE to it; the calculation the employer should have run is frequently the reconstruction the expert runs after the burn.

Area Sample. A sample collected from a fixed location rather than a worker's person. Area samples characterize a space, not a dose; a recurring litigation issue is whether an area sample has been passed off as evidence of what an employee actually breathed.

Bayesian Decision Analysis. A statistical approach that combines sparse monitoring data with prior professional judgment to estimate an exposure distribution and the probability that it exceeds a limit. Increasingly used in defensible reconstructions because it makes uncertainty explicit; when offered by an opposing expert, the priors and their justification are the proper target of the audit.

Bias (Exposure Estimate). A systematic lean in an exposure estimate, as distinct from random uncertainty. Bias arises when the discretionary choices in a reconstruction — which samples, which analogous group, which model inputs — all tilt in the same direction, toward the answer the retaining party needs. A documented directional bias across every judgment call is a recognized ground for challenge under the reliable-application prong of Rule 702.

Breakthrough. Saturation of a sorbent sampling medium during overlong sampling, allowing contaminant to pass through uncaptured. Breakthrough biases the result low and is one reason analytical methods prescribe maximum sampling times as well as minimums.

Breathing Zone. The hemisphere forward of the shoulders within a radius of roughly six to nine inches of the worker's nose and mouth: the region OSHA's sampling guidance directs a sampler to occupy, and the region the exposure standards actually regulate. Consensus practice in industrial hygiene draws the zone more broadly, as a hemisphere of about thirty centimeters centered on the face; individual protocols are more specific still, the asbestos appendices directing that the cassette sit about ten centimeters from the nose and mouth. Which definition governs is a question about the standard cited, not a matter of general usage. Sampler placement outside the breathing zone (on equipment, outside a welding helmet where the standard says inside, or vice versa) is among the most heavily litigated sampling defects.

Calibration (Primary Standard). Verification of a sampling pump's flow rate, or a noise instrument's response, against a reference of known accuracy: expected before and after each sampling session, preferably on site. Because concentration is mass divided by air volume, a missing or unmatched calibration record undermines every number computed from the pump.

Cassette. The filter holder worn at the collar or lapel through which the sampling pump draws air; the filter inside captures particulate (asbestos fibers, silica dust, metal fume) for laboratory analysis.

Ceiling Limit. An exposure limit that may not be exceeded at any moment, as distinct from a full-shift average. Proving or disproving a ceiling violation requires instantaneous or short-interval measurement, not an eight-hour sample.

Chain of Custody. The documented, unbroken record of who possessed a sample from collection through laboratory analysis. Gaps invite the argument that the specimen analyzed is not, or may not be, the specimen collected.

CHMM. Certified Hazardous Materials Manager. See CIH / CSP / CHMM.

CIH / CSP / CHMM. Certified Industrial Hygienist, Certified Safety Professional, and Certified Hazardous Materials Manager: board certifications requiring qualifying education, documented practice, examination, an enforceable ethics code, and recertification. They are the credentials courts and counsel treat as meaningful in this field, in contrast to purchased designations.

Closed-System Transfer Device (CSTD). Engineering control used in hazardous-drug handling that mechanically prevents the escape of drug aerosol and vapor during preparation and administration. Its presence, absence, or misuse is a central control question in hazardous-drug exposure claims.

Competent Person. In OSHA construction standards, a person who can identify existing and predictable hazards and has employer authority to correct them. The title is functional, not honorary: trench and scaffold cases regularly turn on whether the designated person had the training and the authority the definition requires.

Cyclone. A size-selective sampling device that separates out larger particles so the filter collects only the respirable fraction: the particles small enough to reach the deep lung. Respirable-dust standards, including silica, can only be measured through such a device.

Desorption Efficiency. The fraction of collected contaminant the laboratory can recover from the sampling medium for analysis. Results must be corrected for it; whether and how the correction was applied is an auditable step in the calculation behind a citation.

Determinants of Exposure. The physical and work factors that govern a worker's exposure: material composition, task and duration, ventilation and enclosure, respiratory-protection use and efficiency, work rate, room volume, and air exchange. A reconstruction supplies a value, measured or assumed, for each; the audit tests each determinant for its source and for how much the final estimate moves when it is varied.

Dosimeter. A wearable instrument that integrates a worker's noise exposure over the shift into a percentage dose. Calibration, placement, and what the instrument was set to hear (threshold, exchange rate, response) are the recurring points of attack.

Engineered Stone. Manufactured quartz-resin slab material with crystalline silica content far higher than natural stone. Its fabrication (cutting, grinding, polishing) has produced severe, fast-onset silicosis in young workers and driven the current wave of silica product-liability and toxic-tort litigation.

Excursion Limit. A short-term boundary within certain standards (asbestos, for example) limiting exposure averaged over thirty minutes, layered on top of the full-shift limit. It requires its own sampling strategy and its own proof.

Exposure Reconstruction. The estimation of a worker's past exposure, after the fact, from the surviving record: product data, task history, ventilation and control information, and analogous measurements. A legitimate and often unavoidable discipline whose admissibility depends on disclosed determinants, defensible analogies, and stated uncertainty; see Chapters 16 and 17.

Field Blank. An unused sampling medium handled and shipped alongside the real samples to detect contamination introduced in handling or transit. Missing or contaminated blanks undercut the laboratory result.

Field Operations Manual (FOM). OSHA Instruction CPL 02-00-164: the agency's own script for how inspections are opened, conducted, and documented and how citations are developed. Deviations between what the manual requires and what the compliance officer did are facts the defense is entitled to develop.

General Duty Clause. Section 5(a)(1) of the OSH Act, requiring employers to keep the workplace free of recognized hazards likely to cause death or serious harm where no specific standard applies. It is the enforcement vehicle for unregulated hazards, heat prominently among them, and carries a distinct, heavier proof burden for the Secretary.

Ground-Fault Circuit Interrupter (GFCI). A protective device that opens a circuit when it detects current leaking outside the intended path, at thresholds of a few milliamps, well below the level that causes fibrillation. On construction sites, 29 C.F.R. § 1926.404 requires GFCI protection for temporary receptacles or, in the alternative, a documented assured equipment grounding conductor program; a ground-fault death on a protected circuit therefore implies the protection was absent, defeated, or not in the path.

Hierarchy of Controls. The ordered preference for controlling hazards: elimination, substitution, engineering controls, administrative controls, and personal protective equipment last. It is the organizing framework of standard-of-care testimony: the question is rarely whether anything was done, but whether the employer skipped to the bottom of the hierarchy.

Ipse Dixit. Latin, "he himself said it." An expert conclusion connected to the underlying data only by the expert's own assertion, with no disclosed method or reasoning bridging the gap. *General Electric Co. v. Joiner* authorizes exclusion where too great an analytical gap separates the data from the opinion, making the ipse dixit estimate the paradigm case for exclusion under amended Rule 702.

Limit of Quantification (LOQ). The smallest amount of analyte a laboratory method can measure with acceptable confidence (the limit of detection, LOD, is the smaller amount it can merely detect). Sampling must run long enough to collect a quantifiable mass; a result reported near or below the LOQ cannot carry the weight of a violation.

Medical Surveillance. The program of examinations and biological monitoring that substance-specific standards require once triggers (usually the action level) are met. Surveillance records are discoverable and date-stamped evidence of what the employer knew about exposure and effect.

Minimum Approach Distance. The closest an unprotected worker or conductive object may come to an energized conductor, set by voltage in Table A of the crane standard (29 C.F.R. § 1926.1408), in § 1910.269 for utility work, and in NFPA 70E's shock boundaries. The clearance plan, the spotter assignment, and the locate ticket are the records that show whether the distance was managed or merely hoped for.

NIOSH. The National Institute for Occupational Safety and Health: the research agency whose criteria documents, analytical methods, and hazardous-drug list supply much of the technical authority experts cite. NIOSH recommends; OSHA regulates.

Noise Dose / Table G-16. Under 29 C.F.R. § 1910.95, permissible noise exposure is expressed in Table G-16 as allowable durations at given sound levels, integrated by dosimeter into a percentage of allowable dose. What the instrument was configured to include, and for how long, determines the number.

Objective Data. Exposure information — prior monitoring, industry-wide data, calculations — that certain standards accept in lieu of new sampling to characterize exposure. Its sufficiency is a recurring dispute: representative data for one operation may prove nothing about another.

OSHA 300 Log (with 301 and 300A). The employer's running log of recordable injuries and illnesses (300), the incident detail form behind each entry (301), and the annual summary (300A), kept under 29 C.F.R. Part 1904. In litigation the log is read two ways: for what it recorded — notice, pattern, knowledge — and for what it conspicuously did not.

OSHA Technical Manual (OTM). TED 01-00-015 — the agency's technical guidance for its own personnel, including Section II's procedures for personal air sampling. Together with the standards' sampling appendices, it is the yardstick against which a compliance officer's sampling practice is audited.

Permissible Exposure Limit (PEL). The legally enforceable exposure limit, most set out in the Z tables of 29 C.F.R. § 1910.1000 or in substance-specific standards. Many PELs date to consensus values of the 1960s–70s; the gap between a lawful exposure and a professionally acceptable one is a central theme in civil litigation.

Personal Sample. A sample collected on the worker's person, in the breathing zone, over the work shift: the only kind of sample that directly measures the exposure the standards regulate.

Qualified Person (Electrical). Under 29 C.F.R. §§ 1910.331–.335, a worker demonstrated by training to have the skills to distinguish exposed live parts, determine their voltage, and maintain the required approach distances. The term is functional, not titular: assignment of an unqualified worker to work the standard reserves for a qualified one is a program-level failure provable from the training file alone.

Respirable Fraction / Respirable Crystalline Silica. The portion of airborne dust small enough (roughly 4 microns and below) to penetrate to the gas-exchange region of the lung. Silica standards regulate only this fraction, which is why total-dust measurements cannot prove or disprove a silica overexposure.

Root-Cause Analysis (RCA). A structured investigation methodology — causal trees, barrier analysis, systematic why-analysis — that traces an incident from the immediate deviation through its preconditions to the management-system factors that permitted it. In litigation, the RCA is read twice: as the employer's contemporaneous account of causation, and as proof of what the employer knew and when its corrections closed. Findings written as facts survive that double reading; findings written as verdicts do not.

Sampling and Analytical Error (SAE). The combined uncertainty of the sampling and laboratory method, applied as an adjustment before OSHA may cite: the measured result must exceed the limit even after the error band is subtracted. Recomputing the SAE adjustment from the raw data is a standard expert audit, and a recurring source of citations that fail.

Sensitivity Analysis. The systematic testing of how much an exposure estimate changes when each input assumption is varied across its plausible range. Sensitivity analysis reveals which determinants drive the result and whether the opinion collapses when an unsupported assumption is relaxed; its absence from an opposing reconstruction is itself an audit finding.

Similar Exposure Group (SEG). Workers whose exposure profile is expected to be alike because task, materials, and environment are alike. Whether the sampled workers genuinely represent the group, and whether the injured worker belongs to it, determines how far a handful of samples can be generalized.

Sorbent Tube. A glass tube of adsorbent material (commonly charcoal) through which the pump draws air to capture gases and vapors for laboratory desorption and analysis. Subject to breakthrough if run too long, which is why methods prescribe maximum volumes.

Time-Weighted Average (TWA). The exposure concentration averaged over the eight-hour shift: the form in which most PELs are expressed. A TWA is constructed, not observed: sampling times, unsampled intervals, and the assumptions filling the gaps are all discoverable and testable arithmetic.

Two-Zone / Well-Mixed-Room Model. Physical mass-balance models that estimate airborne concentration from a contaminant generation rate, room volume, and ventilation, either treating the room as uniformly mixed or distinguishing a near-field zone around the worker from a far-field zone. A modeled concentration is only as reliable as those inputs, so a model run at a single favorable set of assumptions without a sensitivity or uncertainty analysis is a recurring vulnerability.

Uncertainty (Exposure Estimate). The range around an exposure estimate that reflects measurement error, model error, and the inherent variability of the exposure. Because different defensible methods applied to the same scenario can yield materially different results, a reliable opinion states its confidence bounds; an estimate reported as a single confident number without them has not been completed and is vulnerable under Rule 702.

Wet-Bulb Globe Temperature (WBGT). The composite heat-stress index (temperature, humidity, radiant heat, air movement) used by the ACGIH TLV and NIOSH criteria to judge heat exposure against work demands and clothing. In heat litigation it is the measurement counsel should look for, and usually finds no one took.

Z Tables. Tables Z-1 through Z-3 of 29 C.F.R. § 1910.1000, listing PELs for hundreds of air contaminants not covered by substance-specific standards. Table G-16 serves the analogous role for noise.

APPENDIX C

Table of Key Decisions

Citations reflect the manuscript as of July 2026, supplemented with standard reporters for canonical decisions; several verdicts and Commission decisions remain subject to appeal, remand, or settlement. Verify current posture and controlling authority in the relevant jurisdiction before relying on any entry. Trial-court verdicts carry their decision dates so that the perishable entries can be told at a glance from settled appellate authority: a jury verdict establishes no rule, and several of the entries below were on appeal when this book went to press.

Decision	Forum / Citation	Why It Matters	Ch.
Daubert v. Merrell Dow Pharms., Inc.	509 U.S. 579 (1993)	Established the trial court's gatekeeping duty over expert testimony; reliability and fit are threshold questions.	1, 2, 4, 5, 14, 16, 17
Gen. Elec. Co. v. Joiner	522 U.S. 136 (1997)	Abuse-of-discretion review; a court may exclude an opinion connected to data only by the expert's ipse dixit.	2, 4, 5, 14, 16, 17, 18, 19
Kumho Tire Co. v. Carmichael	526 U.S. 137 (1999)	Extended gatekeeping to technical and experience-based experts — including safety professionals.	1, 2, 4, 5, 16, 17
Sardis v. Overhead Door Corp.	10 F.4th 268 (4th Cir. 2021)	Verdict vacated where a safety expert's untested opinions were admitted; gatekeeping is not optional.	1, 2, 19, 27
Amorgianos v. Nat'l R.R. Passenger Corp.	303 F.3d 256 (2d Cir. 2002)	Reliability applies to each step of the analysis; a flaw in one link can sink the opinion.	1, 19, 27

Decision	Forum / Citation	Why It Matters	Ch.
Tamraz v. Lincoln Elec. Co.	620 F.3d 665 (6th Cir. 2010)	Etiology opinion excluded as speculation dressed as diagnosis; hypothesis is not causation.	1, 14, 18
Milward v. Acuity Specialty Products Group	639 F.3d 11 (1st Cir. 2011)	Weight-of-evidence causation reasoning can be reliable where rigorously applied.	1, 14
Milward v. Rust-Oleum Corp.	820 F.3d 469 (1st Cir. 2016)	The *Milward* sequel: a general-causation win undone by exclusion of the specific-causation opinion.	1, 14
Pluck v. BP Oil Pipeline Co.	640 F.3d 671 (6th Cir. 2011)	Differential-diagnosis opinion excluded where exposure and alternative causes were not reliably addressed.	1, 14, 18, 20
OSHA Inspections, Warrants, and Evidence			
Marshall v. Barlow's, Inc.	436 U.S. 307 (1978)	Employers may insist on a warrant for OSHA inspections; administrative probable cause governs.	7
Marshall v. Horn Seed Co.	647 F.2d 96 (10th Cir. 1981)	Complaint-based warrants require underlying indicia of reliability.	7
Donovan v. Sarasota Concrete Co.	693 F.2d 1061 (11th Cir. 1982), aff'g 9 BNA OSHC 1608	Evidence from an invalid warrant suppressed before the Commission.	7
Smith Steel Casting Co. v. Brock	800 F.2d 1329 (5th Cir. 1986)	Suppression applied in enforcement proceedings for warrant violations.	7
United States v. Leon	468 U.S. 897 (1984)	Good-faith exception; frames the limits of suppression arguments.	7

Decision	Forum / Citation	Why It Matters	Ch.
Accu-Namics, Inc. v. OSHRC	515 F.2d 828 (5th Cir. 1975)	Technical violations of inspection procedure without prejudice do not void a citation — the boundary of procedural attacks.	7
Trinity Indus., Inc. v. OSHRC	16 F.3d 1455 (6th Cir. 1994)	Complaint-triggered full-scope warrants held invalid — a plan keyed to an employee complaint is not the neutral administrative plan Barlow's requires — but the citations were preserved under the good-faith exception; the limits of the suppression remedy.	7
Gilles & Cotting, Inc.	3 BNA OSHC 2002 (1976)	Employee exposure is proved by reasonable predictability of access to the zone of danger, not certainty of contact.	8, 9
Atlantic Battery Co.	16 BNA OSHC 2131 (1994)	States the Secretary's four prima facie elements in their canonical form.	9
Astra Pharm. Prods., Inc.	9 BNA OSHC 2126 (1981), aff'd in part 681 F.2d 69	Companion statement of the burden structure; origin of the four-element formulation.	9
Precision Concrete Constr.	19 BNA OSHC 1404 (2001)	Reasonable diligence comprises inspection, anticipation of hazards, and preventive measures.	9
Frank Swidzinski Co.	9 BNA OSHC 1230 (1981)	Constructive knowledge turns on what a diligent inspection program would have revealed.	9
Williams Enters., Inc.	13 BNA OSHC 1249 (1987)	Willfulness as intentional disregard or plain indifference; the state-of-mind	9

Decision	Forum / Citation	Why It Matters	Ch.
		showing a documented program undercuts.	
Elements, Defenses, and Knowledge			
ComTran Grp., Inc. v. U.S. Dep't of Labor	722 F.3d 1304 (11th Cir. 2013)	A supervisor's own misconduct is not automatically imputed to the employer; the Secretary must prove knowledge.	8, 9
W.G. Yates & Sons Constr. Co. v. OSHRC	459 F.3d 604 (5th Cir. 2006)	Limits on imputing supervisory misconduct; foreseeability matters.	8, 9
Quinlan v. Sec'y, U.S. Dep't of Labor	812 F.3d 832 (11th Cir. 2016)	Knowledge imputed where a supervisor knew of a subordinate's violation — the boundary of the Yates/*ComTran* line.	8, 9
Valdak Corp.	17 BNA OSHC 1135 (1995)	Unpreventable-employee-misconduct defense demands a genuinely enforced program, not a paper one.	8, 9, 25
Dover Elevator Co.	16 BNA OSHC 1281 (1993)	Employer knowledge and constructive knowledge standards before the Commission.	8, 9
Sampling and Measurement Before the Commission			
Southern Scrap Materials Co.	OSHRC Docket No. 94-3393	Sampler placement relative to a welding helmet resolved by how the standard defines exposure — the controlling placement precedent.	10
Bonney Forge Div., Gulf & Western Energy Prods. Grp.	14 BNA OSHC 1968 (No. 79-4053, 1991)	Calibration and instrument-reliability defects weighed against the Secretary's noise proof.	10

Decision	Forum / Citation	Why It Matters	Ch.
Collier-Keyworth Co.	13 BNA OSHC 1208 (No. 80-2848, 1987), vacated per stipulation (1st Cir. Mar. 22, 1989)	Measurement-validity attack on the exposure evidence behind the citation.	10
Multi-Employer Worksites and Control			
Solis v. Summit Contractors, Inc.	558 F.3d 815 (8th Cir. 2009)	Controlling-employer citation policy upheld against a general contractor.	11, 19
Acosta v. Hensel Phelps Constr. Co.	909 F.3d 723 (5th Cir. 2018)	Fifth Circuit accepted controlling-employer liability, abandoning its prior contrary rule.	11
Melerine v. Avondale Shipyards, Inc.	659 F.2d 706 (5th Cir. 1981)	OSHA standards and negligence per se — who the standards protect.	11
Summit Contracting Group, Inc.	No. 18-1451 (OSHRC May 10, 2022)	Controlling-employer duty is reasonable care in a secondary role, not strict liability; citation vacated on that proof.	11
StormForce of Jacksonville, LLC	No. 19-0593 (OSHRC Mar. 8, 2021)	Applied the reasonable-care framework to vacate a controlling-employer citation.	11
Elsner v. Uveges	34 Cal. 4th 915 (2004)	Cal-OSHA standards admissible to define the standard of care in third-party negligence actions.	11
Whistleblower and Retaliation			

Decision	Forum / Citation	Why It Matters	Ch.
Whirlpool Corp. v. Marshall	445 U.S. 1 (1980)	Upheld the right to refuse work under imminent-danger conditions; foundation of § 11(c) protection.	13
Reich v. Hoy Shoe Co.	32 F.3d 361 (8th Cir. 1994)	Employer knowledge of protected activity; suspicion-based inference addressed.	13
Clark Cnty. Sch. Dist. v. Breeden	532 U.S. 268 (2001)	Temporal proximity alone must be very close to support causation — the causal-chain audit's anchor.	13
Privilege, Work Product, and the Safety File			
Upjohn Co. v. United States	449 U.S. 383 (1981)	Corporate attorney-client privilege extends to employee communications made for legal advice — the charter of the internal investigation.	21, 22, 26
Hickman v. Taylor	329 U.S. 495 (1947)	Work-product doctrine; protects litigation-preparation materials, including structured witness statements.	21, 22, 26
United States v. Kovel	296 F.2d 918 (2d Cir. 1961)	Privilege can extend to consultants engaged to assist counsel — the basis of counsel-retained expert engagements.	26
United States v. Adlman	134 F.3d 1194 (2d Cir. 1998)	"Because of" litigation test for dual-purpose documents.	26
Harding v. Dana Transp., Inc.	914 F. Supp. 1084 (D.N.J. 1996)	Placing an investigation at issue can waive its protection.	22, 26

Decision	Forum / Citation	Why It Matters	Ch.
Martin v. Bally's Park Place Hotel & Casino	983 F.2d 1252 (3d Cir. 1993)	Consultant's safety report protected as work product against OSHA subpoena — engagement structure decided the outcome.	22, 26
Dowling v. Am. Hawaii Cruises, Inc.	971 F.2d 423 (9th Cir. 1992)	Routine pre-accident safety-committee minutes not work product — the boundary of protection for ordinary safety records.	22, 25, 26
The Written Program as Evidence			
Wal-Mart Stores, Inc. v. Wright	774 N.E.2d 891 (Ind. 2002)	Internal safety rules admissible but may exceed — and do not define — the objective standard of ordinary care.	25
Grimshaw v. Ford Motor Co.	119 Cal. App. 3d 757 (1981)	Punitive damages built from the defendant's own engineering and cost documents — the internal file read aloud.	25
Canipe v. National Loss Control Serv. Corp.	736 F.2d 1055 (5th Cir. 1984)	Safety consultant's § 324A duty of reasonable care in inspection services undertaken for the employer.	25
Pepperidge Farm, Inc.	17 BNA OSHC 1993 (1997)	General Duty Clause hazard recognition established in substantial part from the employer's own records and responses.	25
Take-Home, Household, and Community Exposure			
Kesner v. Superior Court	1 Cal. 5th 1132 (2016)	Employers and premises owners owe a duty to prevent take-home asbestos exposure, limited to members of the worker's household.	20

Decision	Forum / Citation	Why It Matters	Ch.
Olivo v. Owens-Illinois, Inc.	186 N.J. 394, 895 A.2d 1143 (2006)	Landowner's duty extends to the spouse who laundered contaminated work clothing; the duty is derivative of that owed the worker.	20
Satterfield v. Breeding Insulation Co.	266 S.W.3d 347 (Tenn. 2008)	Foreseeability rather than relationship governs; duty recognized to a worker's daughter.	20
Schwartz v. Accuratus Corp.	225 N.J. 517, 139 A.3d 84 (2016)	Take-home duty may reach beyond the spouse; no categorical limit, and the employer's knowledge at the time fixes the scope.	20
In re N.Y.C. Asbestos Litig. (Holdampf)	5 N.Y.3d 486, 840 N.E.2d 115 (2005)	Foreseeability alone does not create a duty to an employee's household — the leading no-duty decision.	20
Schneider Electric USA, Inc. v. Williams	No. 2023-SC-0436-DG (Ky. Mar. 19, 2026)	Compensation coverage cannot convert non-occupational exposure into employer immunity; a bounded take-home duty recognized.	18, 20
Kamuda v. Sterigenics U.S., LLC	No. 2018-L-010475 (Ill. Cir. Ct. Cook Cnty. 2022) — $363M	The first ethylene oxide community-emissions verdict, and the one that opened the docket: $363 million ($38M compensatory, $325M punitive); a modeled ambient exposure carried the causation case. Approximately 870 remaining Illinois claims settled for $408 million in January 2023.	20, 28
Rutherford v. Owens-Illinois, Inc.	16 Cal. 4th 953 (1997)	Substantial-factor contribution to aggregate dose and risk; the counterweight to every-exposure theories.	14, 18

Decision	Forum / Citation	Why It Matters	Ch.
Bostic v. Georgia-Pacific Corp.	439 S.W.3d 332 (Tex. 2014)	Proof of some exposure is not enough; the plaintiff must establish an approximate dose.	14
Spoliation and Preservation			
Silvestri v. Gen. Motors Corp.	271 F.3d 583 (4th Cir. 2001)	Dismissal for failure to preserve the vehicle; duty extends to plaintiffs and pre-suit.	21, 22
Flury v. Daimler Chrysler Corp.	427 F.3d 939 (11th Cir. 2005)	Dismissal where the central physical evidence was disposed of.	21, 22
Micron Tech., Inc. v. Rambus Inc.	645 F.3d 1311 (Fed. Cir. 2011)	Document destruction under a retention policy can be spoliation once litigation is reasonably foreseeable.	21
Zubulake v. UBS Warburg LLC	220 F.R.D. 212 (S.D.N.Y. 2003)	Litigation-hold obligations for electronic records; the modern e-discovery baseline.	21
Stevenson v. Union Pacific R.R. Co.	354 F.3d 739 (8th Cir. 2004)	Adverse-inference standards for pre-litigation destruction.	21
QBE Ins. Corp. v. Jorda Enters., Inc.	277 F.R.D. 676 (S.D. Fla. 2012)	Corporate-representative deposition obligations — the 30(b)(6) discipline.	26
Third-Party Liability and Verdicts			
Valdivia v. Brock Indus. Servs., LLC	No. C-731101 (La. 19th Jud. Dist. Ct.) — $411M verdict Feb. 28, 2025, appeal and coverage	Dropped-object case; the record-driven verdict examined from both chairs.	6, 23

Decision	Forum / Citation	Why It Matters	Ch.
	litigation pending		
Privette v. Superior Court	5 Cal. 4th 689 (1993)	Frames the hirer's liability for a contractor's employee's injury; doctrinal parent of *Hooker* and *Sandoval*.	11, 19
Hooker v. Dep't of Transportation	27 Cal. 4th 198 (2002)	Hirer liability for retained control requires affirmative contribution to the injury.	19
Sandoval v. Qualcomm Inc.	12 Cal. 5th 256 (2021)	Refined retained-control liability; the modern California framework.	19
Schmid v. Milwaukee Elec. Tool Corp.	13 F.3d 76 (3d Cir. 1994)	Sanction proportionality where an expert altered product evidence; exclusion of the testimony reversed as an excessive sanction.	19
Koger v. Costco Wholesale Corp.	No. 3:20-cv-08759 (N.D. Cal. 2023)	Falling-merchandise claim; the retail dropped-object record.	6
Exposure Torts: Silica and EtO			
Isaacks v. Terumo BCT Sterilization Servs.	No. 2022CV031124 (Colo. Dist. Ct., Jefferson Cnty.) — defense verdict Mar. 14, 2025	Complete defense verdict for the operator of a sterilization plant: the jury found no negligence and so never reached causation or damages — the compliance-record defense, tried successfully.	28
Jordan v. Cambria Co., LLC	No. 2024CV31180 (Colo. Dist. Ct.,	First Colorado engineered-stone silicosis verdict. No design defect; the lead manufacturer not negligent but liable for	28

Decision	Forum / Citation	Why It Matters	Ch.
	Denver Cnty.) — $17,450,000 verdict Apr. 30, 2026, on appeal	misrepresenting the product's character, a second liable for negligence and misrepresentation; sixty-three percent of fault to the employing fabrication shop.	
Reyes-Gonzalez v. Caesarstone USA, Inc.	No. 22STCV31907 (L.A. Super. Ct.) — $52,437,366 gross verdict Aug. 7, 2024; JNOV as to one of the three defendants Oct. 9, 2024; amended judgment Mar. 18, 2025; on appeal	Engineered-stone silicosis products litigation against slab manufacturers. The gross figure is not the exposure: the jury apportioned only twenty-seven and a half percent of fault to the three trial defendants, with seventy percent to parties not on the verdict form — and, after the post-trial ruling removed one of the three, twenty-five percent across the two that remain.	28
Sommerville v. Union Carbide Corp.	No. 24-1491, 149 F.4th 408, 2025 WL 2383496 (4th Cir. Aug. 18, 2025), cert. denied sub nom. Union Carbide Corp. v. Sommerville, No. 25-919	Reversed exclusion of an ethylene oxide dispersion-modeling expert, holding that attacks on an expert's factual basis go to weight rather than admissibility, and recognized Article III standing for medical monitoring absent present disease. The counter-trend to the post-2023 Rule 702 cases; discussed at Chapter 2 §I.	2, 28

Decision	Forum / Citation	Why It Matters	Ch.
	(U.S. May 26, 2026)		
Walker v. C.R. Bard, Inc.	(Ga. State Ct., Gwinnett Cnty.) — $20 million compensatory verdict May 2, 2025, on appeal	Ethylene oxide community verdict against a sterilization plant: negligence and public nuisance found, the punitive phase mistried on polling, and the compensatory award and the scope of any retrial on appeal.	28
Administrative Law and the Enforcement Machinery			
Atlas Roofing Co. v. OSHRC	430 U.S. 442 (1977)	Upheld civil-penalty adjudication before the Commission without a jury — the holding *Jarkesy* now pressures.	8, 28
SEC v. Jarkesy	603 U.S. 109 (2024)	Seventh Amendment jury right for penalty adjudications of common-law character; the cloud over agency enforcement.	12, 28
Loper Bright Enters. v. Raimondo	603 U.S. 369 (2024)	Ended Chevron deference; standard interpretations now reviewed independently.	8, 28
Allstates Refractory Contractors, LLC v. Su	79 F.4th 755 (6th Cir. 2023)	Nondelegation challenge to OSHA standard-setting rejected — for now.	28
Gade v. Nat'l Solid Wastes Mgmt. Ass'n	505 U.S. 88 (1992)	OSH Act § 18 preemption of state occupational safety regulation absent an approved plan, together with the state-plan exception and the § 4(b)(4) carve-out	12

Decision	Forum / Citation	Why It Matters	Ch.
		that preserves state tort law — the state-plan boundary.	

APPENDIX D

Standards and Directives Matrix

Regulatory citations and directive numbers are as cited in the chapters, current as of July 2026; directives are periodically renumbered and standards amended.

Standard / Directive / Authority	What It Governs and Why It Is Cited	Ch.
The Act and Core Procedural Regulations		
OSH Act § 5(a)(1) — General Duty Clause	Recognized hazards without a specific standard; the enforcement vehicle for heat and other unregulated hazards.	4, 6, 8, 12, 23, 25, 26, 28
OSH Act § 11(c); 29 C.F.R. § 1977.12	Retaliation protection and the refusal-to-work doctrine.	13
29 C.F.R. Part 1903 (§§ 1903.6, 1903.7(e))	Advance notice prohibition; inspections conducted in a reasonable manner — the procedural floor for the walkaround.	7
29 C.F.R. Part 1904 (§§ .5, .7, .32, .33, .35, .37, .41)	Injury and illness recordkeeping: recordability, the 300/301/300A forms, five-year retention, employee access, electronic submission.	12, 13, 18, 21, 22, 23, 24, 25, 26
29 C.F.R. § 2200.34	Commission pleading rules; affirmative defenses raised in the answer on pain of waiver.	8, 9
29 C.F.R. Part 1902; Part 1952; § 1953.5(a)	State plan approval, coverage, and the obligation to conform to a new federal standard within six months.	12

Standard / Directive / Authority	What It Governs and Why It Is Cited	Ch.
Fed. R. Evid. 407	Subsequent remedial measures: the bar, and the impeachment, feasibility, and ownership exceptions that carry most of the traffic.	22
General Industry Health and Safety Standards (29 C.F.R. Part 1910)		
§ 1910.95 & Table G-16	Occupational noise: permissible durations, dosimetry, the hearing-conservation action level.	10
§ 1910.1000 & Tables Z-1–Z-3	PELs for air contaminants; the TWA, ceiling, and computation formulas every exposure citation is built on.	10, 15
Substance-specific standards: § 1910.1001 (asbestos), § 1910.1025 (lead), § 1910.1028 (benzene)	Sampling appendices, action levels, medical surveillance, and excursion limits — the mandatory protocols against which the compliance officer's sampling is audited.	10, 16
§ 1910.1020	Access to employee exposure and medical records; thirty-year retention — the deepest discovery well in exposure litigation.	4, 12, 19, 25, 26, 28
§ 1910.1047	Ethylene oxide: PEL, excursion limit, monitoring, and surveillance.	28
§ 1910.1053	Respirable crystalline silica (general industry): PEL, action level, Table 1 controls, medical surveillance.	10, 28
Construction Standards (29 C.F.R. Part 1926)		
§ 1926.100 / § 1926.451(h)	Head protection; scaffold falling-object protection — the regulatory floor in dropped-object claims.	6

Standard / Directive / Authority	What It Governs and Why It Is Cited	Ch.
§ 1926.501 (Subpart M)	Fall protection duties by work surface and height trigger.	24
OSHA Directives and Agency Methods		
Field Operations Manual — CPL 02-00-164 (as amended July 14, 2025)	The inspection script: opening conference, walkaround, interviews, citation development. The yardstick for procedural audit of the government's case.	7
Multi-Employer Citation Policy — CPL 02-00-124	Creating, exposing, correcting, and controlling employer categories; two-step citation analysis.	11
Whistleblower Investigations Manual — CPL 02-03-011	Intake, elements, and investigative procedure for § 11(c) and the adjacent statutes.	13
Silica enforcement — CPL 02-02-080; Silica NEP — CPL 03-00-023	Compliance guidance and targeting for the 2016 silica standards.	10, 28
Heat NEP — CPL 03-00-024	Inspection targeting and heat-hazard documentation while the rule remains stalled.	28
Warehousing NEP — CPL 03-00-026 (term expired July 13, 2026; check status)	The ergonomics-adjacent enforcement frontier in warehousing and distribution.	28
Severe Violator Enforcement Program — CPL 02-00-169	Follow-up inspections and public designation — the economics behind self-audit.	24

Standard / Directive / Authority	What It Governs and Why It Is Cited	Ch.
OSHA Technical Manual — TED 01-00-015 (Section II, Ch. 1)	Personal sampling procedure for air contaminants; calibration, media, minimum volumes.	10, 17, and throughout Part III
Machine Safety and Functional Safety (Products and Guarding Cases)		
29 C.F.R. §§ 1910.212, 1910.217, Subpart O	The regulatory floor for machine guarding; performance-worded, employer-directed, and rarely sufficient to resolve a design case.	19
ANSI B11.0 and the B11 series (B11.1, B11.19, B11.20)	Risk-assessment methodology for machinery and performance criteria for safeguarding; the standard against which a manufacturer's design file is measured.	19
ISO 12100	General principles for design and risk reduction; the international companion to B11.0.	19
ISO 13849-1 / IEC 62061	Functional safety of safety-related control systems; Performance Levels and Safety Integrity Levels for guards, interlocks, and light curtains.	19
NFPA 79	Electrical standard for industrial machinery.	19
ANSI/ASSP Z244.1	Control of hazardous energy; the current consensus treatment where § 1910.147 has aged.	19, 22
ANSI/ASSP Z359.2; ANSI/ASSP A10.32	Managed fall protection programs and construction fall protection systems.	Appendix F
Consensus and Professional Authorities		
ACGIH Threshold Limit Values	Current health-based exposure benchmarks; the standard-of-care counterweight to outdated PELs, including the WBGT heat TLV.	Part III

Standard / Directive / Authority	What It Governs and Why It Is Cited	Ch.
AIHA, *A Strategy for Assessing and Managing Occupational Exposures*	The professional framework for SEGs, exposure judgments, and decision-making on limited data.	1, 15, 17
NIOSH criteria documents, *Manual of Analytical Methods*, and the NIOSH hazardous-drug list	Research-agency methods and recommendations experts cite for sampling and for hazardous-drug identification.	Part III; 17
ANSI/ASSP Z10.0-2019; ISO 45001:2018	Safety management systems; the internal-audit obligation that frames mock-audit practice.	1, 24, 25
ANSI/ASSP Z490.1	Criteria for accepted practices in safety, health, and environmental training — the consensus benchmark for the training file's development, delivery, and documentation.	25
ANSI/ASSP A10.33	Multi-employer project safety programs; how responsibility is allocated on paper.	19
ANSI/ISEA Z89.1; ANSI/ISEA 121-2018 and 121-2023	Industrial head protection performance; dropped-object prevention solutions. The 2018 edition governs conduct before mid-2024; the 2023 revision was ANSI-approved July 2, 2024.	6
NFPA 921	Fire and explosion investigation methodology; embeds the evidence-preservation discipline the spoliation chapter builds on.	21
29 C.F.R. § 1904.39	Reporting the fatality (8 hours) and the hospitalization, amputation, or eye loss (24 hours) — the clock that starts the investigation.	22

Standard / Directive / Authority	What It Governs and Why It Is Cited	Ch.
29 C.F.R. § 1910.119(m)	PSM incident investigation: 48-hour initiation, team with process knowledge, findings retained five years.	22

APPENDIX E

Obtaining the OSHA Inspection File

A FOIA and Discovery Playbook for the Technical Case

Nearly every chapter in this book turns at some point on the government's own file: the compliance officer's notes and photographs, the sampling and calibration records, the interview statements, the citation worksheet with its penalty arithmetic. That file is obtainable, and obtaining it early is often the single most valuable act in a citation defense or a related civil case. This appendix assembles what to request, how, and when.

What the inspection file contains

An OSHA inspection generates a case file (often called the C-file) that typically includes the compliance officer's field notes and narrative; photographs and video; sampling data, calibration records, and laboratory reports; the OSHA-1 inspection report and OSHA-1B violation worksheets showing how each item was classified and each penalty computed; employee interview statements; the citation and notification of penalty; and correspondence. In an exposure case the sampling and calibration records and the OSHA-1B worksheets are usually the highest-value documents, because they expose the measurement chain audited in Chapter 10 and the penalty arithmetic contested in Chapter 8.

Two roads to the file: FOIA and discovery

The file can be sought two ways, and the choice matters. A Freedom of Information Act request (5 U.S.C. § 552) to the OSHA regional office is available to anyone and does not require a pending case, which makes it the fast early tool: a request filed the week a citation issues can return substantial material before the contest strategy is fixed. Once a case is docketed before the Review Commission or in civil litigation, ordinary discovery is available and reaches material FOIA exemptions may withhold. The two are complementary: FOIA gets the early look; discovery gets the rest.

The exemptions you will meet

OSHA routinely invokes several FOIA exemptions, and knowing them shortens the fight. Exemption 7 (records compiled for law-enforcement purposes) is the most common, invoked to withhold or redact material during an open inspection or enforcement proceeding, which is one reason some material arrives only through discovery after the citation issues. Exemption 6 protects personal privacy and is used to redact employee identities and interview statements. Exemption 4 protects confidential commercial information and can appear where a third party's trade secrets are in the file. Exemption 5 covers the deliberative process. A denial or heavy redaction is appealable within the agency and, ultimately, reviewable in court; in a contested case, the same material is often more efficiently pursued through a discovery request or the deposition of the compliance officer described in Chapter 7.

What the file is worth as evidence

Obtaining the file and using it are different questions, and a playbook that answered only the first would leave counsel holding a box. The routes into evidence are three, they are not equally reliable, and the strongest of them is the one that depends least on the agency's own conclusions.

The public-records route. Fed. R. Evid. 803(8) excepts from the hearsay rule a record or statement of a public office that sets out the office's activities, or a matter observed while under a legal duty to report, or, in a civil case, factual findings from a legally authorized investigation, unless the opponent shows that the source of information or other circumstances indicate a lack of trustworthiness. The compliance officer's contemporaneous observations, measurements, sampling data, and photographs sit comfortably inside that language, and the burden of showing untrustworthiness falls on the party resisting them. The reach of the exception is wider than practitioners often assume: in *Beech Aircraft Corp. v. Rainey*, 488 U.S. 153 (1988), the Supreme Court held that "factual findings" embraces conclusions and opinions contained in an investigative report, with trustworthiness — timeliness of the investigation, the investigator's skill and experience, whether a hearing was held, and any motivational problem — as the operative check rather than a categorical bar. State analogues to Rule 803(8) vary, several of them materially, and the governing version should be read before the file is offered.

The citation is a different animal, and counsel who conflate the two lose motions. A citation is the Secretary's allegation, and in the civil case it arrives freighted with the authority of the government and the appearance of an adjudication that, in the uncontested case, never occurred. Jurisdictions split. Many admit it as evidence bearing on notice or on the standard of care, subject to a limiting instruction; many exclude it under Rule 403 or its state counterpart, reasoning that its probative value is substantially outweighed by the risk that a jury will treat an allegation as a finding; some exclude it as an impermissible legal conclusion; and several state statutes address the question directly. Section 4(b)(4) of the Act preserves all of this for state law (Chapter 12 §III), and Chapter 11 §V develops the standard-of-care and negligence-per-se side of it. Two features of the file itself bear on the fight. A citation that was withdrawn, vacated, or reclassified in settlement is worth what the settlement says it is worth, and most informal settlement agreements contain a non-admission clause that is itself an exhibit. And a citation the employer contested and defeated is affirmative evidence in the employer's hands, which is one more reason the contest decision of Chapter 8 §I outruns the penalty.

Preclusion, and its asymmetry. Where a citation was contested and decided on the merits before an administrative law judge, a final order of the Review Commission can support issue preclusion in a later civil action in those jurisdictions that give preclusive effect to the findings of administrative tribunals acting in a judicial capacity: the familiar requirements being that the issue was identical, actually litigated, and necessary to the decision, and that the party to be bound had a full and fair opportunity to contest it. The asymmetry is the practical point. An uncontested citation that ripened into a final order by operation of the fifteen-day clock was never actually litigated, and it will not carry preclusive weight; it may not even carry much evidentiary weight once the employer explains why a five-figure item was not worth contesting. Counsel on both sides should therefore ask of any final order what procedural history stands behind it before treating it as anything at all.

The durable route, and the one this book recommends. The most reliable use of the inspection file is not to offer the agency's conclusion but to use the agency's work as the factual substrate of the expert's own. The compliance officer's measurements become the expert's data, audited as Chapter 10 describes and re-derived rather than adopted; the photographs and dimensioned sketches become the expert's exhibits, authenticated through the officer or by stipulation; the officer is deposed against the Field Operations Manual as Chapter 7 sets out, and can be called. An opinion built that way does not rise or fall with the admissibility of the citation, survives the Rule 403 argument because it does not depend on the government's characterization of anything, and satisfies the amended Rule 702 requirement that the expert reason from sufficient facts rather than from another party's conclusion. Two cautions attach. Employee interview statements in the file are hearsay in most postures and arrive redacted under Exemption 6 in any event; they are investigative leads, not evidence, and should be treated as the identities of witnesses to be deposed. And the officer's field notes may be usable to refresh recollection or, if the foundation is laid, as a recorded recollection, but they are the officer's notes and not the expert's, and an expert who adopts them wholesale has imported someone else's judgment into her own opinion: the precise failure Chapter 2 §II.A describes.

Timing, and why it is urgent

The basis of a reliable opinion has a shelf life, and so does the file. File the FOIA request immediately — the week of the citation, not the week of the hearing — because processing takes time, because an early denial leaves room to appeal or to pivot to discovery, and because the file identifies the perishable evidence (equipment, site conditions, sampling media) that Chapter 21 warns disappears on its own schedule. Request the complete file rather than named documents, so that material counsel did not know to ask for is captured; specify the inspection number and establishment; and calendar the agency's response deadline the way the contest deadline of Chapter 8 is calendared.

A short requesting checklist

Identify the OSHA inspection number, establishment name and address, and inspection dates. Request the complete case file, expressly including field notes and narrative; all photographs and video; all sampling data, calibration records, and laboratory analyses; the OSHA-1 and all OSHA-1B worksheets; employee interview statements (accepting privacy redactions); and all correspondence. Direct the request to the correct OSHA regional FOIA office. In a docketed case, serve a parallel discovery request and notice the compliance officer's deposition, matching the document demands to the Field Operations Manual obligations set out in Chapter 7. Preserve appeal rights on any denial, and treat a redaction as the opening of a negotiation, not the end of one.

References

Beech Aircraft Corp. v. Rainey, 488 U.S. 153 (1988).

Fed. R. Evid. 403, 803(8), and 902(1).

Freedom of Information Act, 5 U.S.C. § 552.

Occupational Safety and Health Administration. (2025). Field Operations Manual (CPL 02-00-164, as amended). U.S. Department of Labor.

U.S. Department of Labor. FOIA regulations, 29 C.F.R. Part 70.

APPENDIX F

An Annotated Model Rule 26 Report

The Deliverable the Whole Book Describes

This book argues that the report is the deliverable (Chapter 3), that the report is what the deposition attacks (Chapter 4), that the report is what the gatekeeper reads (Chapter 2), and that the rebuttal report is where the audit findings land (Chapter 17). It has not, until now, shown one.

What follows is a skeleton report in a construction fall-protection matter, annotated in the margin notes marked ▶. The facts are invented and the analysis is compressed; a real report in this matter would run three to five times as long, and the exhibit list alone would be longer than this appendix. What is not compressed is the architecture, and the architecture is the point. The annotations are keyed to the four places safety and industrial hygiene opinions actually fail under Rule 702: **[BASIS]** — insufficient facts or data; **[METHOD]** — no reliable methodology, or none the court can check; **[APPLICATION]** — a reliable method applied unreliably; **[OVERREACH]** — conclusions the basis and method cannot carry, including legal conclusions.

Counsel should read the annotations as a demand list: this is what to insist on before the report is served, and what to look for in the opponent's. The professional should read them as a drafting checklist. Both audiences should note how much of the report's defensive strength comes from things it declines to say.

IN THE UNITED STATES DISTRICT COURT FOR THE [DISTRICT]

MARTINEZ v. KELLAND CONSTRUCTION GROUP, INC., et al.

Civil Action No. [ ]

EXPERT REPORT OF [NAME], PhD, CIH, CSP

Submitted pursuant to Fed. R. Civ. P. 26(a)(2)(B)

[Date]

I. Assignment and Scope

I was retained by counsel for plaintiff Rubén Martínez to evaluate the fall-protection program, hazard assessment, and site-safety practices in effect at the Kelland Construction Group project at 4100 Delacourt Avenue on and before March 12, 2024, and to determine whether the fall-protection measures in place at the leading edge of the third-floor deck conformed to the requirements of 29 C.F.R. Part 1926 Subpart M and to generally accepted construction safety practice.

I was not asked to, and do not, offer any opinion on the cause, extent, or prognosis of Mr. Martínez's injuries; on the reasonableness of any medical treatment; on the design or structural adequacy of the deck; or on the allocation of legal fault among the parties.

> ► *[OVERREACH] The second paragraph is the most under-used paragraph in expert practice. An explicit scope exclusion does three things: it prevents the deposition from wandering into territory the witness has not prepared, it forecloses a Rule 702 challenge premised on opinions the witness never offered, and it signals a discipline the rest of the report will be read against. Draft it before drafting the opinions.*

II. Qualifications

I hold a doctorate in [field] and am a Certified Industrial Hygienist and Certified Safety Professional. I have [N] years of experience in construction safety program development, hazard assessment, and incident investigation, including [N] years as a corporate safety director with responsibility for fall-protection programs on [type] projects, and I have personally conducted or supervised approximately [N] jobsite hazard assessments involving leading-edge and elevated work. My curriculum vitae is attached as Exhibit A.

> ► *[BASIS] Rule 702(a) asks whether the expert's knowledge will help the trier of fact, and qualification is assessed against the specific opinion offered. A CV appended without curation invites the voir dire that establishes the witness has never done the particular thing at issue. The paragraph above foregrounds leading-edge work because leading-edge work is what the case is about. Chapter 3 §VI treats the point at length.*

III. Materials Considered

The materials I considered in forming the opinions in this report are listed in Exhibit B. They include, without limitation: the site-specific safety plan dated [date];

Kelland's corporate fall-protection program (rev. 2022); job hazard analyses for deck placement dated [dates]; daily superintendent logs for February 1 through March 12, 2024; toolbox-talk records and attendance rosters; the fall-protection equipment inventory and issuance log; weekly site safety inspection reports for the project; the OSHA inspection file for Inspection No. [ ] obtained by Freedom of Information Act request, including compliance officer field notes, 38 photographs, and OSHA-1B worksheets; the depositions of [names]; and the site measurements and photographs I took during my inspection of [date].

▶ ***[BASIS]*** *This list is the report's foundation and its first vulnerability. Two disciplines govern it. It must be complete on the day it is served — a document produced at deposition that was considered but unlisted is the cleanest impeachment available. And it should be organized to show the sources of the facts, because Rule 26(b)(4)(C) makes discoverable the facts and assumptions counsel supplied. Where the expert gathered a fact independently, the list should show it. See Chapter 3 §V.*

IV. Site Inspection

I inspected the project on [date] in the presence of [names], under the joint inspection protocol attached as Exhibit C. I measured the deck elevation at the location identified in the record as the point of the fall (18 feet 4 inches from the deck surface to the grade below), the distance from that point to the nearest anchorage point provided (46 feet), and the dimensions of the unprotected leading edge on the date of the incident as reconstructed from photographs [nos.]. Measurements were taken with a [instrument], calibrated [date]; the field notes are Exhibit D.

▶ ***[BASIS] [APPLICATION]*** *Measurements, instruments, calibration, and field notes. This paragraph is the reason the opinion will survive the argument that the expert reasoned from photographs. Note also the joint-protocol reference: Chapter 21 explains why an inspection conducted without notice can convert the best evidence in the case into a motion.*

V. Factual Background

[Recites the sequence of events, with a record citation for every factual assertion.]

▶ ***[BASIS]*** *Every sentence in this section should carry a citation to a document, a deposition page, or the expert's own measurement. A factual background written as narrative rather than as a cited chronology is the single most common structural defect in safety expert reports, and it produces the deposition question the witness* *cannot answer: where does that come from? Facts genuinely in dispute are identified as disputed, with both versions stated and the analysis run against each, see §VIII below.*

VI. Applicable Standards and Authorities

The following authorities govern the conditions at issue:

1. 29 C.F.R. § 1926.501(b)(2)(i) — fall protection required for each employee constructing a leading edge six feet or more above a lower level. Cf. § 1926.501(b)(1), which imposes the parallel requirement for an unprotected side or edge of a completed walking/working surface. The choice between the two subsections is not clerical: the deck was under construction on the date at issue, the guardrail had been struck for a material delivery rather than never installed, and an opinion pleaded under the wrong subsection invites the applicability attack described at Chapter 9 §II — which is why this report identifies the governing provision and states why it, rather than its neighbor, applies.
2. 29 C.F.R. § 1926.502(d) — performance and use criteria for personal fall arrest systems, including the anchorage requirement at § 1926.502(d)(15).
3. 29 C.F.R. § 1926.503 — training requirements, including retraining triggers.
4. 29 C.F.R. § 1926.20(b)(2) — the employer's obligation to provide frequent and regular inspections by a competent person.
5. ANSI/ASSP Z359.2-2017 — minimum requirements for a managed fall protection program (hazard survey, written program, competent-person designation, rescue planning).
6. ANSI/ASSP A10.32-2012 (R2017), fall protection systems for construction and demolition operations.

► ***[METHOD]*** *The standards matrix is the method's spine, and it should be layered explicitly: the regulatory floor, then the consensus standards, then the defendant's own program. Two disciplines matter.* ***Date every authority to the time of the conduct.*** *Citing a 2023 revision against 2019 conduct is the anachronism error Chapter 16 catalogues, and it is fatal on cross. And* ***state the status of each consensus standard*** *— Z359.2 is not independently enforceable, and a report that implies otherwise has handed the defense its first exhibit. It is offered here as evidence of the recognized practice against which the employer's program is measured, which is what it is.*

VII. Methodology

In forming the opinions below I applied the following method, which is the method taught in [authority] and used in the ordinary practice of construction safety assessment:

Step 1. Hazard identification. I identified the fall hazards present at the work location from the site conditions, the task being performed, and the JHA for that task.

Step 2. Standards mapping. I determined which regulatory and consensus provisions were triggered by each identified hazard, applying the versions in effect in March 2024.

Step 3. Hierarchy-of-controls analysis. For each hazard, I evaluated the controls available in descending order of effectiveness — elimination of the exposure, engineering controls (guardrails, hole covers, safety net systems), administrative controls (controlled access zones, safety monitoring), and personal fall arrest — and determined which were present, which were feasible and absent, and which had been considered by the employer.

Step 4. Program audit. I evaluated Kelland's written fall-protection program and its implementation against the six elements of a managed fall protection program under ANSI/ASSP Z359.2 and against the training and inspection requirements of §§ 1926.503 and 1926.20(b)(2).

Step 5. Notice analysis. I reviewed the site's leading indicators — inspection findings, toolbox talks, corrective-action records, and prior incidents — for evidence bearing on what the employer knew, and when, about the condition at issue.

Step 6. Alternative-explanation analysis. I identified the alternative explanations for the incident advanced in the record or apparent from it, and evaluated each against the evidence. That analysis appears at §IX.

> ▶ *[METHOD] This is the section that decides the Rule 702 motion, and it is the section most often missing. It must do three things: name the method, tie it to an authority outside the expert's own head, and be* ***specific enough that another qualified professional could run it and check the result****. That last is the operational meaning of reliability. A "methodology" section that says only that the expert relied on training and experience is not a methodology section — it is the ipse dixit that Joiner rejected, and the amended Rule 702 has made courts considerably less patient with it. Note Step 6: building the alternative-explanation analysis into the method, rather than reacting to it on cross, is what distinguishes a report from an advocacy document. Chapter 18 explains why the same obligation binds a defense expert offering an alternative cause.*

VIII. Assumptions

The opinions below rest on the following assumptions, each of which is identified as such and sourced:

1. That Mr. Martínez was working at the location marked "A" on Exhibit E at the time of the fall. *Source: deposition of [witness] at 44:12–46:3; OSHA field notes at 7. Disputed: [defendant's witness] places him at location "B." The analysis at §IX.C addresses both locations; the opinions in §X are unchanged under either.*

2. That the guardrail system depicted in photograph [no.] had been removed for material delivery on the morning of March 12 and had not been reinstalled. *Source: superintendent log entry 3/12/24; deposition of [witness] at 88:4–91:20.*

3. That the anchorage points identified in Exhibit F were the only anchorage points available on the third-floor deck on that date. *Source: my site inspection; equipment issuance log. Assumption supplied by counsel: none.*

> ► *[APPLICATION] [BASIS] This section is the report's insurance policy, and Chapter 4 explains why. A stated assumption is a limitation the expert owns; an unstated one is a concession the examiner extracts. Three practices distinguish a good assumptions section.* ***Source each assumption****, distinguishing what the expert established from what counsel supplied — Rule 26(b)(4)(C)(ii) makes the latter discoverable regardless, so concealing it purchases nothing and costs credibility.* ***Identify which assumptions are disputed****, and run the analysis under both versions where the record supports two. An opinion that holds under either version of a disputed fact is close to unassailable on that point; an opinion that depends on the plaintiff's version of every disputed fact tells the jury the expert picked a side before doing the work. And* ***state the sensitivity****: if an opinion would change materially were an assumption wrong, say so here rather than conceding it in a deposition.*

IX. Analysis

A. Fall protection was required and was not provided.

[Analysis applying Step 2 and Step 3 to the record.]

B. The program was deficient in three respects.

[Analysis applying Step 4, with the deficiencies tied to specific program elements and specific records.]

C. The location dispute does not change the analysis.

Under either the location described by [plaintiff's witness] or that described by [defendant's witness], the working surface exceeded six feet above the lower level, no guardrail system was in place at either location on the date in question, and the nearest compliant anchorage exceeded the distance permitted by the manufacturer's instructions for the lanyard issued. The requirement, and the deficiency, are the same at both locations.

D. Notice.

The weekly site safety inspection reports for [dates] each record "leading edge — guardrail incomplete" as an open finding, with corrective action recorded as "in progress" on four consecutive weeks and no closure entry. [Analysis.]

> ▶ *[APPLICATION] Note the structure: each analytical subsection applies a named step of §VII to cited record evidence. That structure is what allows a court to verify that the reliable method was reliably applied, which is the precise question the 2023 amendment added to Rule 702(d). Note also §IX.C: an opinion that survives a disputed fact should say so explicitly rather than leaving the court to work it out.*

X. Opinions

Based on the materials considered, my inspection, and the methodology described above, I hold the following opinions to a reasonable degree of professional certainty within my fields of industrial hygiene and safety:

1. Fall protection was required at the leading edge of the third-floor deck on March 12, 2024 under 29 C.F.R. § 1926.501(b)(2)(i), and none of the three compliant systems recognized by that provision — guardrail, safety net, or personal fall arrest — was in place at the location of the fall.

2. Feasible engineering controls — reinstallation of the guardrail system removed for the material delivery, or installation of a compliant safety net system beneath the leading edge — were available, were used elsewhere on the same project, and would have prevented a fall from that location.

Exhibit G documents their use on the second-floor deck on [date].

3. Kelland's managed fall-protection program departed from the requirements of §§ 1926.503 and 1926.20(b)(2) and from the practice described in ANSI/ASSP Z359.2 in the three respects identified at §IX.B.

4. The condition at issue was recorded as an open finding in Kelland's own weekly inspection reports on four occasions before March 12, 2024, and no record produced in this matter reflects its correction.

Limitations. Opinion 2 assumes the guardrail system removed on the morning of March 12 was the system depicted in photograph [no.]; if the system was of a different configuration, the feasibility analysis would require revision, though the availability of the net system would not. I have not been provided with the equipment maintenance records for the [equipment], and I reserve the right to supplement these opinions if they are produced.

> ► *[OVERREACH] Read what the opinions do not say. They do not say the defendant was negligent, that its conduct was willful, that it violated OSHA, or that its violation caused the injury. Those are legal conclusions and jury questions, and an expert who offers them invites exclusion of the opinion, sometimes of the whole report, while gaining nothing the properly framed opinion does not already supply. "None of the compliant systems was in place" gives counsel everything "the defendant violated the standard" would give, and survives the motion. Note also the phrase to a reasonable degree of professional certainty within my fields. The qualifier is not decoration. An industrial hygienist testifying "to a reasonable degree of medical certainty" has stepped across the line Chapter 14 draws, and opposing counsel will make the crossing the theme of the cross-examination. Finally, the* ***Limitations*** *paragraph. Chapter 4 §III.D observes that the concessions every honest expert must give are best given in the report. A limitation stated by its author is a mark of discipline; the same limitation extracted at deposition is a retraction.*

XI. Rule 26(a)(2)(B) Disclosures

Exhibits to be used as a summary of or support for these opinions: Exhibits A through G, attached.

Prior testimony: I have testified as an expert at trial or by deposition in the matters listed in Exhibit H, covering the four years preceding this report.

Compensation: I am compensated at $[rate] per hour for analysis and $[rate] per hour for testimony. My compensation is not contingent on the outcome of this matter or on the content of my opinions.

> ► *[BASIS] These three are treated as clerical and lost as substantive. The testimony list must be complete for the full four years: an omitted case, found by opposing counsel, is offered not as an oversight but as evidence that the witness conceals. The expert should have read every transcript on that list before serving the report; the examiner will have. And the compensation statement is where the ratio Chapter 27 discusses first appears on the record. See Chapter 3 §VI.*

Signed, [Name, credentials] [Date]

A Reader's Checklist

Run any report, your own or the opponent's, against these ten questions.

1. Does the report state what the expert was *not* asked to do?
2. Are the qualifications curated to *this* opinion, or is a CV standing in for an argument?
3. Is the considered-materials list complete as of the service date, and does it distinguish what the expert gathered from what counsel supplied?
4. Are the authorities dated to the conduct, and is the enforceability status of each consensus standard stated?
5. Does the methodology section name a method, tie it to an authority outside the expert's own experience, and describe it specifically enough that a qualified professional could re-run it?
6. Are assumptions labeled as assumptions, sourced, and marked where disputed?
7. Does the analysis apply the named method steps to cited evidence, step by step?
8. Are alternative explanations addressed in the report rather than left for cross-examination?
9. Do the opinions stop short of legal conclusions, and is the certainty qualifier confined to the expert's actual field?
10. Does the report state its own limitations before anyone else does?

A report that answers all ten is not guaranteed to survive a Rule 702 motion. A report that fails four of them will not.

References

American National Standards Institute / American Society of Safety Professionals. ANSI/ASSP Z359.2, Minimum requirements for a comprehensive managed fall protection program; ANSI/ASSP A10.32, Fall protection systems for construction and demolition operations.

Duty to have fall protection, 29 C.F.R. § 1926.501; fall protection systems criteria and practices, § 1926.502; training requirements, § 1926.503; accident prevention responsibilities, § 1926.20(b)(2).

Fed. R. Civ. P. 26(a)(2)(B), (b)(4)(B)–(C); Fed. R. Evid. 702 (as amended Dec. 1, 2023), 703, 704.

Gen. Elec. Co. v. Joiner, 522 U.S. 136 (1997); Sardis v. Overhead Door Corp., 10 F.4th 268 (4th Cir. 2021).

APPENDIX G

An Annotated Exposure Assessment

A note on placement. This appendix is a working document of the exposure-assessment method taught in Part III rather than a reference table, and it is best read against Chapter 15 §VIII, whose eight elements it works out in full. It is collected here with the other appendices for ease of reference, and it keeps its appendix letter, so every cross-reference to "Appendix G" elsewhere in the book points here.

What Chapter 15 §VIII Asks For, Shown

Chapter 15 sets out the elements a defensible exposure assessment must contain and argues that most reports delivered to employers contain about half of them. This appendix shows the other half. The report below is a compressed model of an ordinary, non-litigation exposure assessment: the kind an employer commissions, files, and forgets, and the kind that surfaces four years later as an exhibit.

Two audiences should read it differently. The employer purchasing industrial hygiene services should read it as a specification: this is what the scope of work should require, and a deliverable missing these elements has not answered the question that was asked. Counsel should read it as a discovery target and a cross-examination map, because every element below is a place where an inadequate assessment leaves a mark, and because Chapter 17's audit of an opposing exposure opinion is, in substance, this document run in reverse.

Annotations are marked ▶.

EXPOSURE ASSESSMENT REPORT

Facility: [Plant], [Location]

Agent: Respirable crystalline silica (quartz)

Similar Exposure Group: Finishing Department — grinders and chippers, all shifts

Assessment period: [dates]

Prepared by: [Name], CIH [certification number]

Date: [date]

1. *Purpose and Decision to Be Made*

This assessment was performed to determine whether the exposure of workers in the Finishing Department SEG to respirable crystalline silica is acceptable, and to support decisions regarding engineering controls, respiratory protection, and the medical surveillance obligation under 29 C.F.R. § 1910.1053(i).

The specific question addressed is: **what proportion of full-shift exposures in this SEG exceeds the 50 μg/m³ permissible exposure limit, and with what confidence can that proportion be stated?**

> ► *An assessment begins with a decision, not with a sampling plan. Chapter 15 §I makes the point: "were exposures within acceptable levels" is not a question a measurement can answer, because it does not specify whose exposures, over what period, or to what standard of confidence. A report whose purpose statement is a sentence of boilerplate will produce a conclusion of the same quality. For counsel: the absence of a stated decision is the first sign that a report was produced to satisfy a checklist rather than to answer a question, and it makes every later step unmoored.*

2. *Basic Characterization*

Process. Castings are ground and chipped at eight stations along the north wall using pneumatic tools. Silica sand from the molding operation adheres to castings entering the department. Production averages [N] castings per shift, with seasonal variation of approximately ±30 percent (source: production records, [dates]).

Workforce. Fourteen employees across two shifts. Job titles, tasks, and shift assignments are in Attachment 1.

Materials. Bulk sand supplier certificates of analysis report 92–96 percent crystalline silica (Attachment 2).

Controls. Local exhaust ventilation is installed at stations 1–4 (installed [date], capture velocity last measured [date], results in Attachment 3). Stations 5–8 have no LEV. General dilution ventilation is [description]. Respiratory protection is issued as [type]; the written program, fit-test records, and medical clearances are in Attachment 4.

> ► *The basic characterization is the workplace, the agent, and the workforce described well enough that a professional who has never entered the building could evaluate the judgment. Chapter 16 §II treats it as the foundation of any reconstruction, and*

the reason is visible here: the LEV installation date makes the sampling data separable into pre- and post-control periods, and an assessment that pools across the boundary describes neither. Note also the control-verification measurements. A report that recites the presence of LEV without evidence it functions has described equipment, not a control.

3. Exposure Group Definition and Its Basis

The Finishing SEG comprises the fourteen employees at stations 1–8 performing grinding and chipping. The group was defined on the basis of common process, common material, common task set, and common control configuration, with one qualification: **stations 1–4 and stations 5–8 differ in control status, and are therefore treated as two subgroups** for the purpose of the exposure judgment, notwithstanding their common job title.

Employees whose duties place them in the department for less than 25 percent of a shift (maintenance, quality) are **not** members of this SEG and are addressed separately in [report].

> ▶ *This is the paragraph most often absent, and it is the one Chapter 16 §V and the worked example in that chapter identify as the most productive line of cross-examination available. A SEG defined by job title rather than by exposure determinants is not a SEG; it is an org chart. Note two disciplines: the group is defined by stated criteria rather than by assertion, and the criteria are then applied honestly enough to split the group when a determinant differs. Note also the explicit exclusion — naming who is not in the group prevents the assessment from being applied later to workers it never described, which is exactly how a defensible assessment becomes an indefensible exhibit.*

4. Sampling Data

Thirteen full-shift personal samples were collected on [dates] using [sampler and method: e.g., SKC GS-3 respirable cyclone at 2.75 L/min, NIOSH Method 7500 / OSHA ID-142]. Calibration records are in Attachment 5; chain of custody in Attachment 6; laboratory reports, including blanks and the laboratory's AIHA-LAP accreditation, in Attachment 7.

Sampling days were selected to span [production conditions] and included [N] days at above-average production. Two samples were voided (pump fault; overloaded filter) and are reported here rather than omitted.

Results (8-hour TWA, μg/m³), stations 1–4: 15, 20, 23, 29, 31, 41. Results, stations 5–8: 62, 88, 91, 140, 155.

> ► *Four things a court, an opposing expert, or a careful purchaser will look for, and which most reports supply only partially.* ***The method, named and cited,*** *because it establishes what the number means and what its limits of quantification and sampling-and-analytical error are.* ***Calibration and chain of custody,*** *because Chapter 10 shows what happens to a measurement whose chain has a gap.* ***The sampling-day selection rationale,*** *because samples collected on convenient days answer a narrower question than the assessment claims to answer. And* ***the voided samples, reported.*** *A report that silently drops the overloaded filter has removed its highest result and has done so invisibly; disclosing it is both the honest practice and, in the litigation this book describes, the difference between a witness who is believed and one who is impeached.*

5. *Data Analysis*

Method. Summary statistics are computed on the natural logarithms of the results. The exposure profile is described by the geometric mean and geometric standard deviation; the 95th percentile is estimated as GM × GSD^1.645; and the upper tolerance limit is the one-sided normal tolerance limit for 95 percent coverage at 95 percent confidence, computed as exp(m + k·s), where m and s are the mean and standard deviation of the log-transformed results and k is the tolerance factor for the sample size of that subgroup. Each subgroup is analyzed on its own samples; the tolerance factor is taken from the subgroup n, not from the combined dataset.

Stations 5–8 subgroup (n = 5): geometric mean 102 μg/m³, geometric standard deviation 1.45. Estimated 95th percentile 188 μg/m³. With k(95,95; n = 5) = 4.20, the upper tolerance limit is 488 μg/m³.

Stations 1–4 subgroup (n = 6): geometric mean 25 μg/m³, GSD 1.43. Estimated 95th percentile 45 μg/m³. With k(95,95; n = 6) = 3.71, the upper tolerance limit is 94 μg/m³.

Worked derivation, stations 1–4, so the reader can check the rest. The six results are 15, 20, 23, 29, 31, and 41 μg/m³. The mean of their natural logarithms is 3.226 and the sample standard deviation is 0.355, giving GM = e^3.226 = 25 μg/m³ and GSD = e^0.355 = 1.43. The 95th percentile is e^(3.226 + 1.645 × 0.355) = 45 μg/m³. The tolerance limit is e^(3.226 + 3.71 × 0.355) = 94 μg/m³.

Analytical uncertainty for the method is ±[N] percent at these concentrations (Attachment 7); this has not been propagated into the tolerance limits above, which reflect sampling variability only. No results were below the limit of quantification; had any been, the censoring treatment of Chapter 17 §IV would be stated here rather than assumed. Propagating it would widen the intervals by approximately [N] percent and would not change either conclusion.

▶ *The report states a distribution, not an average, because the decision concerns the proportion of days above a limit and an average cannot answer that. It reports the GSD, which is the parameter that governs how much of the distribution sits above the mean and the parameter most often omitted. It reports a tolerance limit with its confidence level stated, because a 95th percentile without a confidence statement is a point estimate wearing a percentile's clothing. It names the estimator, states the sample size the tolerance factor was drawn from, and shows one derivation in full: three practices that cost a paragraph and foreclose the most damaging question an opposing expert can ask, which is not whether the number is right but how it was obtained. Note in particular that the tolerance factor belongs to the subgroup, not to the pooled dataset: a report that splits a SEG for the exposure judgment and then computes its limits on the combined n has re-pooled in the statistics what it separated in the analysis, and the resulting limits are too narrow in the anticonservative direction. And it states what the uncertainty analysis did not include and what including it would do — which is the discipline Chapter 17 §VI describes as distinguishing uncertainty from bias, and which is worth a paragraph in every report because it is the paragraph that survives cross-examination.*

6. Exposure Judgment

Stations 5–8: Category 4 (exposures exceed the OEL). The estimated 95th percentile exceeds the PEL by nearly a factor of four, and the UTL(95,95) exceeds it by a factor of nearly ten. Every sample collected in this subgroup exceeded not only the action level but the PEL itself. Confidence in this judgment is **high**: the data are internally consistent, the sample size is adequate for the observed GSD, and no plausible variation in production rate would bring the distribution below the limit.

Stations 1–4: Category 3 (exposures are controlled but approach the OEL). The estimated 95th percentile, at 45 μg/m³, sits at ninety percent of the OEL — above the action level, below the PEL, and squarely within the fifty-to-one-hundred-percent band this category describes; the UTL(95,95) is above the PEL. Confidence in this judgment is **moderate**: six samples with a GSD of 1.43 support the central estimate, but the upper tolerance limit is compatible with a meaningful fraction of days exceeding the limit, and the assessment cannot presently exclude that possibility. Additional sampling (a minimum of six further full-shift samples, weighted toward high-production days) would resolve it.

▶ *This is the section Chapter 15 §IV calls the deliverable, and its two features are the ones most reports lack. The judgment is expressed as a* ***category with a stated***

> ***basis**, not as a sentence saying results were below the limit. And it carries an **explicit confidence statement** — high for one subgroup, moderate for the other, with the reason for the difference stated and a specific remedy proposed. Chapter 15 §VIII's central point appears here in operational form: a conclusion without its confidence is half the deliverable, and it is the half that fails under challenge. Note that "moderate confidence" is a legitimate professional finding. What is not legitimate is presenting a moderate-confidence conclusion as a high-confidence one by omitting the qualifier.*

7. Recommendations

Immediate (stations 5–8): Respiratory protection at an assigned protection factor of not less than 10 is required pending engineering controls, together with the written program, medical evaluation, fit testing, and change schedule that the APF presupposes (see 29 C.F.R. § 1910.134). All employees in this subgroup meet the medical surveillance trigger of § 1910.1053(i).

Engineering (stations 5–8): Extend LEV of the design installed at stations 1–4, which has demonstrably reduced the geometric mean exposure by approximately three-quarters in this facility. Estimated cost [N]. Verification sampling within 30 days of installation.

Assessment (stations 1–4): Six additional full-shift samples as described at §6.

Reassessment trigger: Any change in production rate exceeding 30 percent, any change in sand supplier or composition, any modification to the ventilation system, or 24 months, whichever occurs first.

> ▶ *Three notes. The APF recommendation states what the protection factor presupposes, which is the point Chapter 17 §V identifies as the most practically valuable half-page in Part III: an assigned protection factor is a program value, not a device property, and recommending one without the program behind it recommends a number rather than a protection. The engineering recommendation cites in-facility evidence of effectiveness rather than a catalog claim. And the **reassessment trigger** is the element whose absence does the most damage in litigation — an assessment with no expiration is an assessment the employer will still be relying on in year seven, and the plaintiff's expert will ask why.*

8. Limitations

This assessment characterizes full-shift exposures under the production conditions observed during the sampling period. It does not address short-term or task-specific peak exposures; exposures during maintenance, cleanout, or upset conditions; exposures to other agents present in the department; or exposures of employees outside the defined SEG. The judgment for stations 1–4 is of moderate confidence as described at §6 and should not be relied upon as establishing compliance.

▶ *The last clause is the one an employer's counsel will be grateful for and an employer's operations manager will dislike. A report that permits itself to be read as a compliance certification when its data do not support one has created a document that will be offered against the employer for exactly that reading. Stating the limit is not hedging; it is the assessment declining to say something it does not know.*

The Eight-Element Test

An exposure assessment report should contain all eight. Counsel auditing an opposing expert's underlying assessment, and employers reviewing a deliverable, can run the list in five minutes.

1. A stated decision the assessment exists to support.
2. A basic characterization complete enough that an outside professional could evaluate the judgment.
3. An exposure group defined by stated determinants, with its boundaries and exclusions named.
4. Sampling data with method, calibration, chain of custody, day-selection rationale, and voided samples disclosed.
5. An analysis reporting the distribution — central estimate, GSD, and an upper percentile with its confidence level — rather than an average.
6. An exposure judgment expressed as a category, with an explicit confidence statement.
7. Recommendations with owners, verification, and a reassessment trigger.
8. A statement of limitations, including what the assessment does not address.

A report containing all eight will usually be defended successfully. A report containing four of them is, in the language of Chapter 25, an exhibit with the employer's letterhead on it.

References

American Conference of Governmental Industrial Hygienists. Threshold limit values for chemical substances and physical agents.

Bullock, W. H., & Ignacio, J. S. (Eds.). (2006). A strategy for assessing and managing occupational exposures (3d ed.). American Industrial Hygiene Association.

Jahn, S. D., Bullock, W. H., & Ignacio, J. S. (Eds.). (2015). A strategy for assessing and managing occupational exposures (4th ed.). American Industrial Hygiene Association.

National Institute for Occupational Safety and Health. NIOSH manual of analytical methods, Method 7500 (silica, crystalline, by XRD); Occupational Safety and Health Administration, Method ID-142.

Respirable crystalline silica (general industry), 29 C.F.R. § 1910.1053; respiratory protection, § 1910.134; access to employee exposure and medical records, § 1910.1020.

Topic Finder

References are to chapters and appendices rather than page numbers. Cases appear in italics. Subentries are indented under their headword. A chapter reference locates the discussion; an appendix reference locates the entry in the Glossary (App. B), the Table of Key Decisions (App. C), or the Standards and Directives Matrix (App. D).

B

C

R

S

About the Author

Logan F. Martin, PhD, CIH, CSP, CHMM, is a Certified Industrial Hygienist, Certified Safety Professional, and Certified Hazardous Materials Manager with almost 40 years of experience in occupational exposure assessment and reconstruction and in safety program auditing. She serves as a consulting and testifying expert in OSHA enforcement matters, workplace accident litigation, and occupational injury and disease claims.

She is retained by plaintiff and defense counsel alike, providing case analysis and expert testimony in workplace injury and occupational exposure matters from either side, as well as in contract disputes turning on safety expectations and the allocation of safety responsibilities among the parties.

Dr. Martin is the Principal Environmental Health & Safety and Industrial Hygiene Consultant at LF Martin LLC, the practice she founded in 2001. Her work spans general industry, construction, healthcare, hazardous materials and hazardous waste, transportation, process safety, chemical manufacturing, steel erection, and cranes and hoisting.

She holds a PhD in Occupational Safety and Health from Capitol Technology University, an MS in Occupational Safety and Health Management from Indiana State University, an MBA from Plymouth State University, and a BS in Geology from Winona State University, and is also a Certified Geologist in Maine and a Project Management Professional. For more than a decade she has served as faculty for several universities, teaching and leading safety and industrial hygiene programs at the undergraduate, graduate, and doctoral levels.

Dr. Martin is the editor of the National Safety Council's Supervisors' Safety Manual (Eleventh and Twelfth Editions), has published on exposure modeling, risk management, and leading health metrics, and served as President of the Board of Certified Safety Professionals (2018–2019). She is a peer reviewer for Safety Science and the Journal of Safety Research, and her work has been recognized with the Board of Global EHS Credentialing (formerly the American Board of Industrial Hygiene) CIH Impact Award (2021) and the National Safety Council's Marion Martin Award (2018).

Dr. Martin can be reached at **lfmartinphd@gmail.com**.

www.ingramcontent.com/pod-product-compliance
Ingram Content Group UK Ltd.
Pitfield, Milton Keynes, MK11 3LW, UK
UKHW051118290926
15471UKWH00010B/1744